Astronomy

before the Telescope

Astronomy

before the Telescope

Edited by Christopher Walker
With a foreword by Patrick Moore

Published for the Trustees of the British Museum by
British Museum Press

Published by British Museum Press
A division of The British Museum Company Ltd
46 Bloomsbury Street, London WC1B 3QQ

First published 1996

British Library Cataloguing in Publication Data
A catalogue record for this book is available from the British Library

ISBN 0-7141-1746-3

Designed by Andrew Shoolbred

Printed in Great Britain by Butler & Tanner Ltd, Frome

ILLUSTRATION ACKNOWLEDGEMENTS

Copyright owners not named in the captions are as follows. Figs 2, 6: photos by Chris Jennings. Fig. 3: photo by Clive Ruggles. Fig. 5: Office of Public Works, Dublin. Fig. 12: Bildarchiv Preussischer Kulturbesitz, Berlin. Fig. 27: reproduced by permission of Gerald Duckworth & Co. Ltd, London. Fig. 32: reproduced by permission of the Committee of the Egypt Exploration Society. Figs 43, 44: reproduced by permission of Charles Scribner's Sons, New York. Fig. 48: reproduced by kind permission of the Trustees of the Chester Beatty Library, Dublin. Figs 83, 84, 89, 90: Colin Ronan. Fig 85: photo by Tony Sizar. Figs 87, 91: reproduced by permission of Cambridge University Press/Needham Research Institute, Cambridge. Fig. 94: photo by A. Aveni. Fig. 95: rollout photograph © Justin Kerr 1980. Fig. 100: photo by H. Hartung. Fig. 101: reproduced by permission of the University of Texas Press. Fig. 114: reproduced by permission of the Royal Society, London. Col. Pl. I: English Heritage/Skyscan Balloon Photography. Col. Pl. II: photo Victor Boswell Jr, © National Geographic Society. Col. Pl. VII: photo Scala, Milan.

The maps (Figs 1, 7, 20, 41, 47, 59, 82, 93, 104, 107 and 109) were drawn by Ann Searight. Fig. 86 was drawn by E. Robson. Hieroglyphs by Compuglyph (all rights reserved).

Frontispiece The Istanbul observatory around the year 1577. The director of the observatory, Taqi l-Dīn, is holding an astrolabe; the other astronomers are involved with various instruments, of which only the terrestrial globe and the mechanical clock are European imports. See further chapter 9. (University Library, Istanbul, MS Yıldız 1404)

Contents

The Contributors

Anthony F. Aveni is Russell B. Colgate Professor of Astronomy and Anthropology at Colgate University, Hamilton, N.Y., and a specialist in the ancient astronomy of the New World.

John Britton is an investment manager living in Boston, Massachusetts. He has a PhD in the history of astronomy from Yale University.

J. V. Field is a Visiting Research Fellow at the Royal Institution, London.

Alexander Jones is an Associate Professor of Classics at the University of Toronto. He specialises in the history of the ancient and medieval exact sciences, in particular Greek astronomy during the Roman and Byzantine periods.

David A. King is Professor of the History of Science at the Johann Wolfgang Goethe University, Frankfurt am Main. He has worked on the primary sources for the history of Islamic science, manuscripts and instruments for over twenty-five years. Many of his publications are reprinted by Variorum in three volumes, *Islamic Mathematical Astronomy, Islamic Astronomical Instruments* and *Astronomy in the Service of Islam*.

Patrick Moore is a former President of the British Astronomical Association. His main research has been in connection with lunar cartography. Among his publications dealing with the history of astronomy is *The Planet Neptune* (1996).

Wayne Orchiston is Executive Director of the Carter Observatory, the National Observatory of New Zealand. A former President of the Astronomical Society of Victoria (Australia), he is a Fellow of the Royal Astronomical Society and a member of the International Astronomical Union. His research interests lie mainly in variable stars, meteoritics and the history of Australian and New Zealand astronomy, and he has over 100 publications to his name.

Olaf Pedersen was Professor of the History of Science at Aarhus University from 1956 to 1990, Vice-President of the International Union of History and Philosophy of Science from 1981 to 1985, and President of the Académie Internationale d'Histoire des Sciences from 1985 to 1989. He is also an Associate of the Royal Astronomical Society. Among his publications are *A Survey of the Almagest* (Odense University Press, 1974) and *Early Physics and Astronomy* (2nd edn, Cambridge University Press, 1993).

David Pingree is Professor of the History of Mathematics and of Classics at Brown University, Providence, R.I. His interests lie in Mesopotamian, Greek, Indian, Islamic, and Western and Byzantine astronomy, astrology and astral images, and in their interrelations.

T. W. Potter is Keeper of the Department of Prehistoric and Romano-British Antiquities at the British Museum.

Colin Ronan was President of the British Astronomical Association and director of its historical section for more than twenty years. He also served on the council of the Royal Astronomical Society, and was a prolific author. He died in 1995.

Clive Ruggles is Senior Lecturer in the School of Archaeological Studies at the University of Leicester. He is editor of *Archaeoastronomy*, the supplement to *Journal for the History of Astronomy*.

F. Richard Stephenson, formerly Resident Research Associate at the Jet Propulsion Laboratory, Pasadena, is Professorial Fellow in Physics at the University of Durham. He is co-author or co-editor of several books on various aspects of applied historical astronomy.

N. M. Swerdlow is Professor of Astronomy and Astrophysics and of History at the University of Chicago. His research is in the history of the exact sciences, particularly astronomy, from antiquity to the seventeenth century.

G. J. Toomer, formerly E. P. Warren Praelector in Classics at Corpus Christi College, Oxford, and for many years Professor of the History of Mathematics at Brown University, Providence, R.I., is now an Associate of the History of Science Department at Harvard University. He has published a translation of Ptolemy's *Almagest* and an edition with translation of the part of Apollonius's *Conics* which exists only in Arabic.

G. L'E. Turner is Visiting Professor of the History of Scientific Instruments at Imperial College, London. His most recent books are *Gli Strumenti: Storia delle Scienze, Vol. I* (Einaudi, 1991) and *Scientific Instruments and Experimental Philosophy* (Variorum, 1991).

Christopher Walker is Deputy Keeper in the Department of Western Asiatic Antiquities at the British Museum.

Brian Warner is Professor of Astronomy at the Department of Astronomy, University of Cape Town, South Africa, and has published extensively on nineteenth-century astronomy in South Africa.

Ronald A. Wells was a Fulbright scholar in Egypt at the University of Cairo and at Helwan Observatory in 1983–4, and again at the Institute of Archaeology, Egyptology Division, University of Hamburg, in 1987–8. He is currently a computer resource specialist at the University of California, Berkeley.

Acknowledgements

This book was first conceived some five years ago on a Hellenic cruise, in a conversation between Patrick Moore, Tim Potter and George Hart of the British Museum. Having inherited the project from them, I am especially grateful to Patrick Moore for his continued interest, for his advice on contributors and for writing the Foreword. The Director of the Museum, Dr Robert Anderson, has also given me constant advice and encouragement, and personally invited all the authors to contribute.

It is with regret that we record that Colin Ronan died on 1 June 1995. I am grateful to Dr F. R. Stephenson for advice on the editing of his chapter.

I am indebted also to Dr John Britton and Patrick Moore for assistance in compiling the Glossary, and to Ann Searight for drawing the maps.

The editorial staff of British Museum Press, at first Celia Clear and later Teresa Francis, have as always been supportive, patient and determined in steering the book through to completion.

Christopher Walker
Department of Western Asiatic Antiquities
British Museum

PATRICK MOORE

Foreword

There can be little doubt that astronomy is the oldest science in the world. The skies are all around us; the sun is dominant by day, the moon and the stars by night. They do not stand still; they move slowly but inexorably from one horizon to the other, while the moon shows its phases from night to night and the planets wander around against the starry background. The earth, our home, must initially have seemed to be all-important, with the celestial bodies installed for our benefit, presumably by some divine agency. But how much could the people of Antiquity find out?

First, we must decide just what is meant by 'Antiquity' in an astronomical context. By convention, the Middle Ages begin with the fall of the Roman Empire, and end with the Renaissance, though obviously there can be no hard and fast limits. In astronomy, it may be said that there are three distinct periods. There is the 'dawn', when people became aware of the phenomena in the sky but made no effort to interpret them, or even to observe them in any but the most superficial manner; until the art of writing was developed, there could be no records other than oral tradition. Next comes the long period during which scientific observations were made, and mathematics became something more than simple counting; the stars were divided up into constellations, the movements of the bodies of the solar system were worked out in detail, and it was even possible to obtain some knowledge of the vastness of the universe. Moreover, astronomy became of everyday importance, notably in regulating the calendar, and eventually in map-making and navigation. Then, in the early seventeenth century, came the first astronomical telescopes. The invention of the telescope was claimed by a Dutchman, Hans Lipperhey. Galileo, hearing of the new invention, made his own instrument and turned it skyward with results that changed the whole course of astronomy.

Earlier, everything had to depend upon the naked eye alone, which meant that there was no 'physical astronomy'; all that could be made out were the light and dark patches on the moon, and occasional sunspots. In fact astronomy was positional only, and all interpretations had to be made against this limitation. Under the circumstances we can only marvel at what our ancestors managed to achieve.

One immediate problem is that in very early times there is so little to guide us. Written records are fragmentary until we come to the era of Egypt, Babylon and Greece, and before the eighth century BC we have virtually nothing, apart from occasional allusions to startling phenomena such as comets and eclipses. Chinese records are among the earliest; they tell us about 'guest stars', supernovae and, of course, eclipses of the sun, which were attributed to attacks by a hungry dragon. (The remedy was to make as much noise as possible, in order to scare the dragon away. It always worked!) Even many of the later records have been lost, and there can be no compensation for the loss of the Alexandrian Library. It is fortunate that Ptolemy's *Almagest* has come down to us; without it, we would know far less about ancient Greek science than we actually do. Our debt to Ptolemy is great indeed. Spasmodic attempts to belittle him, and to claim that he was a copyist at best and a fraud at worst, have been singularly unsuccessful, and his nickname of 'the Prince of Astronomers' is not inappropriate. But Ptolemy's knowledge was confined to the Mediterranean, with occasional reference to Babylonian observations, and he could learn nothing about peoples further away from him.

To draw all the strands of the early development of astronomy together into a coherent picture is not an easy task, because the various pieces of the jigsaw are so diverse, and we also have to cover a great span of time. It is a popular belief that there was a brilliant beginning in Egypt, with the Pyramid-builders, and then a explosion of science in Greece which changed astronomy abruptly from mysticism to science. Nothing could be further from the truth. Progress in Greece was slow and incremental. For example, Thales of Miletus, first of the great philosophers of Classical times, lived from about 625 to 547 BC; Ptolemy, the last, died about AD 175. This is an interval of almost eight hundred years, so that in time Thales was as far removed from Ptolemy as we are from the Crusades. Thales believed the world to be flat, and to be floating on water. Ptolemy not only knew it to be a globe, but also had a very shrewd idea of its size. He also drew a map which was based upon astronomical observations.

It was the flat-earth theory which was the first to be discarded by the Greeks, and in the third century BC Eratosthenes, long before Ptolemy, had made a remarkably good estimate of its size. The next step was to dethrone the earth from its proud central position, and this was something which few of the Greeks were able to accept, partly on religious or philosophical grounds, because of the influence of Aristotle, but mainly because there was no good scientific evidence in favour of it. It is probably true to say that the acceptance of the heliocentric theory was far from general right up to the time of Newton. And in countries far away from the sphere of Mediterranean culture, the old ideas persisted for even longer.

Religion had a great deal to do with it. In the time of Pericles, Anaxagoras had to flee from Athens to avoid a charge of impiety, for daring to teach that the sun is a red-hot stone larger than the Peloponnesus. So persecution of this sort is not uniquely a Christian phenomenon, even though it was in Christian countries that it reached its highest pitch and ended less than four centuries ago. The Copernican theory had to deal with attacks from men of the calibre of Martin Luther, while Galileo was brought to trial and forced into a hollow and utterly meaningless recantation. Indeed it was not until 1992 that the Vatican finally conceded that Galileo had been justified in all he said! But we have no record of persecution on these grounds in other cultures, and indeed there was no need for it, since until relatively modern times the question did not arise among, say, the Aborigines or the Maori.

Of course, every culture has its creation myths – even the Book of Genesis is hardly helpful from a purely scientific point of view. To the Egyptians, the sky was formed by the arched body of the goddess Nut, and the sky was solid – a belief which continued for a surprisingly long time. Creation myths in different countries do not have a great deal in common with each other, which in view of the lack of long-distance communication is not surprising. Neither do the constellation patterns agree. If we had followed, say, the Egyptian or the Chinese pattern instead of the Greek, our star charts would look very unfamiliar even though the stars themselves would, of course, be exactly the same.

There were various objects of special attention. The Egyptians regulated their calendar by the heliacal rising of Sirius, which told them when the flooding of the Nile was due – and it was upon this that their whole economy depended. Venus was used to regulate the calendar of the Maya, while among certain African tribes it was Canopus which fulfilled this role. The Pleiades, too, were widely regarded as significant. They are referred to by Hesiod and in the Bible, and also by the Aztecs and the Polynesians.

There were some early races whose work was confined almost completely to positional measurements. The Egyptians had no real concept of the nature of the universe, but their precision was amazing by any standards, and there is no doubt that the Pyramids were astronomically aligned, even if their exact role is still a matter for debate. In the case of Stonehenge and other later stone circles in Britain we remain uncertain of their full significance, and though claims that they may have been primitive 'computers' have often been exaggerated something of the sort cannot be entirely ruled out. Major stone structures arose also in the Far East, as at Kyongju in Korea in AD 647; whether we will ever learn the full truth about how they were used is doubtful, but that they played a part in astronomical observation cannot be doubted.

The development of mathematics played a vital part in astronomy, as in everything else, and some cultures concentrated mainly upon it; in Indian astronomy, for example, observation played a relatively minor role, while the Greeks made every effort to reconcile observations with mathematical theory. Aristarchus's method of measuring the relative distances of the moon and sun were perfectly correct in theory, but through no fault of his own his observational data were insufficiently accurate. Neither could his heliocentric theory be proved observationally, which was one reason why it did not prevail in the ancient world. (Even when the Copernican theory had been published, so many centuries later, the tables produced from it were initially no more accurate than those compiled on the basis of the old geocentric system.) Yet the Greeks knew quite well that the movements of the planets in the sky could not possibly be explained by assuming that they moved round the earth at a steady rate in perfect circles, and it was for this reason that the scheme of deferents and epicycles was developed. Though it has an artificial air, it gave an apparent fit to the observational data available at the time. Ptolemy brought it to its highest degree of perfection, and it prevailed for well over a thousand years after his death – not only in the Mediterranean area, but also elsewhere.

Mathematical theories were developed in medieval times by the Arabs, from around the ninth century AD. They had a reputation as excellent observers; it is true to say that the Muslims excelled in every branch of scientific knowledge, and they also built measuring instruments which were far better than any of their predecessors. Not that early attempts had been inadequate; how else could Hipparchus, around 130 BC, have detected the precession of the equinoxes? But it was the Arabs who paved the way for the great pre-telescopic observatories such as Ulugh Beg's at Samarqand and those built by Jai Singh in India in the early eighteenth century. On the other hand, it is also true that the main concern of the Arabs was astrological, and here we come to another obstacle in the way of real progress.

Asking which came first, astronomy or astrology, is rather like asking whether the chicken preceded the egg or vice versa. Almost all early astronomers were also astrologers, and it was not until the time of Bishop Isidore of Seville, in the seventh century AD, that a sharp distinction was drawn. European-style astrology did not apply to remote lands such as the Americas and Australasia, but here also there was a general acceptance that the stars controlled the destinies of people. For example, the fear of comets seems to have been more or less universal. Eclipses also caused alarm (though the old legend that two Chinese court astrologers, Hsi and Ho, were executed for failing to predict one is certainly apocryphal). In one famous case, an eclipse actually affected the course of human history. The Athenians lost their fleet in the course of their disastrous expedition to Sicily, during the Peloponnesian War, because a lunar eclipse induced the commander, Nicias, to delay evacuation for 'thrice nine days' – by which time it was too late. But for this, the Peloponnesian War might well have had a very different ending, and Sparta would not have replaced Athens as the most powerful city-state in Greece. Christopher Columbus was much more far-sighted during his voyage to the New World; the local inhabitants refused to supply him with provisions, but Columbus knew that an eclipse was due, and threatened to make the moon 'lose its light'. When the eclipse began, the local inhabitants were so terrified that Columbus received everything that he needed.

Not all early civilisations contributed a great deal to astronomy. The Japanese did relatively little, and their astronomy came largely from China. The Romans were not innovators, but it was of course they who carried out a particularly effective reform of the calendar, instigated by no less a person than Julius Caesar – which is why his name is now attached to a crater on the moon. (The name was given by the Jesuit astronomer Riccioli, who drew up a reasonably good lunar map in 1651. However Riccioli was no believer in the heliocentric system, and in allotting names to craters he admitted that he 'flung Copernicus into the Ocean of Storms'!)

For obvious reasons, physical astronomy could make very little advance before the invention of the telescope, but some old ideas sound quaint today; the Greek philosopher Heraclitus believed that the diameter of the sun was about one foot and that it was renewed every day. However, Ptolemy gave the distance of the moon as about 59 earth-radii (close to the actual distance), so that at least he understood that the universe is a large place. The stars were assumed to be lamps fastened to the invisible crystal sphere which surrounded the earth, and it was not until later medieval times that the idea of stars as suns began to gain general acceptance. However, even from early times the distinction between planets and stars was evident. The Roman writer Plutarch believed the moon to be 'earthy', with mountains and ravines. In the second century AD a Greek satirist, Lucian of Samosata, produced the first science-fiction story about a lunar voyage; he entitled it the *True History*, though he was careful to add that he had chosen this title because the story was made up of nothing but lies from beginning to end!

Our knowledge of the astronomy of other lands is less extensive. That the Maya developed astronomical observation to a high degree of perfection is a major recent discovery, and their Venus calendar was very elaborate; as a patron of warfare, Venus was as important to the Maya as was Mars in the Old World. Sadly, the Spanish cleric Diego de Landa decided to destroy their records and was so devastatingly successful that only four Mayan codices survive intact today. In the Americas only the Incas and, to a lesser extent, the Aztecs developed true mathematics; others, such as the Hopi Indians, were avid sun-watchers, but of their beliefs we know next to nothing.

From Africa there are no indigenous astronomical records going back before the last century, apart from those of Egypt and Ethiopian calendrical works largely based on Greek traditions. So we must beware of the infiltration of knowledge when discussing original African beliefs. For example, there have been curious claims that the Dogon people of Mali knew about the tiny white dwarf companion of the star Sirius, which is as massive as the sun though smaller than a planet such as Uranus or Neptune. It is impossible to take this seriously; if the Dogon did indeed know about the Companion of Sirius, they certainly learned it from Western sources at a very much later date.

Just how far the astronomy of one early civilisation was influenced by others is a fascinating question. There can be no doubt about the links between Babylonia, Greece and India, but how far did Chinese influence spread beyond Japan and Korea? Were America and Australasia completely isolated scientifically? Of course, Arab astronomy came to Europe by way of Spain, but that was much later, when communications had been vastly improved.

We can learn much from the old records. We have, for example, observations of comets, supernovae and eclipses; indeed, eclipse timings have been used to check on the slowing-down of the earth's rotation. Perhaps above all, a knowledge of ancient astronomy helps us to look back through the ages and see how the minds of our predecessors worked. In some cases astronomy rose to a crescendo and then declined; China is the obvious example, but Greek astronomy suffered a long hiatus after Ptolemy, and progress in Arab astronomy virtually ended with the assassination of Ulugh Beg.

When the change in outlook did come, at the end of medieval times, it was rapid. Copernicus revived the heliocentric theory with the publication of his *De revolutionibus* in 1543; telescopes were in use from 1609; in 1687 Newton published his *Principia*, and astronomy entered what may be termed its modern phase. These developments transcended all that had gone before, but all in all, we cannot do anything but express our profound admiration for the achievements of the men of early times. Without them, we would have had no foundation upon which to build our knowledge.

This book was first dreamed up on a Hellenic cruise, as a means of putting these thoughts on early astronomy into a more detailed and comprehensive form. It was designed to complement the range of the British Museum's own archaeological and historical collections, and to look beyond mathematics and trigonometry to the contemporary cultural milieux and the surviving material remains. Most of the chapters have been written by scholars actively involved in research into the ancient source material, much of which has only come to light within the last few decades. I hope that it will be apparent that looking back into the records of the past can generate just as much excitement as looking at today's satellite pictures of the solar system or listening to the echoes of the Big Bang.

CLIVE RUGGLES

Archaeoastronomy in Europe

The best-known manifestation of ancient astronomy in North-west Europe is the phenomenon of the solstitial alignment of Stonehenge. Yet despite strong popular belief, fuelled by many speculative publications on the subject, Stonehenge did not incorporate precise astronomical alignments and did not function as an ancient 'observatory' in any sense that would be meaningful to a modern astronomer. The challenge facing the archaeoastronomer, who is interested in the true nature of astronomical practice in preliterate cultures, is to separate what was *possible* – something usually deduced from considering, explicitly or implicitly, the ways in which a modern astronomer might be able to use an ancient site to perform astronomical observations – from what actually took place.

What took place depended upon what ancient people were actually motivated to do; while their intellectual capabilities were indistinguishable from ours, the courses of action that were important to them were almost certainly very different. This principle is best illustrated by a modern example. The Mursi are a small group of cultivators and herders in South west Ethiopia (see p. 306). An anthropologist describes how on one occasion a man, who had been wearing a knotted cord round his ankle for several weeks, announced to a group of bystanders that 72 days had elapsed between the planting and first harvesting of his sorghum crop. He had been keeping track of this interval by successively knotting a piece of cord for every day that passed. The other members of the group treated this information as a curiosity without relevance to their daily lives, not as a 'discovery' to be added to their total stock of knowledge about the world. Their main reaction, in fact, was one of mild surprise that anyone should have taken the trouble to record such a trivial fact, and it was, without doubt, quickly forgotten.

The actions of prehistoric people depended upon their perception of the world, expressed in systems of belief and ritual. Astronomical practice depended upon a range of cultural parameters, many of which we can only guess at through analogy, and many of which are completely inaccessible to us. One thing we can do is to look for repeated trends within the archaeological record, to suggest general principles that influenced

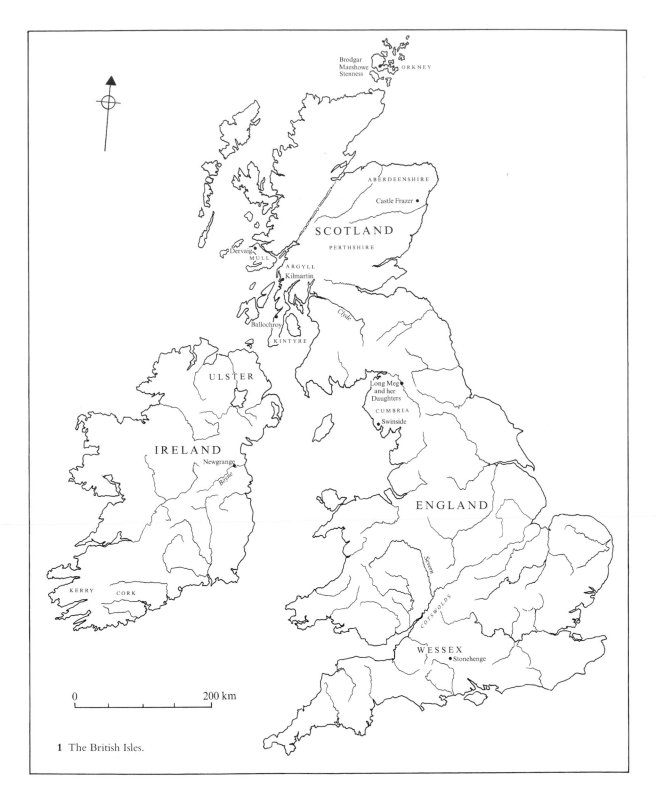

1 The British Isles.

people's actions within certain geographical areas at certain times. Another is to try to achieve some understanding of those people's perception of their environment, both through the cultural context suggested by the broader evidence from the archaeological record and by the careful use of analogy with other cultures, about whom other types of evidence may be available than just the current disposition of archaeological remains.

The clearest statistical evidence from prehistoric North-west Europe relates to small groups of sites, similar in form and confined to relatively small geographical areas, that show consistent orientation trends. One such group is the recumbent stone circles of North-east Scotland. These are rings of standing stones distinguished by the presence of a heavy recumbent slab flanked by two upright pillars (Fig. 2). They are confined to an area of about 400 square km, centred to the west of Aberdeen. Most appear to belong to the Early Bronze Age, which in this part of Scotland dates to the late third and early second millennia BC. The archaeological evidence suggests that the recumbent stone circles, none of which is significantly more grandiose than the others, were the local ritual centres for groups of subsistence farmers living in territories of around 10 square km. The recumbent and flankers seem to be of central importance to the ritual function of the sites. The heavy recumbent was carefully chosen, often transported from some distance and positioned with great care.

About 100 recumbent stone circles are known, over half of which are in a sufficiently good state of preservation to enable reliable measurements to be made. Their most remarkable property is that their orientations are highly clustered in azimuth: viewed from the ring centre towards the centre of the recumbent stone, they fall without exception within a 90° range centred upon SSW. Fieldwork during the 1980s has shown that the sites were deliberately placed where there was at least a moderately distant horizon behind the recumbent stone. Viewed from across the site, the recumbent and flankers demarcate a stretch of horizon which, in the majority of cases, contains a prominent hilltop. The declinations of these hilltops fall between −32° and −17°. This corresponds closely (though not with any great precision) to the theoretical range −30° to −19° that represents the range of possible values of the southerly limit of the moon's monthly motions at different points in the 18.6-year cycle of the nodes. Equivalently, it represents the range of possible setting positions of the full moon nearest midsummer. In the majority of cases the value is within a couple of degrees of the major standstill limit itself, at one end of this range. The major standstill is the time when the monthly swing of the moon's motions is at its greatest. Major standstills occur at intervals of 18.6 years. At monthly intervals around this time the moon comes close to the southernmost (and, in the north, the northernmost) position it can ever reach. Alignments upon the standstill limit would have required organised lunar observations over a period of at least 20 years. However, the precision of these alignments is sufficiently low that the observations involved are not arduous: occasional observations of the full moon nearest the summer solstice would have sufficed.

2 A recumbent stone circle at Castle Frazer, Aberdeen, Scotland.

The evidence that both conspicuous hilltops and lunar orientation were important, and wherever possible were correlated, is repeated in the case of another well-defined group of similar sites, the short rows of up to five standing stones found in West Scotland (Fig. 3). These are similar in number to the recumbent stone circles to the east, and apparently date to a similar period. Fieldwork over the last 20 years has shown that significantly many of these sites, whose orientation (when the site is sufficiently well preserved) is well defined, are oriented upon the southerly limit of the rising or setting moon. There is some evidence of a practice of primary and secondary orientation in two major areas of concentration: Northern Mull and the Kilmartin Valley area of Argyll. In these areas, sites are often found in pairs close together, and when this happens one is oriented upon the moon close to the major standstill limit and the other upon the moon at some other point in the 18.6-year cycle. Isolated sites are oriented in the primary direction. Recent work on the island of Mull has indicated that terrestrial features such as prominent hilltops may also have played a key role in the symbolism underlying the siting and orientation of these sites.

South-west Ireland represents an exciting area where this line of research has been extended in recent years. Counties Cork and Kerry are extremely rich in free-standing megalithic monuments, including over eighty rows of three to six standing stones, similar to those in West Scotland, over ninety 'axial stone circles' similar to the recumbent stone circles in North-east Scotland, and in addition about 100 stone pairs, similar to a group found in Perthshire, Central Scotland, and about 100 burial monuments known as 'wedge tombs'. The morphological similarities between the South-west Irish sites and the Scottish sites reinforce the idea of some form of linkage, direct or indirect, between the two areas. In fact, continuity of tradition is evident in the case of the stone rows: examples are found scattered down through Ireland and there is a further concentration in mid-Ulster. Furthermore, in South-west Ireland, unlike in Scotland, the short stone rows and stone circles are found in close association and evidently are closely related elements of a single tradition. And finally, the orientation trend amongst these sites is quite overwhelming. With the exception of only two stone pairs out of almost 400 sites of different types, the orientations of the South-west Irish sites are strongly concentrated about NE–SW, getting close to N–S or E–W in just a few limiting cases.

Initial results from the Cork and Kerry stone rows shows an interest in prominent hills and the moon similar to that found in Scotland, but there are differences. According to the published evidence on the four- to six-stone rows, the directionality of the sites – as evidenced both from the form of the sites themselves (stone mass and height gradation) and the 'indicated' horizons (distribution of horizon distance with azimuth and presence of prominent hills) – is as often north-east as south-west. This is not only different from the properties of similar sites in West Scotland, where the directionality is apparently

3 The five-stone row at Dervaig N., Mull, West Scotland. Three of the stones are now fallen.

invariably south-west, but also raises the intriguing question of why the builders' interest in marking the moon was confined to its rising position in the north and its setting position in the south. Evidence indicating solar orientations in one direction and lunar in the other would make more obvious sense: the association between the setting sun near to a solstice and the simultaneously rising full moon in the opposite direction is well attested from ethnographic examples such as the Sun Dance of North American Plains Indians. This is not, however, what we find amongst the Irish stone rows.

Notwithstanding these unanswered questions, it is clear that the astronomical symbolism associated with various relatively small and relatively late types of megalithic site from different parts of the British Isles was predominantly lunar. The combined results from a sample of published data on the Scottish recumbent stone circles, the stone rows of Mull, West Scotland, and the four- to six-stone rows of South-west Ireland, are shown in Fig. 4. The preference is evident for declinations within the range −30° to −19° that represents the southerly limit of the moon's monthly motions at different points in the 18.6-year cycle, and around the northern major standstill limit at +28°. Within this context, it seems probable that the famous stone row at Ballochroy on the Kintyre peninsula of West Scotland, which is aligned in the south-west upon the midwinter setting sun and has attracted a good deal of attention in the past as a possible high-precision solar 'observatory', was actually intended by the builders to incorporate a symbolic alignment upon the midsummer full moon, the alignment being set up around the middle of the node cycle. Other claims for high-precision astronomy,

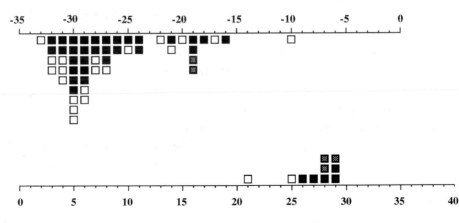

4 Declinations of hill summits indicated by prehistoric stone rows and recumbent stone circles: combined results from published data on the Scottish recumbent stone circles, the stone rows of Mull and the four- to six-stone rows of South-west Ireland. Each square represents a single hill summit, plotted to the nearest degree. The solid black squares represent the most reliable data. The grey tinted squares represent Irish sites where the directionality of the row is unclear (i.e. it is not clear whether the hill in question is in the direction of directionality of the row). The unshaded squares denote hilltops in the opposite direction to the directionality of the row at Irish sites, or hilltops more than 3° from the centre of the recumbent stone in the case of Scottish recumbent stone circles.

including the idea that distant horizon features such as notches could be used to pinpoint the motions of the sun or moon to a few minutes of arc, have now been discounted on the grounds that they can be explained away through data selection effects. The argument for high-precision alignments at prehistoric sites also conflicts with recent evidence from modern astronomy, which indicates that day-to-day variations in atmospheric refraction are much greater than was previously thought. This means, for example, that the idea of horizon notches being used to pinpoint solstitial sunrise and sunset to the very day now seems untenable.

There was, however, solar symbolism. It is found predominantly amongst the megalithic monuments dating to earlier phases of prehistory in areas where an organised, hierarchical society developed. The roots of such development, preceding the construction of the stone rows by some 1,500 years, were in the fourth millennium BC, when Early Neolithic farmers built the earthen mounds and small megalithic tombs that form the longest surviving conspicuous evidence of the activities of our prehistoric ancestors. Even these tombs show definite orientation trends, varying from group to group and area to area: examples are the predominantly north-easterly orientation of the long mounds of the Clyde region; the predominantly south-easterly orientation of the Cotswold-Severn long mounds; and, possibly at the end of this tradition, the predominantly south-westerly orientation of the Clava Cairns of Aberdeenshire.

In certain areas, such as the Boyne Valley of Ireland, the Orkney Islands in Scotland, and Wessex in Southern England, the inconspicuous Early Neolithic burial places were eventually replaced by gigantic monuments such as Newgrange, Maeshowe and the rings of Brodgar and Stenness (Orkney), and Stonehenge in its later phases, reflecting the development of more complex society and more extensive trading networks. The passage grave at Newgrange (Fig. 5) incorporates one of the most famous – and also, from the point of view of understanding the nature of prehistoric astronomy, one of the most enlightening – examples of solar symbolism. The site forms part of a rich complex of prehistoric monuments built in the fourth and third millennia BC amidst fertile agricultural land along the northern banks of the River Boyne. The tomb itself is situated on a long, low ridge with a commanding view over the valley. Around it are a scattering of what appear to be 'satellite' structures: three smaller passage graves along the ridge to the east, and various barrows, standing stones and enclosures down towards the river. Newgrange itself is some 60 m in overall diameter, with a 19 m-long passage stretching from an entrance on the south-east side to three internal chambers where the bones of the dead were placed.

Sir Norman Lockyer had noted in the 1900s that the passage at Newgrange was approximately aligned upon the rising sun at winter solstice. However, the true nature of the interplay between the light from the rising solstitial sun and the architecture of the tomb was only realised more than 60 years later, when it was witnessed at first hand by the excavator Michael O'Kelly in 1969 following the restoration of the site. Just after

5 Newgrange, Co. Meath, Ireland: the sun shining along the passage at the midwinter solstice.

sunrise for a few days around the midwinter solstice, the light of the sun penetrates into the passage through a 'roof-box' above the entrance, and eventually shines along the entire length of the passage where it would have lit up the chambers containing the bones of the dead.

Newgrange contains some valuable indications about the true nature of the astronomical alignments incorporated by prehistoric people into their monuments. It was not intended as a calendrical instrument to be used by living observers. Not only is it quite absurd to think of a priest having to go and sit amongst the bones of the dead in order to determine the date of midwinter solstice: this idea is refuted by the archaeological evidence. At some time after the original construction of the tomb, the entrance was blocked with a large stone. The roof-box seems to have been built precisely so as to allow the light of the midwinter sun to enter after living people could no longer do so. The perceptions and beliefs in the minds of Newgrange's builders that associated death with the solstitial sun are something we can only guess at. That such a symbolic association existed, and was of very great importance – as evidenced by the tremendous effort that was made in order to incorporate it in the architecture of a colossal tomb – is a fact that we can discern from the archaeological record.

The use of monumental stone architecture to express astronomical alignments is itself a strong argument that such alignments were symbolic, rather than intended for any use that would seem to us 'practical'. There are many indicators, astronomical and otherwise, that could be used to determine the time of year quite accurately enough for agricultural purposes. Observations of the heliacal rising or setting of certain stars, for instance, provide an indicator accurate to within a week or so, and require no permanent structure to be built, since they can be made from any convenient spot whence the dawn or dusk sky is visible. Even if the time of year is tracked using the rising or setting position of the sun along a distant mountainous horizon, as is known from many ethnographic examples worldwide, there is no need for anything more than an object to mark a fixed observing spot: observers amongst the Mursi, for example, use a favourite tree. Such observations, while doubtless undertaken by prehistoric people, are not discernible from the archaeological record.

What *is* discernible is the solar symbolism associated with rituals of life and death. The solar alignment at Newgrange is reflected at the great passage grave of Maeshowe in Orkney, erected in the later third millennium BC. Here the entrance is to the south-west, and the slab that blocked the entrance was shorter than the entrance itself, leaving a slit at the top through which the midwinter sun shone just before setting. During the third millennium BC, a number of large stone circles were built in Cumbria (Fig. 6). The circles appear to be carefully designed, with stones placed systematically in the cardinal directions. They also feature single stones placed just outside the circle. At the huge ring of Long Meg and Her Daughters, an outlying stone is placed beyond the circle entrance in the direction of midwinter sunset. At other rings, outlying stones indicate sunrise or sunset at intermediate dates such as the beginning of February and the beginning of May. Concentrations of these dates when the evidence from many sites is put together led the engineer Alexander Thom to postulate a 16-month 'megalithic calendar'. What we may, however, be seeing is the earliest evidence for a ceremonial significance being placed upon the half-quarter days that later became the Celtic festivals of Imbolc, Beltane and so on. The sites themselves are not calendrical instruments; it is much more likely that by incorporating alignments upon a certain calendrical date, they express the

6 The large stone circle at Swinside, Cumbria.

symbolism associated with the ceremonies taking place on that date, and thereby reinforce those ceremonies.

The prehistoric archaeologist has to unscramble changes through time as well as space. The astronomical trends are not always simple. The earliest Neolithic tombs may contain a mix of solar and lunar symbolism, and in many cases no discernible astronomical symbolism at all. In East Scotland, the predominantly south-westerly orientation trend of the Clava cairns carried over into the later recumbent stone circles, and lunar symbolism seems to have held sway throughout. In Wessex, on the other hand, there appears to have been a shift from lunar to solar symbolism as development progressed from the Neolithic into the Bronze Age. This is reflected in the azimuth distributions of the burial cairns from each period, and also in the apparent shift in the axis of

Stonehenge from lunar alignment in the earlier phases to later solar alignment (Col. Pl. I). Despite popular belief it is not clear whether the solstitial alignment of the axis was intended to be towards midsummer sunrise in the north-east or towards midwinter sunset in the south-west – or even both. It was certainly not precise: the distant horizon is flat and featureless, and the Heel Stone, often proposed as a foresight (but which, it is now known, originally had a companion), is much too close to be of any use in pinpointing the event. Other solar and lunar alignments at the site may or may not be coincidental – they are certainly not precise.

Can we use analogy with other cultures to gain any insights into astronomy in the prehistoric British Isles? In the past there have been brave attempts to draw parallels between Mayan astronomer-priests and a supposed astronomical elite in Late Neolithic Britain, responsible for the high-precision observational astronomy envisaged by Alexander Thom. The twin problems with this approach were first the very different cultural characteristics of the Classic Mayan culture and that of Late Neolithic Wessex, and second the fact that Thom's evidence for high-precision astronomy did not stand the test of reassessment. However, an analogy of great potential interest has emerged recently, from far south in the Basque Country. Here there are many examples of what appear to be eight-stone rings. These *sarobe* were constructed by transhumant shepherding people in historic times, and in some cases they were still in use at the beginning of the twentieth century. This means that we have both first-hand accounts and extensive documentary evidence relating to their purpose and function. This evidence shows that the *sarobe* were actually perceived by the builders as stone octagons rather than stone rings. Legal records specify their design, construction and celestial orientation. Each site was laid out using standard units of length and aligned with the cardinal and inter-cardinal directions. Linked to the theme of cosmic order, it acted both as a seat of government and a centre for religious rites. The *sarobe* functioned within a cosmological network of social practices and beliefs rather than merely at an instrumental level.

The *sarobe* are the material remnants of a system of the social organisation of space dating back to at least the early Middle Ages, and possibly much earlier. This system is also reflected in constructs and concepts in the Basque language. This language is pre-Indo-European, which provides evidence that Basque culture was not ruptured by the arrival of Indo-European speakers, so that cultural continuity may be postulated right back to prehistoric times. In addition, it is interesting to note that the Basque standard unit of measurement relates to ancient units used to lay out traditional land holdings in France and possibly in many parts of the British Isles. These observations do not, of course, prove that cultural practice in the Basque Country in historic and modern times was in any way related to that in the Neolithic and Bronze Age British Isles; they do, however, provide a strong motivation for studying the Basque Country further as a useful analogy for ancient cultural practice elsewhere in Europe, and such investigations are well under way.

Archaeoastronomical investigations are still in their infancy throughout much of mainland Europe, although a number of case studies have begun to emerge in recent years. Many of these relate to later prehistoric times where the material evidence features artefacts rich in art and iconography rather than just the orientations of ceremonial architecture, and where contact with contemporary literate cultures provides documentary evidence of astronomical mythology and practice. Excavations in Armenia, for instance, have uncovered numerous bronze and clay artefacts dating to the middle and later part of the first millennium BC that contain symbolic representations of the sun and possibly also the moon and planets. Greek writers refer to the Armenian sun god. We know also that ancient Armenians prayed towards the rising sun, a custom adopted by the early Armenian church and reflected in church orientation and burial orientation practices that survive to the present day.

Studies of grave orientation within prehistoric cemeteries are producing evidence of strong orientation preferences throughout Southern Europe, from Hungary in the east to Spain in the west and the Mediterranean islands in the south. There are indications of solar and lunar orientations in some cases, but a thorough comparative study is still needed.

Evidence for a lunar calendar of very early date comes from Bulgaria, where a model furnace was discovered during excavations of a fifth-millennium BC burnt lodging in the village of Slatino. The model furnace is an exact replica of the real thing, a practice repeated on several artefacts from that period, and thought to relate to a cult of home and fireside. On the base of the model, possibly deliberately hidden from general view, is a table which appears to represent a luni-solar calendar. Other finds from this period feature similar calendars. It seems that these cult objects, while linked with religious concepts and rituals, also served a practical purpose.

The earliest recording of astronomical observations is evidenced by sequences of marks on stone and bone objects dating back more than 10,000 years to the hunter-gatherers of Upper Palaeolithic times. Alexander Marshack has asserted that these represent a luni-solar calendrical notation, a conclusion arrived at by counting sequences of marks, undertaking meticulous microscopic examination, and studying the cognitive processes that could have led to their production. While especially intriguing because it promises to reveal the very beginnings of the use of astronomical observations to keep track of periodic time and hence, perhaps, to regulate cultural and ritual activities, this work remains highly controversial.

One of the most challenging aspects of archaeoastronomy as a whole is that it represents an area of enquiry where the physical scientist and the archaeologist or anthropologist have conflicting, but both undeniably relevant, points of view on the nature of admissible evidence, and on the conclusions that can reasonably be drawn from that evidence. Many authors, and particularly those trained in the humanities, would conclude (and have concluded) that any rigorous statistical approach, since it can ultimately only isolate overall trends and necessarily excludes most of the cultural and functional diversity represented in the material record, is of very limited value indeed. Others, however, and particularly those trained in the numerate sciences, would say the same of any investigation which fails, through the *lack* of such an approach, to attempt to distinguish between deliberate design features and chance occurrences. Ironically, both camps end up by arguing that not to accept their viewpoint would be to open the floodgates to unabated speculation. In striving for ways of reconciling these two approaches, archaeoastronomy is in a unique position not only to reveal facts about astronomical practice in preliterate contexts but also to reveal sounder general ways of tackling problems on the very boundary between C. P. Snow's two cultures.

To summarise the results from the north-west of Europe, where by far the most extensive archaeoastronomical investigations have been carried out, lunar and solar symbolism is found in a range of prehistoric monuments, from some of the earliest Neolithic tombs through to the relatively late Bronze Age stone rings and rows. There is no evidence of the use of astronomical observations for practical purposes such as the determination of the time of year. This is not to say that such observations did not take place, which would be counter to extensive evidence from a wide range of simple farming communities; it is simply that such activities would not tend to leave a trace in the archaeological record. What *has* been bequeathed to us is a set of more esoteric – and perhaps ultimately far more interesting – associations between the ceremonial architecture of life and death and the sun and moon, associations that manifest aspects of otherwise largely unfathomable systems of prehistoric ritual and thought.

Bibliography

Archaeoastronomy, Annual supplement to *Journal for the History of Astronomy* (ed. M.A. Hoskin), Science History Publications, Cambridge.

Archaeoastronomy, Journal of the Center for Archaeoastronomy, Maryland.

Aveni, A.F. (ed.) 1989. *World Archaeoastronomy*. Cambridge University Press.

Burl, Aubrey 1993. *From Carnac to Callanish: The Prehistoric Stone Rows and Avenues of Britain, Ireland and Brittany*. New Haven and London: Yale University Press.

Heggie, D.C. (ed.) 1982. *Archaeoastronomy in the Old World*. Cambridge University Press.

Ruggles, C.L.N. 1984. *Megalithic Astronomy*. Oxford: British Archaeological Reports (BAR British Series 123).

Ruggles, C.L.N. (ed.) 1988. *Records in Stone*. Cambridge University Press.

Ruggles, C.L.N. (ed.) 1993. *Archaeoastronomy in the 1990s*. Loughborough: Group D Publications.

Ruggles, C.L.N. and Saunders, N.J. (eds) 1993. *Astronomies and Cultures*. Niwot, Colorado: University Press of Colorado.

Ruggles, C.L.N. in press. *Astronomy in Prehistoric Britain and Ireland*. New Haven and London: Yale University Press.

RONALD A. WELLS

Astronomy in Egypt

Historians often underestimate, while others frequently exaggerate, the capabilities of the ancient Egyptians. The wider truth is that they were a practical people. The development of science and mathematics in Egypt, such as it was, lies rooted in that practicality. More than six millennia ago in the Nile Valley and Delta, man's primal gleanings from the night sky crystallised into a variety of myths that formed the basis of Egyptian religion. Since its principal deities were heavenly bodies, the priesthood mastered the ability to predict the time and place of their gods' appearances. This skill, a rudimentary form of positional astronomy, was responsible for a time unit of 365 days; the division of night and day into twelve segments each; and the formation of a relatively sophisticated lunar calendar, which was used to determine the times of feasts and offerings. Eventually, economic progress forced the simplification of this religious calendar into one that the ordinary Egyptian could more easily manipulate for business transactions. This civil calendar of twelve 30-day months with a special 5-day unit added to bring the total to 365 is the earliest predecessor of our modern Western calendar.

Many of these achievements were already in place before the unification of Upper (Nile Valley) and Lower (Delta) Egypt around 3000 BC, a remarkable feat because of the lack of a well-developed system of writing. However, the spread of writing throughout Egypt after the time of the First Dynastic king Menes (c. 2920 BC, named on contemporary inscriptions as Narmer) certainly had an important effect on the Egyptian calendar because it led to the adoption of the civil version (c. 2800–2900 BC). The introduction of writing and the civil calendar can therefore serve as the milestone in our discussion of the development of astronomy in Egypt. Prior to that, events in the Predynastic (before 3100 BC) and Early Dynastic (Dynasties 1–3, c. 3100–2613 BC) Periods will be sketched. Afterwards, we will touch on developments in the Old (Dynasties 4–8, c. 2613–2125 BC), Middle (Dynasties 11–14, c. 2125–1650 BC), and New Kingdoms (Dynasties 18–20, c. 1550–1070 BC) with a brief discussion concerning the Ptolemaic Period (c. 305–30 BC).

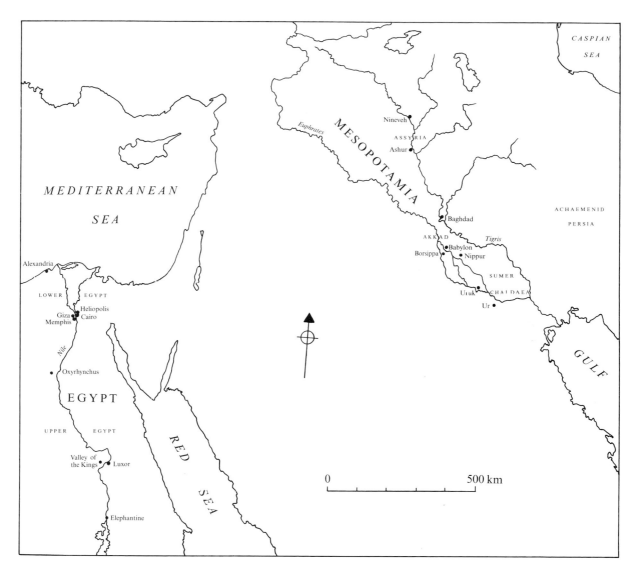

7 Egypt and Mesopotamia.

Predynastic and Early Dynastic Periods

Because the sun god Ra was the pre-eminent god, the annual solar motion along the horizon at sunrise was keenly observed by the early Egyptians who noted its north-ernmost and southernmost turning points, called the solstices. Almost all of Egyptian astronomy and religion are ultimately derived from this simple horizon movement.

One of the greatest of Egyptian legends, the mythology of the sky goddess Nut giving birth to Ra, catalysed both time-keeping and calendar development, endowed the concept of divine royalty, and instituted the matrilineal inheritance of the throne.

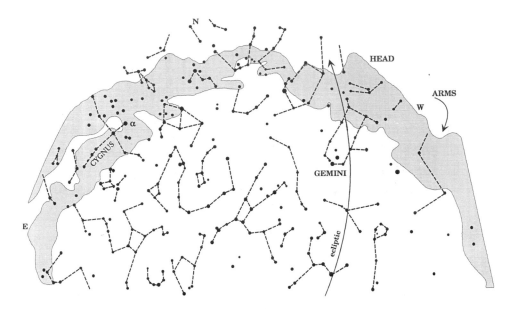

8 Modern representation of the Milky Way for northern latitudes. Shown are the bifurcation at Cygnus (left) forming the legs, with the bright star Deneb (α Cygni) marking the birth canal exit; the star clouds at Gemini (right) forming the head (face downwards); and the distribution of the brighter stars in the constellations along and within the Milky Way, reminiscent of the stars shown along the body of Nut (Col. Pl. II). Note also that the ecliptic passes through the head's mouth. (Adapted from Skymap 4, in D. H. Menzel and J. M. Pasachoff, *A Field Guide to the Stars and Planets*, 2nd edn, Houghton Mifflin, Boston, 1983, p. 48)

Nut was portrayed as a naked female stretched across the sky. The sun is shown entering her mouth, passing through her star speckled body, and emerging from her birth canal. Common throughout Egyptian history, one of the best preserved variants of this panorama from a New Kingdom Ramesside tomb is illustrated in Col. Pl. II. The astronomical basis for the scene is given in the following three diagrams.

Figure 8 illustrates how the faint outer arm of the Milky Way galaxy was perceived as the goddess. The legs are formed by a bifurcation at the cross shape of Cygnus, while the head consists of swelling star clouds in the vicinity of Gemini. Primitive figurines often show the female genitalia as a cross in the appropriate spot. The similar location of Cygnus, with its principal star Deneb marking the birth canal exit, provides the female identification. The path of the sun in the sky passes directly through Gemini, where the concave shape of the Milky Way could be interpreted as the head's mouth. But since the path does not pass near Cygnus, further clarification of the astronomical relationships is needed.

About 1.25 hours after sunset on the spring equinox when the Milky Way is discernible, Fig. 9 indicates that the head of Nut can be seen passing below the horizon face upwards with her mouth open at, or very close to, the position where the sun had

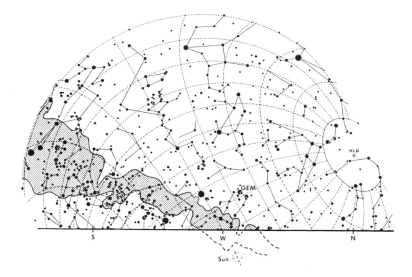

9 The appearance of the western horizon at Cairo at 7:15 pm on the spring equinox (19 April 3500 BC Julian). The solar depression is −16° (position of the sun below the horizon shown dashed), just dark enough for the head of Nut as part of the Milky Way to be seen on the horizon. The left twin of the constellation Gemini is at the eye socket and the right twin is at the mouth of the goddess, located close to where the sun disappeared below the horizon at sunset. (Wells 1992, p. 317, fig. 6)

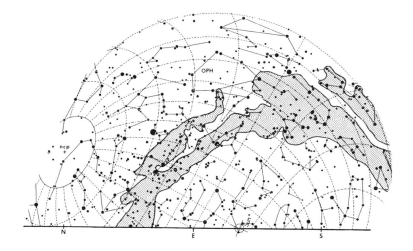

10 The appearance of the eastern horizon at Cairo at sunrise (7:05 am) on the morning of the winter solstice (16 January 3500 BC Julian). Although the stars are no longer visible, note that a line drawn from the north celestial pole (ncp) through Deneb (α Cygni) and intersecting the horizon at the sunrise point forms a great circle. This configuration of the Milky Way showing the lower half of Nut's body began to rise with the left foot shortly after midnight. The constellation of Cygnus was fully risen with Deneb elevated 8° above the horizon by 5:00 am. The solar depression at that time was −26°; at 5:30 it was −19° and at 6:00 am it was −13°. The wisps of the Milky Way would have disappeared after 5:45 am, although Cygnus and Deneb would have remained visible until shortly after 6:00. (Wells 1992, p. 313, fig. 4)

earlier set. A more evocative portrait of a faintly glowing head on the horizon having consumed the sun could hardly have been visualised by these early sky watchers.

The final act in this drama occurs 272 days later on the morning of the winter solstice. Figure 10 shows that the lower half of the goddess, visible above the horizon for only a few hours, has reached a point at sunrise where a great circle drawn from the north celestial pole through Deneb intersects the horizon at exactly the spot where the sun rises. The great circle marks the shortest path that the infant Ra would follow after exiting the birth canal to the point of its appearance on the horizon at sunrise, reminiscent of the ancient Egyptian method of birthing exemplified by the hieroglyph 𓀼. This geometry is valid only on the winter solstice since the sunrise point is further north on other dates. Perhaps the most remarkable aspect of these events, a convincing tie to actual astronomical observations, is that the number of days between the spring equinox and the winter solstice is the period of human gestation!

These correlations explain two related aspects of Egyptian culture, one religious, the other connected with royalty. The Egyptians considered Ra to be a self-creating god. Indeed, the celestial picture shows that Ra enters Nut at sunset on the spring equinox, at which time the goddess presumably conceives. Nine months later, she gives birth to him on the winter solstice. The implied method of conception is oral, but that is not an unusual belief in a primitive society.

Although Ra creates himself, he cannot perform the task without the assistance of the goddess. This crucial point later governed the behaviour of Egyptian kings. The inheritance of the throne in Egypt was matrilineal. The eldest prince had to marry his eldest sister in order to lay legitimate claim to it. It did not always happen in this manner if the appropriate sibling was lacking; but the ancient Egyptians went to remarkable lengths to maintain this relationship as far as possible. This incestuous brother–sister marital relationship of Egyptian rulers, disturbing to many, was clearly the result of the queen acting out the role of the goddess Nut while the pharaoh became the god Ra incarnate. The royal couple passed on their roles to their eldest daughter and son; and in this manner the institution of royalty became divine.

Another legend related to time-keeping, the passage of Ra through the region on the other side of the horizons called the Duat, gave rise to a variety of stories connected with the safety of his voyage. These have been collected together into several so-called 'books', chief among them the *Book of Gates*. They consist mostly of spells that the deceased must correctly recite for a safe transition through the underworld, a later manifestation of Ra's performance to guarantee that his solar barque would make the nightly journey without mishap.

The *Book of Gates* usually consisted of twelve portals through which the solar barque had to pass. Each gate, corresponding to the 12 hours of night, was in the charge of a demonic gatekeeper and several attendants (see panels of Col. Pl. II). Ra had to recite correctly the names of the gate, the gatekeeper and his assistants in a spell before passing on to the next one. Later, the deceased pharaoh had to deliver the recitations to be safely united with his father Ra. And still later, the ordinary Egyptian would have to learn the recitations if he or she expected to reach the Elysian Fields.

Although we do not have the predynastic form of the *Book of Gates*, its source is relatively easy to reconstruct. It would originally have been a primer for telling the time at night by using stars to predict the time of sunrise. The latter was important for making daily offerings to the sun god. A particularly bright star would have been noticed rising a given interval ahead of the sun in sufficient time to permit the necessary preparations before sunrise. A single star would, of course, be an inadequate herald of sunrise because of the daily increasing interval between its appearance and sunrise produced by the 4-minute differential rotation rate of the earth with respect to the sun and stars. In order for this star clock to work for the whole year with 12 equal hours (arbitrarily defined as 60 minutes each in this example), a series of 24 bright stars would be required. Each in turn would serve 15 days as the marker of the hour before sunrise, then 15 days marking the previous hour, etc., until it returned to its duty as the herald of sunrise again a year later.

One rising bright star is difficult to distinguish from another, especially if they all appear near the same horizon point. However, if the bright star were part of three or four other fainter ones that formed a distinctive pattern rising a few minutes before the principal star, then they would forewarn the latter's rising. The early Egyptians would have identified the fainter stars as the attendants, the brighter star as the gatekeeper, and the place of their appearance as the gate itself. The recitation requiring the names of these deities was in the beginning merely a mnemonic device ensuring that each star group would be recognised. Only the particular group just rising would be at the 'horizon' gate marking the current hour. Others would be distributed overhead towards the western horizon while the remainder would still be in the underworld. As each group dominated the horizon gate, the latter would be associated with the stories that characterised the order of the sequence. Later, when the Egyptians wrote down these narratives, they depicted each group separately associated with its own gate. Since twelve star groups came out of the underworld during the night, then the underworld would be illustrated as having twelve gates. Actually, variants of the *Book of Gates* have eight, ten or other numbers of gates. The different numbers reflect the number of patterns which were used in a particular star clock and indicate that the first 'hours' were of unequal length, depending on the sizes of the star configurations.

The astronomical origins of these two principal myths provide insight into the development of the Egyptian calendars. A calendar is measured against a base period whose fundamental unit of time is normally the earth's orbital period. The sacrosanct cycle for the ancient Egyptians was clearly the length of time it took Ra to return each year to his birthplace on the south-eastern horizon at the winter solstice. The accurate determination of this period as 365 days by a simple counting procedure, probably using wooden stakes, cords and shadows, can be approximately dated to sometime around 4500 BC because the mouth area of Nut was aligned more to the west about 1,000 years earlier than seen in Fig. 9.

The determination of when the all-important birth of Ra festival would occur led to a calendar used primarily in Lower Egypt, where the god's chief cult centre was later established at Heliopolis. This calendar was derived from tallies of phases of the moon, the other astronomical body keenly observed by the ancient Egyptians. It usually contained 12 observationally determined 29 or 30-day months, the first beginning after the festival had occurred. Because the calendar averaged only 354 days, whenever the first day of a new year occurred within 11 days after a winter solstice feast, the new year would have to contain an intercalary thirteenth month placed at its beginning to keep the festival always within the last month. Intercalations occurred about once every 2 or 3 years.

A similar calendar arose in the south. However, because of the agrarian dependence of Upper Egypt on the floods of the Nile, its regulating factor was the heliacal rising of Sirius, its first appearance at dawn just prior to sunrise after an absence of about 70 days. Sirius rose heliacally in 4500 BC near the time when the Nile itself began to rise, around the beginning of June. The annual rising of the star, which the Egyptians called *Peret Sepdet*, the 'Going Forth of Sothis', was highly revered because of this association with the onset of the flood. The cult centre from which the influence of Sirius spread over the country was on the island of Elephantine near modern Aswan, the fabled source of the River Nile according to ancient Egyptian tradition. The lunar calendar of Upper Egypt therefore used the festival of *Peret Sepdet* as the principal feast in its last month governing the intercalation scheme.

The lunar calendar from which we have recorded dates, however, was the product of the amalgamation of Upper and Lower Egypt. The relationships between the northern 'lunisolar' and the southern 'lunistellar' calendars made their coalescence simple. The chief objects responsible for the religious festivals regulating the calendars were theologically related (Sirius was equated with Isis, the daughter of Ra), and they rose on their festival days at nearly the same place on the horizon at the beginning of the First Dynasty. The azimuthal difference was only around 2° between midwinter sunrise and *Peret Sepdet*, which occurred 6 months later near the summer solstice.

At some point during the coalescence of Upper and Lower Egypt, the slowly increasing use of the southern calendar as it spread northwards usurped many of the functions of the northern calendar. Although the amalgamated lunar calendar was regulated by Sirius, the name of its last month, almost always eponymously derived from its principal festival, instead retained the old northern name in deference to the 'birth of Ra' (*Mswt Ra*, later Graecised into *Mesore*).

The Egyptian lunar calendar that was ushered in by the pharaonic period was a workable one; but it was also impractical in terms of the economy. A year in which there were sometimes 12, sometimes 13 months, all the beginnings of which had to be determined by observation, was very difficult for the general public to handle. In fact, a special class of priests, the Overseers of the Hours, took charge of time-keeping, regulating the calendar, and determining the beginning of each lunar month. But as the ability to write also spread through the country during the first three dynasties, the necessity of keeping dated records, letters and inventories of stocks and commodities, especially perishable goods, demanded a simpler method of keeping track of months and days.

The desire to have a simpler system that did not require such an effort to regulate led to the development of another calendar. The simplification was based on rounding the numbers of the lunar calendar to 12 months × 30 days for a total of 360 days. A regular pattern of 12 months of even days, none beginning observationally, provided a powerful economic incentive for its adoption. Like the lunar calendar, which was never abandoned but continued to regulate the observance of religious feasts, the new civil calendar was divided into three 4-month seasons corresponding to the periods of growth, harvest and inundation. A special interval of 5 days, known as the 'epagomenal' days, preceded the onset of each year, making its total equivalent to the revered 'cycle' of Ra. The ancient Egyptians called them 'the five days upon the year'.

The Old Kingdom

The Old Kingdom, sometimes called the Age of Pyramids, was the beneficiary of a dual calendar system, a robust economy and religious orders, all essentially derived from observations of the sun, moon and stars.

The Fourth Dynasty pyramids at Giza reflect the astronomical/religious zeal of the Old Kingdom pharaohs. Their shape was dictated by the manner in which clouds and airborne dust scattered sunlight into broad swaths forming stairways to heaven. In fact, they were stone pathways to the gods, the conduit by which the soul of the deceased pharaoh could reach the immortal ones, the northern circumpolar stars. The latter were called *ikhemu-sek*, literally 'the ones not knowing destruction', an indication that the Egyptians recognised that these stars never rise or set. The entrances to these pyramids are all in their north faces, and the corridors are sloped downwards at an angle such that the north circumpolar stars could easily be seen from them. This reverence explains the peculiar diagonal orientation of the three pyramids with respect to each other. They were simply offset away from the river by one full width so that their north faces did not block each other's view of the circumpolar stars and in particular Thuban (α Draconis), the pole star of the era.

Their orientation was also governed by the fact that the Egyptians believed the entrance into the underworld lay due west, the point on the horizon where the sun set into the mouth of Nut on the spring equinox. The pharaoh was expected to pass safely through the underworld before joining the immortal gods.

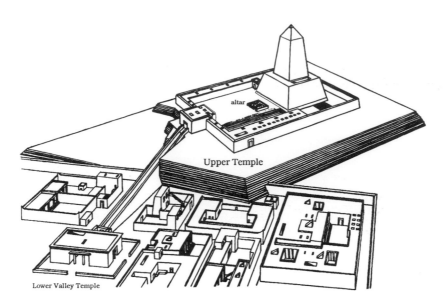

11 The sun temple city of Neuserre (sixth king, Fifth Dynasty, 2445–2421 BC) as reconstructed by L. Borchardt from excavations in the 1890s. The main temple area consists of an obelisk on a podium (representative of the sun's rays), an altar in front for sacrificial animals, an animal slaughter area on the north side, and enclosed walkways leading from the temple gate. A causeway connects the entrance gate of the upper temple with the lower 'valley' temple located in the middle of the priests' town. The latter had an orientation of N46°E towards a series of axial-crossing stars that could have been used to mark the hours at night and especially the pre-dawn hour before sunrise. The basic design, determined by the excavations of H. Ricke and G. Haeny, of Userkaf's sun temple, first in the series of six, was similar except that the lower temple in the priests' town had an orientation of N53°E toward a different series of stars which included Deneb as the brightest. The remaining four have not yet been discovered. (Adapted from L. Borchardt in F. W. von Bissing (ed.), *Das Re-Heiligtum des Königs Ne-Woser-Re. Band I: Der Bau*, Berlin, 1905; Wells 1993, p. 311, fig. 2)

The worship of Ra reached its zenith during the Fifth Dynasty when six of the nine kings of that era built huge temples expressly to honour him. These kings were among the first to adopt the title Son of Ra as part of their name in official recognition of their divinity as set forth by the mythology of Nut. The Palermo stone records that King Userkaf sacrificed two oxen and two geese daily at his sun temple. The temples were important not only for pious reasons but also because administration of the vast temple estates provided for the bulk of the state economy. Moreover, they had a special architectural design, illustrated in Fig. 11, which facilitated the measurement of the night hours for the prediction of sunrise. Two of the six temple complexes have been excavated. The lower so-called 'valley' temples in the priests' towns, located below the main sun temples, were each oriented towards separate series of stars on the north-eastern horizon that could have been used in the manner described for the *Book of Gates* to mark the hours at night. Analysis of them has shown that they marked unequal segments of the night, but that the average of these lengths was close to 60 minutes. In

12 Early surveying instruments. The notched palm rib (a precursor of the theodolite), called *bay* by the Egyptians, was used for surveying with the smaller L-shaped instrument, the *merkhet*, which had a cord attached to a plumb-bob for marking the local vertical. After determination of an axial line, the *bay* could also be used to observe stars crossing it. The *merkhet* could also be used as a shadow-clock to determine day hours. These instruments belonged to an Overseer of the Hour priest of the Twenty-sixth Dynasty named Hor, who, according to an inscription on the *merkhet*, 'know(s) the motion of the two discs [i.e. sun, moon] and every star to its place'. The *bay*, 34 cm in length, but possibly longer in its original form, has an inscription which says, in part, '. . . for indicating the commencement of a feast, and for placing every person in his hour'. The latter may be a reference to priests in charge of particular hours. (Egyptian Museum, Charlottenburg, Berlin, 14084/14085)

addition, Userkaf's temple, the first one of the six built, was associated with a series containing Deneb as its brightest star, appropriately the one from which Ra himself was born (Figs 8 and 10).

The specific orientations of these lower subsidiary temples permitted their roofs to be used as observing platforms to monitor the axial crossing stars used as the hour markers. The Overseers of the Hours would have used an instrument called the *bay*, a palm rib with a notch cut into one end, illustrated in Fig. 12. The same instrument together with another called the *merkhet*, essentially a plumb-bob for determining the local vertical, would also have been used to ascertain the orientation of the building itself and the axial line on the roof. Similar devices were probably used to determine the orientation of pyramids and to maintain it as construction progressed.

The Middle Kingdom

Our principal knowledge of Middle Kingdom astronomy comes from a series of wooden coffins, primarily from the Ninth and Tenth Dynasties, whose lids contain painted scenes on the inside, which are actually tables of rising stars. They consisted of 36 rising stars, each successively marking an 'hour' for an interval of 10 days. These hours were, however, only of 40 minutes ($24/36 \times 60$) duration. Referred to as the 'decans', these stars were located in a belt south of the ecliptic, all having the same invisible interval of 70 days prior to their heliacal rising. This new method of denoting the night hours arose by combining only those stars which behaved like Sirius with 10-day weeks of the civil calendar. Although 18 decans marked the period from sunset to sunrise, 3 were assigned to each interval of twilight leaving 12 to mark the hours of total darkness. The 12-unit division of the night may therefore have originated in this combination of the decanal stars with the civil calendar decades.

Although named, the only one of the decans unambiguously identified today is Sirius (*Sepdet*). The constellation of Orion (*Sah*) is also identifiable though association of names with its individual stars is indeterminate. The coffins also provide the only other certain constellation identification from ancient Egypt, and the only example in which stars are actually drawn in approximately the known configuration. Not part of the decanal system or the star tables, one of the accompanying scenes depicts the seven stars of the Big Dipper (Ursa Major) in the form of the foreleg of an ox and is named *Meskhetiu*.

Scholars usually refer to the tables as 'diagonal star clocks' because a given star appears one line higher marking an earlier hour in adjacent decade columns. Given one of these tables of decanal stars and the appearance of the night sky at any moment, one could tell the time by noting the tabular position of a given star for the particular date. The tables included an extra entry to account for the epagomenal days in the civil calendar. But they did not account for the fact that 365 days do not accurately return the sun to the same star. Every 4 years the tables would have been one day in error. After 120 years, they would be off by a whole month. Evidently, the Egyptians attempted to solve this problem by shifting the star names by the appropriate amounts to reset the clock with the civil calendar. But this procedure appears to have been abandoned by the time of the New Kingdom.

The New Kingdom

New Kingdom astronomy is characterised by a variety of tomb paintings and by the introduction of the water-clock. Two new elements appear in the tomb scenes. One is the arrangement of zodiac-like representations of star groups in the form of animals and deities; the other depicts men facing the viewer and seated in front of grid patterns containing stars and star names to the side.

These 'zodiacs', in contrast to their Ptolemaic counterparts, have defied attempts to decipher them apart from the already known examples of *Sepdet*, *Sah* and *Meskhetiu*. While certain vague relationships may be deduced from them, it is not likely that all or most of the elements will ever be correctly explained unless independent, contemporary texts are found with similar descriptive elements.

The priests seated before star grids in the tombs of Ramesses VI, VII and IX,

13 Outflow type of water-clock used by the Egyptians from the New Kingdom onwards. The decorations on the exterior of the vase refer to the months of the year. The baboon seated over the position of the outlet represents Thoth, god of measurement and time reckoning. Monthly scales on the interior indicate the hour as the water level drops. (Oriental Institute, Chicago, 16875)

however, represent the final stage of telling time with stars used by the Egyptians. The new procedure involved any stars transiting the meridian and several adjacent lines rather than rising stars. It was derived directly from earlier tables of transiting decanal stars found in the cenotaph of Seti I and the tomb of Ramesses IV. Whereas the earlier method used 12 decans in 36 tables that transit only the local meridian measuring equal 'hours' during the night, the final stage exhibited in the Ramesside tombs was not limited to decanal stars, consisted of 13 stars (one marking the beginning of night) in 24 tables, used transits over 3 lines parallel with the meridian on either side in addition to the meridian, and resulted in unequal 'hours'.

Two priests were obviously involved in the observation of the stars, one seated facing north who performs the function of the *bay*, the other seated facing him who could see the stars behind. The viewer of the figure in the tomb scene, of course, was intended to be the spirit of the pharaoh who would then be able to determine the time of night. The inscriptions list particular stars for the beginning and the 12 hours of the night and indicate that they can be seen 'over the left shoulder', 'over the right shoulder', 'over the left ear', 'over the right ear', etc. The vertical lines in the grid denote these same positions before and after the meridian transit. The horizontal grid lines denote the hours of night.

Analysis of the tables has indicated that water-clocks were probably required to regulate the choice and use of these transiting stars. The first direct evidence of water-clock usage in Egypt comes from an inscription of a prince, Amenemhet, under Amenophis I about 1520 BC. The water-clock was shaped like a vase (Fig. 13) having a scale on the inside denoting hours and a hole into which a finely bored plug fitted; the diameter of the bore was no larger than that of a hypodermic needle. The device was filled with water which escaped through this small-bore outlet.

During daylight hours, time was measured by shadow-clocks, the forerunners of medieval sun dials. Although many later examples exist, the device must date back to the earliest periods of telling time. The extant examples all resemble the *merkhet*, which was probably a dual instrument. The plumb-bob would have permitted its quick levelling by hand when facing the sun so that the L-shaped upright arm, or a small cross–beam attached to it, would accurately cast its shadow to the hour mark on the long arm.

A text from the cenotaph of Seti I describes its use, from which it is deduced that it could measure 4 hours before and after noon, that there was an hour after sunrise and another before sunset during which it could not be used (presumably because shadows would be too long, making the instrument unwieldy), and that there was then an hour of twilight before sunrise and again after sunset. These numbers divide the day–period including twilight into 12 hours. Hence, the division of the day into twelve segments is probably at least as old as the division of the night into twelve segments which we traced to the diagonal star clock tables of the Middle Kingdom.

The Ptolemaic Period

The character of Egyptian astronomy changes significantly with the rise of the Greek Ptolemies as the rulers of Egypt (305–30 BC). Both Greek and Babylonian influences are now visible in surviving temples, monuments and papyri. The Graeco-Babylonian zodiac is incorporated into Egyptian stylistic art and appears on temples and monuments. Greek and Demotic papyri contain astrological horoscopes and Babylonian-like omens. However, there is unfortunately nothing in purely Egyptian or even Ptolemaic records remotely resembling Babylonian observations of planets and lunar and solar eclipses, the latter of which have proved so valuable in the determination of the slowing of the rotation rate of the earth.

The legacy of Egypt

Historians of science concede only two items of scientific significance bequeathed to us by the ancient Egyptians: the civil calendar of 365 days used by astronomers even as late as Copernicus in the Middle Ages, and the division of the day and night into 12 hours each. These fundamental contributions may see meagre to many, engineering of the pyramids and surviving temples notwithstanding. But we must also include in the legacy that which they bequeathed themselves, namely why and how they used their concept of astronomy to regulate both their religious and secular lives. It has given us a more realistic insight into a society which lasted more than 3,000 years.

Bibliography

Ancient Egypt, Discovering its Splendors. 1978. Washington, DC: National Geographic Society.

Cottrell, B., Dickson, F. P. and Kamminga, J. 1986. Ancient Egyptian water-clocks: a reappraisal. *Journal of Archaeological Science* 14, 31–50.

Edwards, I.E.S. 1985. *The Pyramids of Egypt*, rev. edn 1993. Harmondsworth: Penguin.

Edwards, I.E.S. 1994. Do the Pyramid Texts suggest an explanation for the abandonment of the subterranean chamber of the Great Pyramid? In *Hommages à Jean Leclant*, eds C. Berger, G. Clerc and N. Grimal, pp. 159–67. *Bibliothèque d'Étude* 106/1, Institut Français d'Archéologie Orientale.

Edwards, I.E.S. 1994. Chephren's place among the Kings of the Fourth Dynasty. In *The Unbroken Reed: Studies in the Culture and Heritage of Ancient Egypt in Honour of A.F. Shore*, eds C. J. Eyre *et al*, pp. 97–105. Egypt Exploration Society Occasional Publications 11, London.

Mengoli, P. 1988. Some considerations of Egyptian star-clocks. *Archiv der Geschichte der Naturwissenschaften* 22/23/24, 1127–50.

Neugebauer, O. 1951. *The Exact Sciences in Antiquity*. 2nd edn 1967; reprinted 1969. New York: Dover Publications.

Neugebauer, O. 1975. *A History of Ancient Mathematical Astronomy*, 3 vols. Berlin, Heidelberg and New York: Springer-Verlag.

Neugebauer, O. and Parker, R.A. 1960. *Egyptian Astronomical Texts*. I. *The Early Decans*. 1964. II. *The Ramesside Star Clocks*. 1969. III. *Decans, Planets, Constellations and Zodiacs*. Providence: Brown University Press.

Parker, R.A. 1950. *The Calendars of Ancient Egypt*. Oriental Institute of Chicago, Studies in Ancient Oriental Civilization, no. 26. University of Chicago Press.

Quirke, S. 1992. *Ancient Egyptian Religion*. London: British Museum Publications.

Wells, R.A. 1985. Sothis and the Satet Temple on Elephantine: a direct connection. *Studien zur Altägyptischen Kultur* 12, 255–302.

Wells, R.A. 1990. The 5th Dynasty Sun Temples at Abu Ghurab as Old Kingdom star clocks: examples of applied Ancient Egyptian astronomy. *Beiheft zu Studien zur Altägyptischen Kultur, Akten des Vierten Internationalen Ägyptologen Kongresses München 1985*, 4, 95–104.

Wells, R.A. 1990. Sothis and the Satet Temple on Elephantine: An Egyptian Stonehenge? *Beiheft zu Studien zur Altägyptischen Kultur, Akten des Vierten Internationalen Ägyptologen Kongresses München 1985*, 4, 107–15.

Wells, R.A. 1992. The mythology of Nut and the birth of Ra. *Studien zur Altägyptischen Kultur* 19, 305–21.

Wells, R.A. 1993. Origin of the Hour and the Gates of the Duat. *Studien zur Altägyptischen Kultur* 20, 305–26.

Wells, R.A. 1994. Re and the Calendars. In *Revolutions in Time: Studies in Egyptian Calendrics*, ed. A.J. Spalinger, pp. 1–37. Van Siclen Press.

JOHN BRITTON AND CHRISTOPHER WALKER

Astronomy and Astrology in Mesopotamia

Apart from a few fragments of lists of the names of stars or constellations and occasional references to eclipses in contemporary documents or later traditions, there is no record of astronomy in Mesopotamia (Babylonia and Assyria) before the eighteenth century BC. The Sumerians, who invented the cuneiform writing system shortly before 3000 BC and thereby laid the foundations of most of what is characteristic of Mesopotamian civilisation, were plainly the first to give names to the constellations, names still familiar to us in some cases – the Bull (Taurus), the Lion (Leo), the Scorpion (Scorpio). Long after the decline of Sumerian culture these names continue in use in their Sumerian form through the rise and fall of Assyria and down to the last days of Babylonian astronomy. But so far as we can tell it was the Babylonians who began regular observation of the moon and the planets.

The earliest Babylonian texts, already couched in an astrological format, forecast the possible consequences of lunar eclipses and record the intervals between successive first and last visibilities of Venus during the reign of King Ammiṣaduqa (1702–1682 BC). The Venus observations have survived, by some historical accident, incorporated in the later Babylonian astrological compendium *Enūma Anu Enlil* ('When the gods Anu and Enlil . . .'), and to them are appended a later scheme for the visibility of Venus (8 months 5 days visibility, 3 months invisibility, 8 months 5 days visibility, 7 days invisibility) which finds a remarkable echo two millennia later in Mayan astronomy (see p. 276).

Whatever impetus may lie behind the growth of astronomy in other parts of the world, there can be no doubt that in Mesopotamia the requirements of astrology were all important. The development of a satisfactory civil or cultic calendar played a limited role. The science of divination in all its forms, whether involving the making of astronomical observations or the inspection of a sheep's liver or the flight of birds or the pattern of oil poured on water, was for the Babylonians and Assyrians the pre-eminent science. As with other branches of learning in Mesopotamia it is characterised by three features: the making of lists, respect for tradition, and extrapolation from observed experience *ad absurdum*. Thus some quirk of tradition, perhaps the result of no more

14 A Late Assyrian diviner's manual concerning the nature of terrestrial and celestial omens. Unusually, the text comments that the validity of an omen depends on the month and the watch of the day or night. About 650 BC. (British Museum, WA K2847)

than an ancient scribal error, results in astrological forecasts being made for lunar eclipses not only on the 14th, 15th or 16th days of the lunar month (on which days lunar eclipses may well occur in a lunar calendar) but also on the 19th, 20th and 21st days of the month, and this for every month of the year. These astronomically impossible forecasts were recopied by one scribe after another for centuries.

The Mesopotamian attitude (and in cultural matters Assyria largely reflects Babylonia) is summarised in a Babylonian diviner's manual of the seventh century BC (Fig. 14): 'The signs on earth just as those in heaven give us signals. Sky and earth both produce portents; though appearing separately, they are not separate (because) sky and earth are related.' (Translation by A.L. Oppenheim, 1974). The portents are not causative. Neither the planets nor a sheep's liver have any influence of themselves, but they forecast future possibilities and give an indication of the intention of the gods. If in the past an eclipse of the moon occurred on the 14th of the month Du'uzu and began and cleared up on its southern quarter, and if this eclipse was followed by the death of a great king, then the next time such an eclipse occurs there is a risk that another great king will die. The risk may be removed or mitigated by the appropriate ritual.

Some of the constellations were referred to as 'gods of the night'. The sun, moon and Venus were certainly gods, but the Babylonians were able to maintain a distinction between a god as such and his or her celestial manifestation. Astral deities form only a part of the amalgam which was Mesopotamian religion. Many gods appear to have had no astral connection whatever. Other gods have an astrological link to a wide variety of stars or constellations and vice versa. Even where the link between a deity and, say, a planet is secure, the Mesopotamian astronomer–astrologers normally used a specific name for the planet rather than the name of the god in recording observations.

The texts available to us strongly suggest that at all periods down to the late fifth century BC (when the first personal horoscopes or nativities appear) astrology and astronomy was the concern solely of the king and his court and, in Southern Meso-potamia, of the temples. Commoners had no apparent concern with the subject. The apodoses of astrological omen texts are primarily concerned with affairs of state, the economy and the person of the king. It was the king who was responsible for the welfare of the country as a whole, for reading the signs and for averting the dangers, and it was the king who was responsible for regulating the calendar. Only from the fifth century BC onwards, perhaps as a result of the Persian kings not being resident in Babylonia, was responsibility for the appointment and maintenance of astrologers transferred to senior temple officials.

At no time was any practical distinction made between astronomers making observations and astrologers making predictions. In the last few centuries BC a specific group, known as 'scribes of *Enūma Anu Enlil*', performed both functions. But earlier, under the Assyrian kings of the eighth–seventh centuries BC, a variety of cultic officials are responsible for the whole range of work involved.

It is from the period of the Late Assyrian empire that we get our best contemporary picture of astronomer–astrologers in action. The Assyrian kings obviously relied heavily on their astrologers. The surviving archive from Nineveh, which covers the period from 674 to 648 BC, contains letters and reports to the king not only from his local experts but also from a variety of other centres in Assyria and Babylonia. These cover the whole range of phenomena in which the Assyrians were interested. Some letters were brief: 'We kept watch on the 29th day; we did not see the moon'. Others are more detailed and offer advice:

> To the king, my lord: your servant Nabu-ahhe-eriba. Good health to the king my lord. Since this is a gloomy day, I did not send the introductory blessing. The eclipse swept from the eastern quadrant and settled over the entire western quadrant of the moon. The planets Jupiter and Venus were present during the eclipse, until it cleared. With the king, my lord, all is well; it is evil for the Westland. Tomorrow I shall send the king, my lord, a full report on this eclipse of the moon.
>
> Translation by S. Parpola, 1993, 75.

Apart from occasional (but partly predictable) phenomena such as eclipses, the principal concerns of the astronomers were with the observation of the moon at the beginning and end of the month (for the purpose of determining the beginning of the month), the heliacal rising and setting of the planets, and the positions of the planets among the constellations. The perceived appearance (colour and brightness) of the planets and stars were remarked on, and frequently also the weather. Such concerns remain characteristic of Mesopotamian astronomy to the end. It is often suggested, by people who have never worked in Iraq, that the Mesopotamian astronomers were assisted by the clarity of their night sky. In practice this is far from true. The phenomena of most importance, first and last visibilities of the moon and planets, take place near to the horizon, and the horizon of Mesopotamia is all too often obscured by dust and rain storms. The Assyrian reports frequently remark that when the new moon was visible it was high in the sky, implying that the new crescent would have been observed the preceding evening but for the weather conditions; in consequence the date of the beginning of the month should be adjusted. Later Babylonian astronomical diaries are also full of reports of weather conditions and dust storms.

The calendar

The Mesopotamian calendar was luni-solar, with the month beginning on the evening when the lunar crescent was first visible after conjunction. It seems probable that if 30 days had passed since the previous first visibility of the lunar crescent a new month was begun even if weather conditions prevented the visibility of the crescent. There is no recorded case of a 31-day month. By the fifth century BC the beginning of the month was determined by computation.

Various month names were used in various areas, but the month names of the calendar of the city of Nippur prevailed in Babylonia from the second millennium BC onwards and also in Assyria in the first millennium BC. The names are in part familiar from the Jewish calendar of the last centuries BC: Nisannu, Ayaru, Simanu, Du'uzu, Abu, Ululu, Tashritu, Arahsamnu, Kislimu, Tebetu, Shabatu, Addaru.

The lunar calendar was kept in line with the solar year by the addition of an intercalary month at irregular intervals. In the third millennium BC a variety of months were intercalated, but from the early second millennium BC onwards only the months Ululu (the sixth month) and Addaru (the twelfth month) were used for intercalation, except in Assyria in the late second millennium BC, where quite different arrangements applied.

The decision to intercalate a month was taken by the king, as is shown by a letter of Hammurabi, king of Babylon 1848–1806 BC:

Tell Sin-iddinam, Hammurabi sends you the following message, 'This year has an additional month. The coming month should be designated as the second month Ululu, and wherever the annual tax had been ordered to be brought in to Babylon on the 24th of the month Tashritu it should now be brought to Babylon on the 24th of the second month Ululu.'

Ideal rules for intercalation are summarised in the diviner's manual already referred to:

> Twelve are the months of the year, 360 are its days. Study the length of the year and look (in tablets) for the timings of the disappearances, the visibilities [and] the first appearances of the stars, (also) the position(?) of the *Iku* star at the beginning of the year, the first appearance of the sun and the moon in the months Addaru and Ululu, the risings and first appearances of the moon as observed each month; watch the 'opposition' of the Pleiades and the moon, and (all) this will give you the (proper) answer, thus establish the months of the year (and) the days of the month, and do perfectly whatever you are doing.
>
> Translation by A.L. Oppenheim, 1974 (Fig. 14)

The rules are spelled out in more detail in the Assyrian astronomical compendium, *MUL.APIN*, 'The Plough', which in its present form may date to the centuries around 1000 BC (Col. Pl. III). Two points must be made. First, the rules use a schematic calendar of 12 months to the year, 30 days to the month, and thus 360 days to the year. This calendar, which is in some sense the ancestor of the later system of zodiac signs and of our division of the circumference of a circle into 360°, harks back to the administrative calendar attested in the Ur III period (twenty-first century BC) and perhaps even in the proto-literate texts of the Uruk III period (*c.* 2900 BC). It will have been understood by the Babylonians that such a calendar did not actually match the luni-solar year. Second, it is not possible to demonstrate that the rules enunciated were ever applied. During the late third millennium BC the variety of calendars in the cities of Sumer and the plethora of data concerning intercalation defy any consistent astronomical interpretation. During the following Old Babylonian period (2000–1600 BC) the frequency with which we find two or even three years in succession having an intercalary month contradicts any idea of a precise astronomical basis for the calendar. While the available data on intercalation for the period 700–500 BC might eventually be amenable to astronomical analysis this work has not yet been done.

It is only from about 500 BC onwards that we begin to see signs of a consistent and predictable cycle of 7 intercalations in 19 years, corresponding to the astronomical realities. The invention of this cycle is traditionally attributed to the Greek astronomer Meton of Athens in the late fifth century BC, but it was almost certainly known in Babylon by the beginning of that century.

Time-keeping

The water-clock (Babylonian *dibdibbu*) is referred to in both astronomical and mathematical texts. In early times its primary function seems to have been to demarcate (probably for civil purposes) the three 'watches' into which the night was divided. The water-clock would have been filled at the beginning of the watch, and the ending of the watch would have been marked by its being empty. Since the length of the night

15 A Late Babylonian table for the water-clock, giving two alternative values for the weight of water which corresponded to a watch of the night at 5-day intervals throughout the year. The tablet is signed by the scribe Nabu-apla-iddin. About 500 BC. (British Museum, WA 29371)

varies throughout the year different amounts of water were needed at different seasons. The existing tables for the water-clock, which again assume an idealised year of 360 days, range from a primitive Old Babylonian scheme which changes the amounts of water only four times a year, through the scheme of the compendium *MUL.APIN* which has intervals of 15 days, to a Late Babylonian table which has intervals of 5 days (Fig. 15). The range of values in the latter text suggest that we are no longer dealing with a water-clock filled once and left to empty, but with a water-clock having a constant head, and thus a constant outflow, which is what would have been needed accurately to time astronomical observations.

Time during the day might have been measured both by the water-clock and by the gnomon or shadow-clock. *MUL.APIN* again has a simple table for the shadow-clock giving the time intervals corresponding to shadows of varying lengths on the two equinoctial and the two solstitial days in the idealised calendar. However, in the already-mentioned Late Babylonian water-clock table the identical comment is written against every 5-day interval, 'One cubit shadow, $1\frac{2}{3}$ double-hours day'. This appears to mean that after $1\frac{2}{3}$ double-hours of daylight the shadow of the gnomon has a length of 1 cubit. Astronomically it is impossible that this statement can be correct throughout the year –

unless the length of the gnomon was systematically varied, for which we have at present no other evidence. It appears that at least one Babylonian scribe had no practical knowledge of or use for the shadow-clock. Two fragmentary copies of a text datable to about the second century BC contain what have been described as 'instructions for the construction of a gnomon-like instrument', but while the apparent complexity of the terminology implies that something more than a simple gnomon is involved the text has defied detailed interpretation.

A number of Late Assyrian observations and of Late Babylonian eclipse reports are timed in relation to the meridian passage of one of a group of stars known as *ziqpu* stars. There are two lists of these stars, identifiable as stars culminating at or near the zenith at Nineveh and in Babylonia respectively. The *ziqpu* stars are also occasionally used in religious contexts to determine the correct time for various rituals. The more common convention, however, in the Late Babylonian reports of eclipses gives time in units of *uš* or $\frac{1}{360}$ of a day measured from or to sunset or sunrise.

Cosmology, constellations and zodiac

There are limited and conflicting accounts of Mesopotamian cosmology. One describes a six-level universe with three heavens and three earths: two heavens above the sky, the heaven of the stars, the earth, the underground waters of the Apsu, and the underworld of the dead. The so-called Epic of Creation (*Enūma ēliš*) recognises a threefold division: heaven, earth and Apsu. Some traditions holds that the earth was created by the god Marduk as a raft floating on the Apsu. The gods were divided into two pantheons, one occupying the heavens and the other in the underworld. The only Babylonian world map depicts the inhabited world as a circle surrounded by waters on the outer edge of which are seven *nagû*, perhaps cosmic mountains. North is indicated by the remark, 'where the sun is not seen'.

It is the compendium *MUL.APIN* which again provides us with the basic framework within which to identify the Mesopotamian constellations. The first section of the text originally listed sixty stars or constellations, distributed across the 'paths of Enlil, Anu and Ea', among which the five planets and six circumpolar stars were later interpolated. The 'paths' of the gods Enlil, Anu and Ea are now understood to correspond to sectors of the eastern horizon over which the constellations rose, the path of Enlil extending north from about $+17°$, of Anu between $+17°$ and $-17°$, and of Ea south from about $-17°$. Within these 'paths' the constellations are listed from east to west. The task of identifying the constellations is helped by the subsequent sections of *MUL.APIN* which specify (in the idealised 360-day calendar) the dates of heliacal rising, simultaneous risings and settings, and time intervals between the dates of heliacal risings of various constellations. Another group of texts, misleadingly described by Assyriologists as 'astrolabes', list constellations in groups, month by month. The *ziqpu* lists are another guide to the approximate location of individual stars.

In Late Babylonian observation texts a group of thirty-one stars in the zodiac belt, known to Assyriologists as 'Normal Stars', are regularly used as reference points for the movements of the moon and the planets. This allows their certain identification in all but a few cases; there remains some uncertainty in the region of Sagittarius.

Details of how the individual constellations were visualised come from three sources. An incomplete text from Ashur describes several of the constellations star by star, for instance:

> The Little Twins, two clothed (humanoid) figures, wearing beards. Two [stars] are drawn [at their heads]. The front figure carries a whip in his right hand, [. . .] holds the whip. The rear figure – [his right hand] holds lightning in front of him, his left hand holds [. . .].

Another Babylonian text recently identified describes lines of stars in the sky. The significance of these lines is disputed (meridian or azimuth lines?), but the text has, for instance, given the first clear indication that the star Sirius, named by both Sumerians and Babylonians as 'the Arrow', is visualised as fired by a humanoid (having a foot and an elbow) from the Bow at Orion (Sumerian, 'the faithful shepherd of Anu'; Babylonian, 'he who is hit'). Graphic representations are limited to a fragmentary Assyrian 'plani-sphere' with schematic drawings of six constellations and two tablets from Seleucid Uruk with drawings of the Pleiades, moon and Taurus in one case, and Jupiter, Hydra, Leo, Mercury and Virgo in the other.

The integration of these various strands of evidence leads to identifications of the Mesopotamian constellations of which we can be fairly confident in the zodiac belt, but which are increasingly speculative as we move north or south. Comparison with Ptolemy's Catalogue (see p. 87), which it must be admitted probably served as an essential background to the task of identification, suggests that the names and delimiting of many of the Greek zodiac constellations are of Mesopotamian origin, although we know nothing of the Mesopotamian picture of Aries ('the Hired Man') and Pisces is split into 'the Swallow' and 'the Fish-tails'.

As to this zodiac belt, MUL.APIN specifies seventeen (sometimes interpreted as eighteen) constellations lying in the path of the moon, and these constellations (not zodiac signs) are used as reference points by the Assyrian astronomers. Progress towards the eventual system of zodiac signs is indicated by a Babylonian text of about the fifth century BC which lists the 12 months (ignoring the intercalary month) and their associated constellations, but assigns both Pleiades and Taurus to month II, both Orion and Gemini to month III and both Pegasus and Pisces to month XII. The final system of twelve zodiac signs of 30° first appears around the middle of the fifth century BC. It is a reference system for the position of the moon and planets, and the fundamental reference system for all of Babylonian mathematical astronomy. This sidereal zodiac appears to have been fixed so that the longitude of the bright star β Gemini was 90°.

Consequently, the equinoxes and solstices occurred at about 10° of their respective signs in 500 BC, an amount which decreases due to precession by 1.4° per century or to roughly 5° by 150 BC.

Babylonian observations in the first millennium BC

The outline of Mesopotamian astronomy given above sets the background to the major developments in Babylonia from the eighth century BC onwards, the beginning of continuous compilation of records of astronomical observations in chronological sequence and the rise of scientific astronomy.

Ptolemy, writing c. AD 150, remarks that the earliest observations available to him come from the reign of the Babylonian king Nabonassar (Nabu-naṣir, 747–734 BC), and indeed he utilises eclipse records from that reign. This is paralleled by a fragment of a Babylonian table of eclipses and eclipse possibilities which has been tentatively recognised as covering the years 747–745 BC. It may be from this point on that the Babylonian astronomers started keeping the series of daily, monthly and annual 'Diaries' of observations part of which was excavated at Babylon (and perhaps Borsippa) in the 1880s and survives today in the British Museum. The series is very incomplete. Only one year from the seventh century is represented (652 BC), one from the sixth (568 BC), four from the fifth, but thirty-five from the fourth, and thereafter a high proportion of the years down to 61 BC.

The Diaries typically contain for each month: a statement of the length of the preceding month; the time interval between sunset and moonset on the first day of the month; time intervals between sun/moonrise/set in the middle of the month; the time interval between moonrise and sunrise on the morning of the moon's last visibility; the dates on which the moon approached the various Normal Stars (see p. 49) and the watch of the night in which this occurred; and the date and description of lunar and solar eclipses. For the planets they record dates and position among the stars of first and last visibility, direct and retrograde motion and stationary points, and conjunctions with Normal Stars. The relation of the moon and planets to the Normal Stars is expressed in terms of 'cubits' (about 2.5°) or 'fingers' (about 5') above (north), below (south), in front (west) and behind (east); but although the general sense of these statements is agreed, comparison with modern calculations suggests that the positions were often estimated rather than measured. The dates of solstices and equinoxes and of the heliacal rising of Sirius are recorded, but analysis has shown that at least during the Seleucid period (311 BC onwards) these are calculated, not observed. Weather conditions are regularly reported, and occasional astronomical phenomena such as meteors and comets (including Halley's Comet in 164 and 87 BC).

In keeping with the fundamental astrological purpose of this activity, and the aim of forecasting the fortunes of the country as well as of the king, the monthly entries give the positions of the planets within the zodiac signs (not constellations), details of the prices of six basic commodities and of the height of the River Euphrates, and a record of interesting historical events and miscellaneous curious phenomena.

The Diaries were used to compile a number of chronologically arranged tables of planetary and lunar phenomena. The tables of eclipses are of particular interest. The 18-year cycle of lunar eclipses and the succession of eclipse possibilities at 6 or occasionally 5-month intervals was probably already known to the Assyrians. Fragments of extensive tables of eclipses giving regnal year and month give an immediate overview of Babylonian chronology from 747 BC onwards, and it may be this source material which enabled Ptolemy to compile a chronology of the Babylonian, Persian and Seleucid kings.

From at least 236 BC onwards the Diaries were used to compile another group of texts now known as Goal Year Texts. These exploit the known period relations between repeating phenomena of the moon and the five planets: 71 years and 83 years for Jupiter, 8 years for Venus, 46 years for Mercury, 59 years for Saturn, 79 years and 47 years for Mars, and 18 years for the moon. This astrologically based order of the planets was followed in all non-mathematical astronomical texts of the Seleucid period. By looking at the Diary for the appropriate year in the past the scribe could compile a list of the phenomena to be expected in the year ahead (the Goal Year). The surviving series of Goal Year texts ends in 24 BC, with one possible later example from AD 41.

The Goal Year texts may have been at least in part the source from which were compiled two other groups of texts, Almanacs and Normal Star Almanacs. The former include month by month the characteristic lunar and planetary phenomena, planetary positions at the beginning of the month and dates of their entry into the next zodiacal sign, solstices and equinoxes, Sirius phenomena and eclipses. The Normal Star Almanacs add to this information the dates on which the planets pass by, above or below, the Normal Stars. The surviving Almanacs and Normal Star Almanacs cover the periods 220 BC–AD 75 and 281–100 BC respectively. The two groups of texts have been very little studied, but they appear to reflect procedures and computations quite different from those associated with the mathematical ephemerides and related texts.

The mathematical ephemerides represent Babylonia's principal contribution to the science of astronomy, which we shall now describe in detail. They use Babylonian sexagesimal numbers and place-value notation whereby $1, 1, 1; 1, 1 = 1 \times 60^2 + 1 \times 60^1 + 1 \times 60^0 + 1 \times 60^{-1} + 1 \times 60^{-2}$. (Here the punctuation is a modern notation, not used in the original texts.) They are principally published in Neugebauer's *Astronomical Cuneiform Texts*, hereafter abbreviated as ACT (see Bibliography).

Babylonian scientific astronomy

Around 500 BC Babylonian astronomy began a process of transformation which led to the development of radically new techniques for predicting celestial phenomena. These techniques were mathematical in nature, rational in approach, and entailed separating complex phenomena into components which could be described by mathematical functions which could then be combined to predict the phenomena in question. Some of these functions are very elegant, and the process of their discovery and integration into consistent arithmetical 'theories' represents the first true scientific revolution.

The objects of the new astronomy were largely traditional and comprised determining the circumstances of celestial events which were important to cultic practice. For the moon this meant determining the dates of the syzygies (new and full moon), the times and circumstances of eclipses, the lunar visibilities near syzygies, and the lengths of the months. For the planets it meant determining the dates and places of the planetary phases – appearances and disappearances, stations and (for the outer planets) oppositions. As we shall see, planetary theory was less ambitious than lunar theory, which was the outstanding triumph of Babylonian astronomy. Nevertheless, both lunar and planetary theories largely succeeded in solving the problems they addressed, the theory of Venus being a conspicuous exception.

Mathematical tools

PERIOD RELATIONS   The mathematical models which make up Babylonian scientific astronomy rest on a small number of relatively simple concepts and arithmetical methods. The most fundamental of these is the concept of 'period relation' in which Π events of one kind are assumed to correspond exactly to Z events of another kind, where Π and Z are relatively prime integers.

A simple example of this is the 19-year cycle, which was almost certainly known in Mesopotamia before 500 BC (and which is still used to establish the dates of Easter and Passover). This assumes that

(1)   235 synodic months = 19 years,

which is a short way of saying that 235 revolutions of the moon in synodic elongation correspond precisely to 19 zodiacal revolutions by the sun. From this we can compute that

(2a) $$1 \text{ year} = \frac{235}{19} = 12;22,6,18\ldots \text{ months, and also that}$$

(2b) $$\overline{\varDelta\lambda}_{\text{sun}} = 360° \times \frac{19}{235} = 29;6,22,58\ldots \text{ degrees/month,}$$

where $\overline{\varDelta\lambda}_{\text{sun}}$ is the average monthly progress of the sun in sidereal longitude. This period relation is in fact quite good, for on average the sun and moon return to within $-0;11°$ of their initial positions after 235 months. Furthermore, this fact is directly visible, since lunar eclipses frequently recur at 235-month intervals.

Another famous early period relation for the moon is the so-called Saros, which holds that in 223 months the moon returns to its same position relative to the nodes, having passed by one node or the other 38 times, while in the same interval it also completes 239 cycles of lunar velocity or anomalistic months. This implies that

(3) $\qquad$ 223 synodic months $= 38$ eclipse possibilities

$\qquad\qquad\qquad\qquad\qquad = 239$ anomalistic months,

or that the average interval between eclipse possibilities, also known as the eclipse period (EP), is equal to

(4a) $\qquad$ $1 \text{ EP} = \dfrac{223}{38} = 5; 52, 6, 18 \ldots$ months, and also that

(4b) $\qquad$ 1 anomalistic month $= \dfrac{223}{239} = 0; 55, 58, 59 \ldots$ synodic months.

The Saros is an unusual example, because it links three independent phenomena. These were sometimes augmented by the further relationships, mentioned by Ptolemy (but note his $10; 40°$ for $10; 30°$):

(5) $\qquad$ 223 synodic months $= 18$ revolutions of the sun $+ 10; 30°$

$\qquad\qquad\qquad\qquad\qquad = 6,585\tfrac{1}{3}$ days,

which leads to the following approximate parameters:

(6a) $\qquad$ 1 mean synodic month $= \dfrac{6,585; 20}{223} = 29; 31, 50, 18 \ldots$ days,

(6b) $\qquad$ 1 sidereal year $= \dfrac{6,585; 20}{18 + (10; 30/360)} = 365; 15, 35 \ldots$ days,

(6c) $\qquad$ $\overline{\varDelta\lambda}_{\text{sun}} = \dfrac{(18 \times 360°) + 10; 30°}{223} = 29; 6, 19, 22 \ldots$ degrees/month.

These parameters are rougher than those derived from the 19-year cycle, but nevertheless reflect the order of magnitude of each phenomenon.

In the later mathematical lunar theories, these relationships were superseded by more accurate approximations. Thus in place of the 19-year cycle, the System A lunar theory used the slightly more accurate ($\delta\lambda = + 0; 4.5°$ in 235 months) period relation

(7) $\qquad$ 2,783 months $= 225$ years; 1 year $= 12; 22, 8$ months,

while the Saros was replaced by separate and more accurate period relations for its individual components, to wit:

(8a)   1,655 months $= 282$ EP; 1 EP $= 5; 52, 7, 39 \ldots$ months (System A);

(8b)   2,729 months $= 465$ EP; 1 EP $= 5; 52, 7, 44, 30$ months (System B),

and

(9a) $\qquad$ 251 synodic months $= 269$ anomalistic months, whence

(9b) $\qquad$ 1 anomalistic month $= 0; 55, 59, 6, 28 \ldots$ synodic months.

By comparison the corresponding values calculated from modern theory are:

(10a)   1 sidereal year $= 12; 22, 7, 29 \ldots$ months,

(10b)   1 EP $= 5; 52, 7, 44, 45 \ldots$ months, and

(10c)   1 anomalistic month $= 0; 55, 59, 6, 41 \ldots$ months,

showing that each of these period relations is in fact quite good, and also significantly better than its earlier counterpart.

For the planets, period relations take the form: Π synodic phenomena (of a given sort) correspond to Z revolutions of the zodiac, which implies a mean synodic arc $(\overline{\Delta\lambda})$ equal to $360° \times Z/\Pi$. Here again we find relatively short-period approximations reflected in Goal Year Texts, as well as more accurate relationships covering longer intervals in the later mathematical theories. These are collected in Table 1, where they are designated 'GYT' and 'ACT' respectively. Here 'Error', calculated from modern theory, reflects the mean change in sidereal longitude over Π synodic phenomena, while the 'Error/year' shows the quality of the period relation.

Table 1 Planetary period relations

Planet/ source	Π Syn. phen.	Z Revs	Y Years	$\Delta\lambda$	Error $\Pi\,(\Delta\lambda)$	Error/ year
Saturn	28	1	29	$12; 51 \ldots°$	$-5; 46°$	$0; 12°$
GYT	57	2	59	$12; 37, 53 \ldots°$	$+1; 6°$	$0; 1°$
ACT	256	9	265	$12; 39, 22, 30°$	$-1; 21°$	$0; 0, 18°$
Jupiter	11	1	12	$33; 43 \ldots°$	$+4; 35°$	$0; 23°$
GYT1	65	6	71	$33; 13, 50 \ldots°$	$-5; 40°$	$0; 5°$
GYT2	76	7	83	$33; 9, 38 \ldots°$	$-1; 5°$	$0; 0, 46°$
ACT	391	36	427	$33; 8, 44 \ldots°$	$-0; 50°$	$0; 0, 7°$
Mars	15	2	32	$48; 0°$	$+10; 41°$	$0; 20°$
GYT1	22	3	47	$49; 5 \ldots°$	$-8; 19°$	$0; 11°$
GYT2	37	5	79	$48; 39 \ldots°$	$+2; 22°$	$0; 2°$
ACT	133	18	284	$48; 43, 18 \ldots°$	$-1; 12°$	$0; 0, 15°$
Venus						
GYT	5	3	8	$216; 0°$	$-2; 25°$	$0; 18°$
ACT	720	431	1151	$215; 30°$	$+12; 20°$	$0; 0, 38°$
Mercury	104	33	33	$114; 13 \ldots°$	$-2; 10°$	$0; 4°$
GYT	145	46	46	$114; 12, 24 \ldots°$	$+0; 26°$	$0; 0, 34°$
ACT	1513	480	480	$114; 12, 36 \ldots°$	$-0; 28°$	$0; 0, 3.5°$

Excepting Venus, each of the planets' ACT period relations is simply a combination of a small number of shorter periods with offsetting errors. Thus for

Saturn   265 years $= 4 \times 59$ years $+ 1 \times 29$ years,

Jupiter   427 years $= 5 \times 83$ years $+ 1 \times 12$ years,

Mars   284 years $= 3 \times 79$ years $+ 2 \times 47$ years, and

Mercury   480 years $= 9 \times 46$ years $+ 2 \times 33$ years.

For Venus the ACT period relation derives from the assumption that

(11a)   5 syn. phen. $= 8 \times 360° - 2; 30°,$

which accumulates to 1 revolution in 144 '8-year' cycles. This makes

(11b) 144×5 syn. phen. $= 144 \times 3 - 1$ revolution $= 144 \times 8 - 1$ years,

which is the ACT period relation. In all cases the ACT period relations are an improvement over the Goal Year periods. (The relatively poor alternative Goal Year period relations for Jupiter and Mars (GYT1) were used for visibility phenomena, despite their inaccuracies, because in these cases Π synodic phenomena also correspond to an integral number of months.)

It should be noted that none of these composite period relations depends on estimating the overage or shortfall in a Goal Year period to a precision greater than half a degree, which is also consistent with the precision we find reflected in the observational records. Indeed, only in the case of the sun, where the System A period assumes a shortfall in 235 months of a quarter of a degree (i.e. half a lunar diameter) do we find evidence of greater precision. It cannot be sufficiently emphasised that *precise* measurements played no role whatsoever in the development of these parameters, or for that matter in Babylonian astronomy generally.

LINEAR ZIGZAG FUNCTIONS   Period relations define the average relationship between the phenomena involved, and thus implicitly the associated mean motions. To describe variations from average phenomena, Babylonian astronomers relied on two fundamental techniques. The simplest of these is what we call a 'linear zigzag function', where a quantity increases or decreases by a constant amount (d) in successive intervals, reflecting off some fixed maximum (M) and minimum (m) as shown in Fig. 16. If the phenomenon described returns to its initial value after Π occurrences, then it will be true that

(12) $$\frac{2(M-m)}{d} = \frac{2\Delta}{d} = \frac{\Pi}{Z} = P,$$

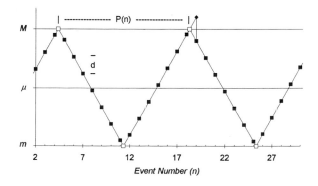

16 Linear zigzag function.

where Z is an integer which counts the number of waves (peak to peak or trough to trough) in Π occurrences, and P is called the period of the function.

The earliest known examples of linear zigzag functions are found in Tablet XIV of the astrological series *Enūma Anu Enlil*, which was compiled towards the end of the second millennium BC, but probably dates from Old Babylonian times. Here we find two functions, one of which describes the amount of time the moon is visible at night over the course of a month, and another which describes the variation of the *increment* of the first function over the course of a year. Both functions give values in time-degrees (1 time-degree $= \frac{1}{360}$ of a day $= 4$ minutes), reflecting an early introduction of this unit, and both use proper sexagesimal place-value notation to express these values.

The first function (Fig. 17a) reflects an idealised month of 30 days, a maximum and amplitude of 3,0° and an increment of 12°. The function repeats after one wave (i.e. $Z = 1$), so its period is simply $6,0/12 = 30$ (days). In a seemingly baroque gloss on the underlying function, the scribe has made the function progress geometrically for the first and last 5 days of the month – presumably to reflect the less than complete visibility of the new and old moon.

The second function (Fig. 17b) reflects a schematic calendar of twelve 30-day months, is tabulated at semi-monthly intervals, and varies between 16° and 8° at the solstices (winter and summer respectively) with a semi-monthly increment of 0; 40° and a mean value of 12° at the equinoxes. Like the first it contains but a single wave, and its extrema reflect an assumed ratio of 2 : 1 for the ratio of the longest to the shortest night.

These simple functions reflect the extreme age of the arithmetical techniques underlying linear zigzag functions. They also comprise the earliest example we possess of the separation of a complex phenomenon into components describable by simple arithmetic functions, each with its own independent variable.

A later and more typical function is Column G of the System B lunar theory,

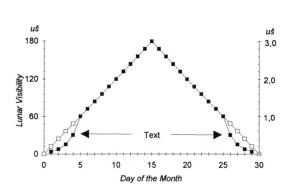

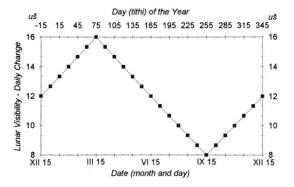

17a *Enūma Anu Enlil*, Tablet XIV. Lunar visibility during the month.

17b *Enūma Anu Enlil*, Tablet XIV. Daily change in lunar visibility during the year.

which describes the variation in the length of the month due to the variation in lunar velocity. The function is tabulated at monthly intervals and reflects the parameters

$$(13) \qquad M = 4, 29; 27, 5 \quad m = 1, 52; 34, 35 \quad \text{and} \quad d = 22; 30,$$

which result in an amplitude $\varDelta = M - m = 2, 36; 52, 30$ and a mean value $\mu = 3, 11; 0, 50$. Once again the units are time-degrees, and the function tabulates the excess over 29 days of the length of successive months. Converted to days the mean value becomes the more familiar

$$(14) \qquad \mu - 1 \text{ mean synodic month} = 29; 31, 50, 8, 20 \text{ days},$$

a parameter which was retained unchanged by Ptolemy (and in fact continued to appear in medieval tables), and which differs only in its last place from the value derived from modern theory for -500: $29; 31, 51, 8, 39.4$ days.

The period of this function, $2\varDelta/d$, reduces to $P = 251/18 = 13; 56,40$, which seems unfamiliar at first glance. However, if we assume that the tabular interval (one month) is slightly more than the period of the underlying phenomenon, then $\varPi$ tabulated occurrences will represent $\varPi + Z$ rather than Z cycles of the underlying phenomenon. The 'true' period thus becomes

$$(15) \qquad p = \frac{\varPi}{\varPi + Z} = \frac{251}{269} = 0; 55, 59, 6, 28 \text{ months},$$

which reflects the (excellent) period relation, noted above, that

$$(16) \qquad 251 \text{ synodic months} = 269 \text{ anomalistic months.}$$

Similar functions, differing only in parameters, comprise the majority of linear zigzag functions encountered in Babylonian astronomy. However, variations also exist, including truncated functions, and at least one instance of a function composed of a summation of the differences between two truncated functions separated in phase. A common element of nearly all these functions is that the independent variable is an integer number of events, rather than an argument such as longitude, or some invariable unit of time. As a result, the argument itself often includes the variation depicted by the function as well other variations. This is illustrated in the preceding example where the function G is tabulated for successive months, which include the variation it describes.

STEP FUNCTIONS   The second and distinctively Babylonian technique for describing periodic variations is based on fixed zones of constant amplitude which vary discontinuously at their boundaries, giving rise to the name 'step functions'. (Strictly speaking, the step function only underlies the more complex resulting function, but it is convenient, if imprecise, to apply the term to both functions.) Here the variable quantity – typically a synodic arc – is expressed strictly as a function of longitude by a procedure which rests on the following facts:

- Any phenomenon, which varies solely as a function of longitude, and which recurs after Π events and Z revolutions of the zodiac (Π and Z being relatively prime), will occur at Π distinct points in the zodiac.

- The average interval between these points will thus be $\bar{I} = 360°/\Pi$, and Z such intervals will equal the mean synodic arc, $\overline{\Delta\lambda} = 360° \, (Z/\Pi) = Z\bar{I}$

- If $\bar{I}$ is now replaced by intervals of different length, then Z such intervals will also vary in proportion to the lengths of the component intervals, without, however, affecting the uniqueness of each point or the periodicity of the scheme.

The Babylonian procedure was to divide the zodiac into a small number of zones (two, three and six are attested) comprising intervals of constant, but different lengths, and to make $\Delta\lambda$ always equal to Z such intervals. Wherever Z intervals fell within a single zone, the resulting synodic arc would be constant as a function of longitude and equal to $w_i = Z I_i$; elsewhere it would vary linearly in proportion to the number of intervals from each zone.

Since each zone comprised an integral number of intervals, the constraints on the function were simply that

$$(17) \qquad\qquad \Sigma v_i I_i = \Sigma a_i = 360° \quad \text{and} \quad \Sigma v_i = \Pi,$$

where a_i is the length of zone (i) comprising v_i intervals of length I_i. This gave the schemes' designer(s) great flexibility, since zones could be divided or added as needed, while even the most constraining two-zone scheme allowed a wide latitude for choice of velocities or their corresponding intervals. The attested schemes also evidence a partiality for 'nice' numbers, and the capacity to indulge such aesthetic constraints underscores the power of the methodology.

A simple example of the construction of a step function is the System A scheme for the longitude of the moon at syzygy, which reflects the zodiacal anomaly in the velocity of the sun. Here $\Pi = 2,783$, $Z = 225$ whence $P = 1$ year $= 12; 22, 8$ months as in (7). The model has two zones of velocity, of which the faster is centred on *Sagittarius* 20° and the slower on *Gemini* 20°. One velocity is clearly chosen for its convenient value, $w_1 = 30; 0$ degrees/month, which corresponds to 'nice' intervals of length $I_1 = 0; 8°$. This leaves the other velocity to be determined, and if the two zones are to be approximately equal in length, satisfy the constraints in (17), and have intervals of convenient length, there are but two choices:

$$w_2 = 28; 7, 30° = \frac{15}{16} w_1, \quad I_2 = 0; 7, 30°, \quad a_2 = 166°, \text{ and}$$

$$w_2 = 28° = \frac{14}{15} w_1, \quad I_2 = 0; 7, 28°, \quad a_2 = 154; 56°.$$

Of these the first seems intuitively preferable, reflecting more nearly equal zones, computationally 'nicer' intervals, and ratios of the velocities as $n : (n + 1)$, where both n and $(n + 1)$ are regular integers. It was also the one chosen.

Schemes of this sort were used to describe the synodic arcs of each of the planets (the scheme for Venus being a still unpublished discovery). While they may appear more primitive than linear zigzag functions, such schemes are in fact capable of considerably greater sophistication, as illustrated in the following example for the synodic arc of Mars.

In this scheme, the period relation is given by $\Pi = 133$ synodic phenomena corresponding to $Z = 18$ revolutions of the zodiac, which is divided into six zones of $60°$, each with a different characteristic velocity. These are described in Table 2, whose last column also illustrates a characteristic of all such schemes, namely that, as in the solar model, the ratios of the velocities are all of the form $p : q$, where both p and q are small regular integers. Indeed, one remarkable aspect of this scheme for Mars is that, if one assigns velocities of $30°$, $60°$ and $90°$ to three of the zones (an obvious choice given the length of the maximal synodic arcs), this scheme is the only one for which the ratios of the velocities meet this condition.

Table 2 Scheme for Mars' synodic arc ($\Delta\lambda$)

Zone	Longitude	v_i	I_i	w_i	$w_i : w_4$
1	$30°$–$90°$	24	$2; 30°$	$45°$	$3 : 4$
2	$90°$–$150°$	36	$1; 40°$	$30°$	$1 : 2$
3	$150°$–$210°$	27	$2; 13, 20°$	$40°$	$2 : 3$
4	$210°$–$270°$	18	$3; 20$	$60°$	$1 : 1$
5	$270°$–$330°$	12	$5; 0°$	$90°$	$3 : 2$
6	$330°$–$30°$	16	$3; 45°$	$67; 30°$	$9 : 8$

The power of this technique is illustrated in Fig. 18, which shows the synodic arcs derived from the foregoing scheme as a function of longitude. Also shown are the

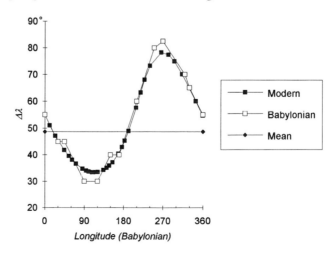

18 Synodic arc for Mars (c. −450).

synodic arcs exhibited by successive first stations of Mars from -500 to -436, adjusted for precession and the norm of the Babylonian zodiac. The agreement between the two is extraordinary, given the amplitude and asymmetry of the variation of the synodic arc, which in reality ranges from roughly $33°$ to $79°$. In comparison, the Mars scheme yields synodic arcs which range from $30°$ to $82; 30°$, and which remain in virtually identical phase with the actual phenomenon throughout the zodiac. In contrast, the linear zigzag scheme for Mars has an accurate maximum ($M = 80; 7, \ldots °$) and mean value but is constrained by its structure to reflect a minimum ($m = 17; 19, \ldots °$), which is far too low, as well as a symmetry which belies the reality.

Theories

It is convenient to speak of the assemblage of functions which describe the behaviour of a particular celestial body as a 'theory', even when our knowledge of the underlying assumptions is limited and indirect. So defined, Babylonian astronomical theories divide into two broad categories – lunar and planetary – which are distinguished by both their objectives and their procedures. Within each, two further classes are distinguished, depending on whether the synodic arc is described by a step function with longitude as its argument, which we call System A, or by a linear zigzag function with event number as its argument, which we term System B.

The theories are expressed in three classes of texts: ephemerides, which tabulate the functions necessary to compute the phenomenon in question for successive phenomena; procedure texts, which describe in compressed fashion the procedures for calculating each function; and auxiliary texts, which tabulate functions which relate to but do not figure directly in the computation of ephemerides. Generally, it is the ephemerides which comprise the clearest expression of each theory.

LUNAR THEORIES   The first goal of Babylonian lunar theory was to calculate the time intervals between syzygies of the same type. This interval is affected by the variable velocities of both the moon and sun, with the latter primarily affecting the distance in longitude between syzygies. One of the great triumphs of Babylonian astronomy was the correct separation of these two effects.

In addition, the calculation of the intervals of visibility near syzygy and the detailed circumstances of eclipses required taking proper account of variations in the length of daylight, in the moon's latitude (or nodal elongation), and in the angle between the horizon and the ecliptic. Thus a successful lunar theory needed to portray satisfactorily the separate effects of five different variables, of which one depended on lunar anomaly, one on nodal elongation, and three on zodiacal position at syzygy.

A typical lunar ephemeris covers a single year and consists of a distinctively wide tablet with thirteen to fourteen rows on a side, one for each month, containing twelve to eighteen columns of functions, calculated for new moons on the obverse and full

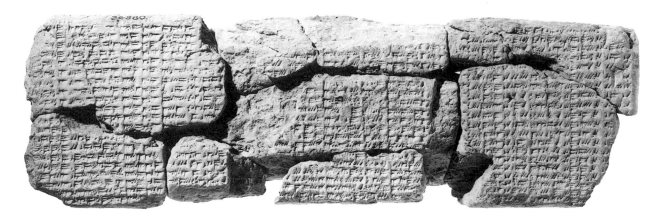

19 An ephemeris in 18 columns for new moons for the years 208–210 of the Seleucid Era (104–102 BC). On the basis of this ephemeris the Babylonian System B was first analysed by F. X. Kugler in 1900. (British Museum, WA 34580)

moons on the reverse (Fig. 19). The functions comprising the two theories – shown in Table 3 together with their letter designations in ACT – are broadly similar, despite individual differences which extend to the details of every function (other than the date). Both systems begin by calculating the arguments and related quantities used in the final calculations, which appear in the same order: date, argument of lunar anomaly (System A only), argument of zodiacal anomaly (longitude at syzygy), length of daylight, lunar latitude and lunar velocity.

After seven or eight columns of arguments both systems proceed to compute the length of the month, beginning with a provisional month-length which reflects only the variation due to lunar anomaly. This is followed by corrections for solar anomaly and (in System A) epoch, which are combined in Column K to arrive at the effective 'month length' from syzygy to syzygy measured in time-degrees from evening epoch in System A and from midnight epoch in System B. Interestingly, all known examples of functions from System A are connectable up to this point in the theory, reflecting a single continuous and consistent ephemeris, whose attested entries extend from −475 to +42. This contrasts with System B whose separate ephemerides frequently contain functions which are not connectable.

From this point on the theories treat the interrelated problems of calculating the visibilities and the dates of the syzygies. This part of the theory is still incompletely understood, due partly to the paucity of preserved material and to the fact that System A omits several functions (N–R) necessary for calculating the visibilities, which System B reflects.

Table 3 Lunar theory: contents

Column		System A	System B
Arguments and related functions			
1	(T)	Date	Date
2	(Φ, A)	Argument of lunar anomaly	Solar velocity (ΔB)
3	(B)	Longitude of syzygy	Longitude of syzygy
4	(C)	Length of daylight	Length of daylight
4b	(D)	–	Half length of night
5	(E)	Lunar latitude	Lunar latitude
6	(Ψ)	Eclipse magnitude	Eclipse magnitude
7	(F)	Lunar velocity	Lunar velocity
Calculation of month lengths (syzygy to syzygy)			
8	(G)	Month length due to lunar anomaly	Month length due to lunar anomaly
9a	(H)	–	Difference in $[J](9)$
9	(J)	Correction for zodiacal anomaly	Correction for zodiacal anomaly
10	(C')	Correction for epoch	–
11	(K)	Month length	Month length
Calculation of dates and visibilities			
12a	(L)	–	Date of syzygy, midnight epoch
12	(M)	Date of syzygy, evening epoch	Date of syzygy, morning or evening epoch
13a	(N)	–	Time: syzygy to sunset/rise
13b	(O)	–	Elongation at visibility
13c	(Q)	–	Effect of horizon angle
13d	(R)	–	Effect of latitude
13	(P)	Duration of visibility	Duration of visibility

A detailed description of the functions comprising these theories is beyond our present scope. Nevertheless, a few of the differences bear mention. In System A, the longitude of syzygy is derived from a step function and thus is strictly a function of longitude; in System B this longitude results from a summation of Column A, which depicts the monthly change in longitude by a linear zigzag function whose argument is the number of months from some starting point. While the practical differences are small, the argument of zodiacal anomaly is less rigorously constructed in System B than in System A.

One of the critical columns in both systems is Column G, which gives the provisional length of the month reflecting only the variation due to lunar anomaly. In System A this function is derived in an ingenious fashion from the argument of anomaly, Column Φ, and accurately depicts a complex variation which is strictly dependent only on lunar anomaly. At the same time its mean value is ambiguous and clearly anticipates further systematic correction. In contrast, the corresponding function in System B (equation 13 above) varies linearly, and thus rather more crudely, about an exceedingly accurate mean value.

These differences are broadly characteristic of the two systems. In general System A is tighter, more rigorously founded, and more internally consistent than System B, while System B reflects more accurate parameters. This is especially true of its most important central parameters: the length of the mean synodic month measured in days, and the lengths of the eclipse period and anomalistic month measured in months. Each of these is virtually as accurate as we can compute them from today's theories, whereas the corresponding parameters in System A are either ambiguous as in the case of the mean synodic month, or slightly less accurate. An exception to this generalisation is the length of the anomalistic month, which equals

(18)
$$27; 33, 16, 30 \text{ days in System A,}$$
$$27; 33, 16, 26, \ldots \text{ days in System B, and}$$
$$27; 33, 16, 30, 34.5 \text{ days by modern computation.}$$

One other curious feature of the two lunar theories is their virtually complete lack of common elements at every level of detail. Except for the canonical assumption that the ratio of longest to shortest day is as $3 : 2$, and a cameo appearance of $12; 22, 8$ months (the System A period for the year) as an implicit and unimportant parameter in Column H of System B, the two theories are relentlessly different in terms of rigour and consistency, the functions they employ, and the parameters these are based on. It is as if two competitors were assigned the same problem, but precluded from using any element of the other's solution.

PLANETARY THEORIES   The goals of Babylonian planetary theory were less ambitious than those of lunar theory and consisted of predicting the longitudes and dates of the principal synodic phenomena – also known as the synodic phases – for each planet. For outer planets these phenomena were (using the Greek letter designations from ACT)

Γ first appearance after invisibility [igi]
Φ first station [uš *mahritu*(igi-*tú*)]
Θ opposition (acronychal rising) [e(-me)]
Ψ second station [uš *arkitu*(*ár-tú*)]
Ω disappearance [šú].

For inner planets they included

Γ first appearance in the east [igi *šá* kur]
Φ station in the east [uš *šá* kur]
Σ disappearance in the east [šú *šá* kur]
Ξ first appearance in the west [igi *šá* šú]
Ψ station in the west [uš *šá* šú]
Ω disappearance in the west [šú *šá* šú],

although ephemerides for Mercury omit the (barely visible) stations.

The first problem for planetary theory was to determine the differences in longitude between successive phenomena of the same kind. These differences, or synodic arcs

($\Delta\lambda$), were described by functions which were characteristic of a given planet and which (with some exceptions) normally applied independently to each of the synodic phenomena of that planet.

As in the lunar theory we find the same phenomena described in parallel theories both by step functions (System A), in which $\Delta\lambda$ is a function of longitude, as well as by linear zigzag functions (System B), in which the variation is a function of event number. These functions are based on the period relations described in Table 1 (p. 54), and unlike the lunar theories, the same period relation often underlies both types of function. System A functions are attested for all planets – although Venus is a special case – and often with several variants. System B functions are unattested for Mercury and Venus, and are known only from a single fragment for Mars.

From the synodic arcs, $\Delta\lambda$, the corresponding synodic time intervals, $\Delta\tau$, are calculated simply as

(19) $$\Delta\tau = \Delta\lambda + C_p + i_p \times 360°,$$

where C_p is a constant, characteristic of each planet, i_p is the number of complete solar revolutions in one synodic interval of that planet, and the resulting synodic times are expressed in *tithis*, where 1^τ equals $\frac{1}{30}$ of a month.

The motivation for this procedure is the following. The sun takes 12 months and a fraction, or $(360 + \epsilon)^\tau$, to complete one revolution of the zodiac, so that its average velocity in degrees per *tithi* is equal to $360°/(360 + \epsilon)^\tau$. During one synodic interval the sun travels $\Delta\lambda° + i_p \times 360°$, which at average velocity corresponds to a synodic time of

(20)
$$\Delta\tau = (\Delta\lambda + i_p \times 360)°/\left(\frac{360°}{(360 + \epsilon)^\tau}\right)$$
$$= \Delta\lambda + \epsilon\left(\frac{\Delta\lambda}{360} + i_p\right) + i_p \times 360,$$

where the first term is the synodic arc and last term simply equals $12i_p$ whole months. If now in the second term we replace $\Delta\lambda$ with the mean synodic arc, $\overline{\Delta\lambda} = 360(Z/\Pi)$, that term will be constant, and the expression for the synodic term becomes

(21) $$\Delta\tau = \Delta\lambda + \epsilon\left(\frac{Z}{\Pi} + i_p\right) + i_p \times 360.$$

This is just what we find in the planetary theories, where in each case $\epsilon = 11;4^\tau$, which is the 'epact' of the solar model in the System A lunar theory.

The approximation introduced in this last step greatly simplifies the computation of $\Delta\tau$, while introducing an error equal to

(22) $$\delta\Delta\tau = \frac{\epsilon(\Delta\lambda - \overline{\Delta\lambda})}{360}.$$

In the case of Mars this error can reach approximately 1 *tithi*, while for the other planets it is insensible. In any event, it is smaller than the error from assuming uniform motion of the sun over the synodic arc, which can reach 6^r for Mars and 2^r for Jupiter. These errors are relatively unimportant for the times of the stations, which are themselves imprecise, and the planetary theories give a satisfactory account of these phenomena.

The times of the visibilities are another matter, for in addition to these errors, no explicit account is taken of variations in either latitude or the angle between the ecliptic and the horizon, which sensibly affect the visibility of Mars, Mercury and especially Venus. In the case of Venus the interval of invisibility at inferior conjunction varies from 1 day to (more than) 15 days, a fact well known to the Babylonians and due to the variation in Venus's latitude. A comparable, if less dramatic inequality is created by the variation of the horizon angle, which shifts the whole interval of invisibility by up to 14 days relative to inferior conjunction. These wide variations appear to have largely defeated the theories for Venus, of which, however, only few examples are preserved.

In the case of Mercury these difficulties are addressed, with mixed success, by making two of the four phases dependent on the other two by means of 'pushes' which are a function of longitude, and by making the functions which describe the independent phases different from each other. In the theory for Mars, where opposition and second station are made similarly dependent on the scheme for first station, no corresponding adjustment is made for the two visibility phases (Γ and Ω). Consequently, these correspond poorly to the actual phenomena. All in all it is doubtful that the planetary theories were competitive with Goal Year techniques in predicting the dates of visibility phenomena.

In general the planetary theories for visibility phenomena appear rough and incomplete in comparison with the lunar theories, where, moreover, the techniques which might have improved them were routinely employed. Thus the shortcomings of the planetary theories seem to have been more a matter of motivation than of knowledge, which may ultimately have reflected a need to balance the labour of computation against the importance of the result.

Historical development

We know almost nothing about the details of the invention of these astronomical theories, including the names of their authors. We assume on slender evidence that the locus of invention was Babylon, but Uruk may have played a more important role in this process than we credit. What we can say with assurance is that these theories require a zodiac, which on evidence from two Diaries appears to have been introduced between −463 and −453, and that the full System A lunar theory is attested in a text covering −318 to 315. In addition a template of the System A Mars theory is preserved on a text which includes a list of summer solstices ending with −331, the penultimate year of the reign of Darius III. Thus we can safely infer that the invention took place within the Achaemenid period in Babylonia, and roughly within the century −450 to −350.

Within these limits the picture is less clear and the evidence more tenuous. The earliest text which exhibits functions from a fully developed theory covers a Saros cycle of solar eclipse possibilities from −474 to −456, and was probably written not too long afterwards. It includes Columns Φ and F from lunar System A, alongside primitive schemes for the longitudes of conjunction and eclipse magnitudes, which reflect no variations for zodiacal anomaly.

Clear evidence of this inequality first appears in a text for −397, which also includes Columns Φ and F from System A along with a primitive step function for the zodiacal anomaly and also a more primitive function for the length of daylight than we find in either later theory. Finally, an eclipse text covering the period −416 to −380, and also probably written not too long afterwards, contains a scheme for eclipse magnitudes which reflects the eclipse period underlying System B, and attempts ingeniously, if unsuccessfully, to depict the effect of this anomaly.

We are led then to the somewhat unexpected conclusion that the problem of lunar anomaly, as developed with great elegance in System A, was already solved by the second half of the fifth century BC, whereas the problem of zodiacal anomaly remained unsolved until sometime between −390 and −330, when the System A theories for both moon and planets were probably completed.

At present, we cannot say how the development of lunar System B related to this time-frame. It is first attested in an auxiliary text covering −257 to −244, nearly a century after the presumed completion of System A. Nevertheless, the evidence of an important System B parameter in a fourth-century text points to a more nearly contemporaneous date of invention, while the superiority of its critical parameters and the covert inclusion of one parameter from System A suggest that System B was developed later.

Conclusion

The Babylonian astronomer–astrologers lingered on in the temple of Bel at Babylon until at least late in the first century AD, the latest datable text being an almanac for the year AD 75. By then, however, a substantial part of their astronomical tradition had already passed to the Greeks, where it profoundly influenced the subsequent development of astronomy. The details of this transmission are obscure, but it seems likely that at least a substantial part took place in the first half of the second century BC and was in any event available to Hipparchus.

The Babylonian influence on Greek astronomy as reflected in Ptolemy's *Almagest* included: many constellation names, the zodiacal reference system, the degree as the basic unit of angular measure, the digit as a unit of eclipse magnitude, the use of sexagesimal fractions; observations, especially of eclipses going back to the beginning of the reign of Nabonassar in 747 BC; and finally, fundamental parameters including the excellent value for the mean synodic month from System B and the problematic value for the length of the tropical year (derived from it and the 19-year cycle), period relations for the moon and planets, and the magnitudes of their principal inequalities, none of which could have been derived from the Greek observational record alone. In addition, recently discovered papyrus fragments have shown that at least some of the detailed Babylonian mathematic theories were known in Roman Egypt, suggesting that its influence on the development of Greek astronomy may have been even stronger and more pervasive than is presently assumed.

Even more fundamental and far reaching, however, was the Babylonian discovery that it was possible to create mathematical models which would yield reliable numerical predictions of complex astronomical phenomena. For it was this outgrowth of the Babylonian compulsion to predict, which – recast into a more general and powerful mathematical format – motivated not only the *Almagest* but virtually all subsequent astronomy and indeed science.

Bibliography

Aaboe, A. 1964. On period relations in Babylonian astronomy. *Centaurus* 10, 213–31.

Galter, H.D. (ed.) 1993. *Die Rolle der Astronomie in den Kulturen Mesopotamiens*. Grazer Morgen-ländische Studien, 3, with bibliography, 407–49.

Hunger, H. 1992. *Astrological Reports to Assyrian Kings*. State Archives of Assyria, VIII. Helsinki University Press.

Hunger, H. and Pingree, D. 1989. *MUL.APIN: An Astronomical Compendium in Cuneiform*. Archiv für Orientforschung, Beiheft 24. Horn, Austria.

Koch, J. 1989. *Neue Untersuchungen zur Topographie des babylonischen Fixsternhimmels*. Wiesbaden: Otto Harrassowitz.

Neugebauer, O. 1955. *Astronomical Cuneiform Texts*, 3 vols. London: Lund Humphries.

Neugebauer, O. 1975. *A History of Ancient Mathematical Astronomy*, 3 vols. Berlin, Heidelberg and New York: Springer Verlag.

Oppenheim, A.L. 1974. A Babylonian diviner's manual. *Journal of Near Eastern Studies* (Chicago) 33, 197–220.

Parpola, S. 1993. *Letters from Assyrian and Babylonian Scholars*. State Archives of Assyria, X. Helsinki University Press.

Sachs, A.J. 1948. A classification of the Babylonian astronomical tablets of the Seleucid period. *Journal of Cuneiform Studies* 2, 271–90.

Sachs, A.J. and Hunger, H. 1988–96. *Astronomical Diaries and Related Texts from Babylonia. Vol. I: Diaries from 652 B.C. to 262 B.C., Vol. II: Diaries from 261 B.C. to 165 B.C., Vol. III: Diaries from 164 B.C. to 61 B.C.* Vienna: Österreichische Akademie der Wissenschaften.

Stephenson, F.R. and Walker, C.B.F. (eds) 1985. *Halley's Comet in History*. London: British Museum Publications.

Waerden, B.L. van der 1966. *Die Anfänge der Astronomie: Erwachende Wissenschaft II*. Groningen. English language edn 1974: *Science Awakening. II: The Birth of Astronomy*. Leiden and New York.

G. J. TOOMER

Ptolemy and his Greek Predecessors

For the early Greeks astronomy was a practical matter, a means to fix the times for performing agricultural operations or religious rites at a period when they had no adequate calendar. This is apparent from the poem on *Works and Days* by Hesiod in the eighth century BC, where we read, for instance: 'When the Pleiades, daughters of Atlas, rise, begin reaping, and when they set, begin ploughing'. That is to say, the rising of the star-group of the Pleiades just before dawn (its heliacal rising, which was about the middle of May at Hesiod's time and location), having been invisible for some 40 days previously, as Hesiod goes on to explain, is the signal for the start of harvesting. Likewise the setting of the Pleiades just before dawn (the 'cosmical setting', then at the end of October in Greece) warns that it is time to begin the autumn ploughing and sowing. This 'agricultural calendar' relied primarily on the risings and settings of prominent stars (Sirius, Arcturus) or star groups (Orion, the Pleiades, the Hyades), and also the summer and winter solstices, which the Greeks called 'turnings' (*tropai*), because at these times the rising sun had reached its northernmost or southernmost point on the horizon and turned to move in the opposite direction. At a later period formal treatments of it, listing the significant events of the whole astronomical year, were composed, and, like modern farmer's almanacs, these also incorporated linked weather predictions. Some of these were done by well-known astronomers, including Meton, Hipparchus and Ptolemy, and some examples have survived, mostly from late antiquity. Although these were not uninfluenced by scientific astronomy (for instance, the dates they give for solstices and equinoxes were based on the observations and results derived from them), they remain essentially a product of practical and popular astronomy.

Although the early Greeks had identified a number of stars and star groups, they had no notion of dividing the whole visible sky into constellations. The existence of planets was recognised, and Venus at least was named as 'evening star' and 'morning star', but they did not yet recognise that these two were the same entity. As the city-states developed, they established formal calendars, assigning names to the 12 months of the year. The names varied greatly from city to city, but invariably the months were (or

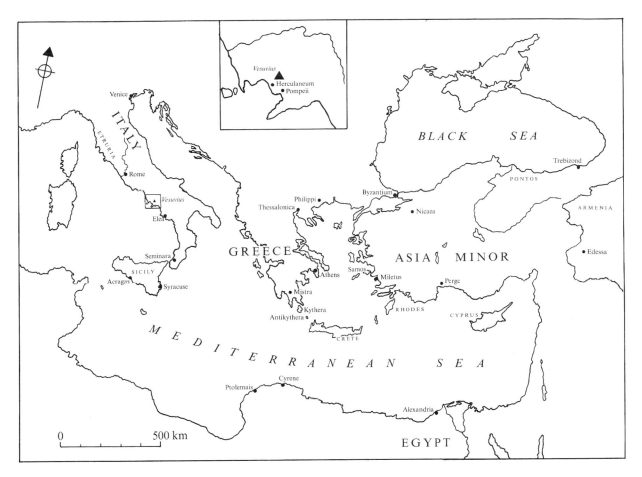

20 The Greek, Roman and Byzantine world.

were supposed to be) true lunar months, and the years true solar years. This entailed great practical difficulties, since the months could be either 29 or 30 days long (called 'hollow' and 'full' respectively), and while 12 such months were considerably less than a solar year, 13 were considerably more. The problem of determining whether a given month should be full or hollow, and of reconciling the years with the months by intercalating a thirteenth month in certain years, was never solved satisfactorily in the official city calendars, and was mostly dealt with by arbitrary action on the part of the responsible officials, which often resulted in alarmingly large deviations of the calendar from the true solar year or lunar month. The farmers continued to rely on the traditional agricultural calendar marked by the stars. However, as scientific astronomy began to develop, the problem did arouse the interest of the professional astronomers, who, as we shall see, proposed solutions (none of which, as far as is known, was ever adopted into a Greek civil calendar).

The pre-Socratic philosophers to Apollonius

Beginning in the sixth century BC there arose, first in the Greek cities of Asia Minor, but later spreading to other parts of the Greek world, the intellectual movement which goes under the name of the 'pre-Socratic philosophers'. Their speculations encompassed a wide range of physical and cosmological matters, including the nature of the visible heavenly bodies, their relationship to the earth and the causes of their apparent behaviour. A sample of these is provided by statements attributed to Anaximander (Miletus, first half of the sixth century BC):

> The stars [i.e. all the heavenly bodies, including planets, sun and moon] are produced as a circle of fire, separated off from the fire in the universe, and enclosed by air. There are certain pipe-shaped passages acting as vents through which the stars are seen; it follows that it is when the vents are stopped up that eclipses take place.

This is a fair representative of the pronouncements of these early thinkers which have been transmitted by later compilers. However, in the course of the next century some of the concepts which became central to Greek astronomy were enunciated by one or more of the philosophers. Parmenides of Elea in Southern Italy (early fifth century BC) mentioned that the earth is spherical, and stated that the moon receives its light from the sun. Empedocles of Acragas in Sicily inferred that solar eclipses are caused by the moon passing before the sun. But even in the later fifth century BC astronomical knowledge was deficient by later Greek standards, even among advanced thinkers: for Democritus the atomist is said to have suspected that there were several planets, but gave neither numbers nor names.

A development of great significance for the future of astronomy in Greece was the transmission of knowledge from Mesopotamia, which in the fifth century BC was much more advanced in this respect. We do not know how or even exactly when this happened. It is possible that the twelve signs of the zodiac, which are of Mesopotamian origin (see p. 49), were known in the Greek world as early as the sixth century BC. It is certain that Babylonian influences were present in the work of the Athenian Meton, who is precisely dated by his observation of the summer solstice in 432 BC. Much of Meton's activity was directed towards the traditional Greek astronomy of the agricultural calendar: he constructed what was perhaps the first 'parapegma', a written or engraved list of the principal astronomical events (risings and settings of stars, solstices, etc.), with holes along the sides into which a movable peg was inserted to mark the current day. However, he may be considered the first 'scientific' astronomer in Greece: first, because he made actual observations (the solstice was determined by means of an instrument constructed for the purpose, perhaps no more than a pillar fixed on a platform to allow measurement of the length of the noon shadow); second, because he attempted to fit the traditional agricultural calendar into a civil calendar. For this purpose it was impossible to use the actual Athenian or any other Greek civil calendar, since that was full of vagaries. Instead, Meton devised an ideal civil calendar, which used the names of the Athenian months, but in which the succession of full and hollow months, and the order of the years in which a thirteenth month was intercalated, were determined not by the whim of a magistrate, but by precise rules which he specified. Underlying this was a 19-year intercalation cycle, which assumed that 19 solar years were precisely equivalent to 235 lunar months. Such a cycle had been in use in Babylon somewhat earlier, and its use by Meton is surely not mere coincidence (see p. 46).

The fifth century BC in Greece, and especially at Athens, was marked by the rapid development of geometry as a deductive system. The success of this as a way of discovering what were considered to be indisputable truths probably led to the use of geometrical models as a way of depicting and explaining astronomical facts. By the early fourth century BC there seems to have been general agreement amongst philosophically inclined Greeks that the universe could be described by the following simple model. At the centre lies the earth, spherical and motionless. At the outer limit is the sphere in which the fixed stars are embedded, which rotates once daily about the earth from east to west, carrying with it all the intermediate bodies. These (the sun, the moon and the planets) have their own spherical motions in the opposite direction, the sun for instance, making a circuit in about a year, and the moon in about a month. Here, however, severe difficulties confronted them, since even crude observation revealed that these motions could not be explained by simply supposing that each body was carried by its own sphere: the planets for instance, while in general maintaining a west-to-east motion against the background of the fixed stars, not infrequently appear to reverse direction and go east to west ('retrograde' motion). Furthermore, although the moon and planets move along approximately the same path as the sun's annual motion (the circle of the ecliptic), they also deviate from it to north and south (they have motions in 'latitude'). Plato is credited with proposing that the goal of astronomy is to explain these apparently irregular motions on the basis of uniform circular motion. Whatever the truth of this attribution, that requirement is characteristic of Greek astronomy from now on, and the first attempt to construct a system embodying it was made by Eudoxus, who lived in the first half of the fourth century BC and was thus a contemporary of Plato.

Eudoxus, who also made profound contributions to the axiomatisation of geometry, proposed a model combining simplicity with great ingenuity. In this each body was carried by one or more spheres rotating uniformly about the earth as their common centre (hence it is known as a 'homocentric' system), but with different poles, these spheres all being connected with one another, so that the motion of the outer was transmitted to that of the inner. Thus the outermost sphere of the fixed stars revolves round the earth once daily, about the poles of the equator, carrying with it the spheres of the sun, moon, etc. The sun, for instance, is fixed on a sphere whose poles are those of the celestial ecliptic, and which rotates in the direction opposite to the daily rotation

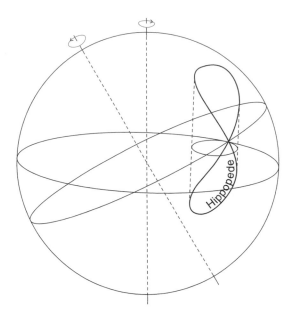

21 Generation of the *hippopede*, the device used by Eudoxus to produce retrogradation on his planetary model. It is the path traced on a spherical surface by a point on the equator of a sphere rotating with uniform velocity, this sphere in turn rotating with the same velocity but about different poles and in the opposite direction.

once annually. The moon required additional spheres to account for its deviation in latitude from the ecliptic. For the planets the great problem was accounting for their retrogradations. Eudoxus discovered that if one examines the motion of a point located on the equator of a sphere rotating with uniform velocity, this sphere in turn being fixed on another sphere with different poles, rotating with the same velocity but in the opposite direction, the combined motion of the point will be a figure-of-eight (in fact the path cut out of the surface of a sphere by a cylinder touching it at one point and intersecting it internally; the Greeks called it a *hippopede*, or horse-fetter, which had that shape – Fig. 21). Accordingly, Eudoxus supposed that each planet had such a combination of two spheres (now visualised as a *hippopede*) superimposed on the equator of the sphere carrying it around the ecliptic. This would in principle produce the variation in speed, and even retrogradation, observable in the planets, as well as a deviation in latitude. The period of rotation on the *hippopede* was necessarily the synodic period of the planet (the time in which it returned to the same position with respect to the sun), while the period of rotation on the sphere carrying it was the sidereal period (the time in which it returned to the same fixed star).

For all its theoretical ingenuity, Eudoxus's system had severe defects, some of which must have been apparent even in the comparatively primitive state of astronomical knowledge in fourth-century Greece. Since the synodic and sidereal periods of the planets are (or should be) fixed by observation (the rather crude amounts which Eudoxus assigned to them are preserved, although his treatise is lost), the only 'free variable' is the size of the *hippopede* (in other words, the distance between the poles of its two constituent spheres). Changing the size of this also changes the planet's deviation in latitude (which is in fact quite independent of its variation in speed). It turns out that for Mars, for instance, no *hippopede*, even the largest possible, can produce retrogradation in combination with even crudely correct figures for the synodic and sidereal period: it is significant that Eudoxus assigned a grossly wrong synodic period to this planet. But above all, no homocentric system could account for the variation in brightness of Mars and other planets, the most obvious explanation for which is a variation in their distance from the earth. Some effort was made by Callippus in the following generation to correct some of the defects by introducing more spheres, and this version of the scheme was adopted by Aristotle (which is the only reason we know anything about it). But neither he nor anyone else could answer this fundamental objection against a homocentric system. Eudoxus's proposal was a dead end, and its chief interest is as the first attempt by a Greek to apply geometry to astronomy, using the principle that any explanatory model must employ uniform circular motion. It is also important as an indication of the state of astronomical knowledge of the Greeks in the early fourth century BC: for instance, the planets enumerated by Eudoxus are the same five, with the same names, which became canonical (and are indeed the only planets known to antiquity), Saturn, Jupiter, Mars, Venus and Mercury. Since these had all been known in Mesopotamia long before, the question of Babylonian influence again arises. This is even more certain in another astronomical work of Eudoxus, a description of the heavens, as visible in Greece, in which he grouped all the fixed stars into the constellations which are still in use today. This work is a combination of traditional Greek nomenclature and mythology with Babylonian elements (most obviously in the twelve constellations of the zodiac). It is impossible to separate Eudoxus's own contributions from earlier elements, but the substance became definitive when it was cast into poetic form by Aratus in the early third century BC: his poem, *Phaenomena*, was immensely popular both in the Greek and later in several Latin versions, and indeed was the principal source of what knowledge of the heavens an educated Greek possessed (Col. Pl. VI).

Very few Greek astronomical works have survived from the period between Aristotle (later fourth century BC) and Hipparchus (mid second century BC). Those that have give only an inadequate picture of the intense interest in both theoretical and practical astronomy which flourished after the conquests of Alexander the Great led to the rapid expansion of Greek culture and Greek settlement to new areas. These included Mesopotamia and Egypt, where Alexander's foundation, Alexandria, became a great centre of intellectual endeavour. Ptolemy has preserved observations made in the third century BC by Timocharis and other Alexandrian astronomers, but we know nothing about the genesis of the two models which were to dominate classical Greek astronomy, the eccentric and epicyclic hypotheses, except that they arose in the period between Callippus (*c.* 330 BC) and Apollonius of Perge (*c.* 200 BC). In the eccentric model the planet or other body is supposed to rotate with uniform motion on the circumference of a circle placed eccentrically to the earth. In the epicyclic model the body rotates

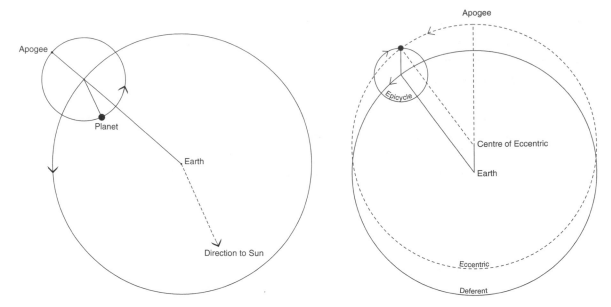

22 (*left*) How the epicyclic model can reproduce retrogradation in a planet. The planet rotates on the epicycle in the same sense as the epicycle rotates about the earth. Depicted is the situation for an outer planet, in which the radius vector from epicycle centre to planet always remains parallel to the direction from earth to sun. When the planet is on the side of the epicycle closest to the earth (near opposition) it will appear retrograde.

23 (*right*) Geometric equivalence of the epicyclic and eccentric hypotheses. In the situation depicted (for the sun), the deferent is the same size as the eccentric, and the radius of the epicycle is equal to the eccentricity. The (angular) speed of the epicycle round the deferent is equal to the speed of the planet on the eccentric, while the speed of the planet on the epicycle is the same, but in the opposite sense. The radius vector from the epicycle centre to the planet always remains parallel to the direction from the earth to the apogee.

uniformly about the centre of a small circle (the 'epicycle') which in turn is carried with uniform motion about a larger circle (the 'deferent') whose centre is the earth. It is obvious that either of those models will produce a variation in the distance of the body, and it is easy to show that each will also, under suitable assumptions (which include the rotation of the centre of the eccentric about the earth), produce retrogradation in a planet (Fig. 22). These two models are in fact fully equivalent in the mathematical sense; the simplest case is illustrated here (Fig. 23). In this the radius of the epicycle is equal to the eccentricity of the eccentric, and the speed of the epicycle about the earth is the same, and in the same sense, as the speed of the body about the centre of the eccentric, while the speed of the body on the epicycle is also the same, but in the opposite sense (this simple model was used by Hipparchus and Ptolemy to represent the motion of the sun). It was no doubt in the context of this kind of mathematical transformation that Aristarchus of Samos (*c.* 280 BC) came to formulate his 'heliocentric hypothesis', famous in modern times, but little regarded in antiquity. According to that the sun is the centre of the universe, and the earth, like all the other planets, revolves about it, while rotating daily on its own axis (the latter rotation had already been envisaged by Heracleides of Pontos in the fourth century BC). While this might have been acceptable from a purely mathematical viewpoint, it was antagonistic to ancient physics, and also entailed that the fixed stars must lie at an unthinkably enormous distance (since their relative positions remained unchanged during the earth's annual orbit). As a result the heliocentric hypothesis was never taken seriously by ancient astronomers.

Although practical astronomy was assiduously pursued in the Greek world during the third century, it seems to have been confined within traditional channels: thus we hear of solstice observations by Aristarchus and Archimedes, and observations of the declinations of fixed stars by Timocharis and his associates. These would have served the established forms of the astronomical calendar and the star-globe respectively. The powerful mathematical tools embedded in the epicyclic/eccentric hypothesis were used merely for theoretical purposes, for explaining how, in principle, the apparently irregular phenomena could be produced by combinations of uniform circular motions. They were not, as yet, used to calculate or predict such phenomena, and indeed some of the mathematical apparatus necessary for that task had not been invented. This theoretical use of the model is apparent in its earliest surviving use by Apollonius of Perge, in a theorem that is preserved by Ptolemy in Book 12 of the *Almagest*. Apollonius addressed the question, assuming the epicyclic or eccentric model for a planet, how can one mathematically determine where its stationary point (the place where it begins or ends its retrogradation) will occur? The answer is simple, although the proof is far from simple by ancient methods. In the epicycle model (Fig. 24), if the planet, P, is at a stationary point, and a line, OPT, is drawn from the earth, O, through P, intersecting the epicycle, then the ratio of the distance of the planet to half of the chord cut off on the epicycle by the intersecting line is equal to the (angular) speed of the planet about the epicycle centre to the speed of the epicycle about the earth, that is

$$\frac{OP}{PT} = \frac{\text{speed of planet}}{\text{speed of epicycle}}$$

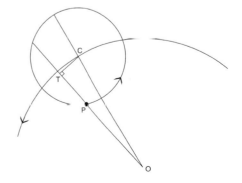

24 The theorem of Apollonius for the stationary point of a planet in the epicyclic hypothesis. If the planet is stationary at P, then OP : PT = velocity of P : velocity of C.

The speed of the planet and the speed of the epicycle are easily derived from the synodic and sidereal periods. But the problem cannot be solved for an actual planet in real numbers without knowing the relative size of epicycle and deferent for that planet, and there is no evidence that Apollonius had determined this or was interested in doing so: he was concerned only with the abstract mathematical problem. The same attitude is displayed in the single astronomical work surviving from this time, Aristarchus's treatise on the distances of the sun and moon. This is a mathematical exercise showing how the limits for those distances can be derived from certain numerical assumptions, about the inaccuracy of which the author appears unconcerned, although it must have been obvious: for instance, the apparent diameter of the moon is assumed to be $\frac{1}{45}$ of a right angle (2°, about four times too great).

Hipparchus

The transformation of Greek mathematical astronomy from a descriptive to a predictive science was the work of Hipparchus, who was born in Nicaea in Asia Minor, but spent most of his working life (*c.* 150 to *c.* 125 BC) in Rhodes. He was an original genius, an innovator in both mathematical and observational techniques, with a judgment detached enough to distrust his own as well as others' results, and the industry to verify and correct them. But for all his remarkable abilities, he could not have achieved what he did without the resources supplied to him by contemporary Mesopotamian astronomy. We have already seen evidence for Mesopotamian influences in the work of Meton and Eudoxus, and other Greeks before Hipparchus had knowledge of individual features of Babylonian astronomy, but Hipparchus is the first Greek (and perhaps in some respects the only one) to have had detailed knowledge both of the contents of the great archive of observational material at Babylon going back to the eighth century BC, and also of the powerful mathematical techniques for calculating and predicting lunar and planetary phenomena which had been developed there in more recent centuries by the astronomer scribes, and which were still being practised in his time (see pp. 52–66). We have no information about how Hipparchus acquired this knowledge (all but one of his many works are lost, and most of our information about him comes from Ptolemy's *Almagest* and other secondary sources), but it is so detailed and well informed that the only plausible explanation seems to be that he was personally instructed, and supplied with translated material, by one of the astronomer scribes at Babylon. Although that city was captured by the Iranian dynasty of the Parthians in Hipparchus's lifetime, it had been accessible and familiar to Greeks since Alexander's conquests.

However he acquired this information, Hipparchus provided the main channel through which it was transmitted to the Greek world. This must have involved a great deal of labour. For instance, he compiled a complete list of lunar eclipses observed at Babylon from the eighth century BC down to his own time, which was not merely a matter of having the relevant Babylonian lists translated, but must have required conversion of the Babylonian lunar and regnal dates into a running calendar usable by Greek astronomers (Hipparchus used, and may have introduced into astronomical practice, the unvarying Egyptian year of 365 days). This wealth of data was fundamental both for himself and for his successor Ptolemy in their reform of Greek astronomy. But observational material was perhaps not the most important thing that Hipparchus derived from Mesopotamia: he also owed a great deal to the mathematical theory that had been developed there. One of the great advantages for mathematical calculations which the Babylonian astronomers possessed was the sexagesimal system (a place-value notation similar to our decimal system, but with base 60) – see p. 51. Hipparchus adopted a version of this (using Greek numerals) to replace the existing clumsy system of unit fractions; without this innovation it is difficult to imagine how he could have constructed astronomical tables. He also adopted the 360° division of the ecliptic and other circles. Even from our inadequate knowledge of the details of his astronomical practice, we can be sure that he also took over Babylonian arithmetical techniques, for instance in his calculations of the positions of both sun and moon. Recent discoveries have shown that parts of the most sophisticated Babylonian lunar ephemerides existed in Greek versions, and Hipparchus is the most probable source of these, too. But above all, it seems, it was the idea of astronomy as a predictive science, using mathematical techniques, and based on precise period relations, which Hipparchus took from Babylon and adapted to existing Greek theory. In so doing he produced a fusion of the two elements which was to prove one of the most successful and lasting achievements of ancient civilisation.

Hipparchus's achievement is best illustrated by the example of his lunar theory, about which we know much from Ptolemy's discussion. He took for granted that the lunar motion could be explained by a simple epicyclic or eccentric model, as his predecessors had done. But in order to use such a model to calculate the lunar position at a given time, one needs to know the size of the epicycle (or, in other words, the amount of the eccentricity), and also some 'mean' position of the moon (i.e. its location at a moment when the epicycle or eccentric had no effect on its position, for instance at the apogee). To determine the lunar eccentricity Hipparchus devised an ingenious method using three lunar eclipses. The moon at these three eclipses is represented by three points, M, N, Q, on the eccentric (Fig. 25). The angles formed by these three points at C, the centre of the eccentric, and at E, the earth or observer, are given. Then by Euclidean geometry one can find the ratio $e : R$ of the eccentricity EC to the radius of the eccentric, and also the direction of the apogee at one of the three eclipses. The angles at E are given from the true position of the moon at the three eclipses; but these are not directly observed: what Hipparchus found from Babylonian or other eclipse records was only the time of mid-eclipse. However, at this moment the moon is directly opposite the sun; so he calculated the solar position (probably using Babylonian arithmetical methods rather than the solar eccentric model which he developed later) at these three times, and added 180° to find the moon's position. The angles at C are

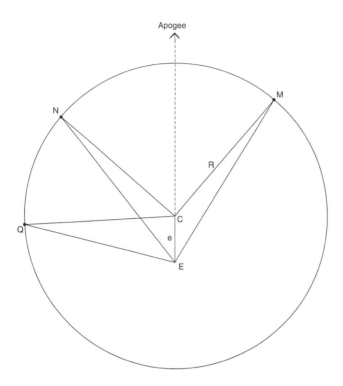

25 Hipparchus's determination of the lunar eccentricity and apogee position from three lunar eclipses. From the angles formed by the moon in the three positions M, N, Q, at the centre C and the earth E, the ratio e : R and the direction of EC can be calculated.

found from the moon's mean motion between the eclipses. This Hipparchus derived from the period relations which he took unchanged from the Babylonian astronomers, which are an essential foundation of his lunar theory. These are attributed to Hipparchus by Ptolemy in the *Almagest*, and are explicit or implicit in the cuneiform lunar ephemerides which were extracted from the site of Babylon in the nineteenth century, the discovery of which initiated an era in our understanding of Greek, as well as Babylonian, astronomy. They are:

- In 251 months the moon performs 269 returns in anomaly. (A return in anomaly is a return to the same speed; in Greek terms, a return to the same point on the epicycle.)

- In 5,458 months the moon performs 5,923 returns in latitude.

- The length of the mean synodic month is 29; 31, 50, 8, 20 days.

To solve this problem (and indeed to carry out the whole programme of producing a computational astronomy using Greek methods), one other tool is needed, namely trigonometry, since a number of triangles have to be solved from given angles and sides. Before Hipparchus this did not exist: in Babylonian astronomy, which operated arithmetically, it was not needed, while Greek astronomy and geometry had hitherto been largely theoretical, and when problems which one would expect to see solved by trigonometry did arise, they were dealt with by approximation methods to establish upper and lower limits, as can be seen in the work of Archimedes and Aristarchus. It is no coincidence that Hipparchus is the first Greek to be associated with a trigonometrical function: he is said to have computed a table of chords. Ptolemy explains in the first book of the *Almagest* how to compute such a table (which is related to the modern sine function), and although Hipparchus's table was probably rather simpler, it served his purposes.

By comparable procedures Hipparchus also established an eccentric/epicycle model for the sun, based on the observed length of the seasons and of the solar year (to this end he devoted much attention to observing the times of equinoxes). From this he was able to proceed to the problem of predicting lunar and solar eclipses. But this in turn involved the question of parallax, which could only be solved if the distances of the moon and sun could be found. A number of Greeks before Hipparchus had speculated about this, and Aristarchus had even devised a theoretical method for computing the distances, which, as we noted, was useless from a practical viewpoint. The procedure which Hipparchus adopted was the first to give a reasonably correct value for the lunar distance. It is based on the fact that the sun and moon appear under the same angle at a certain position of the moon on its orbit (Fig. 26). Measurements of this angle (the 'apparent diameter' of the moon or sun) can be made by means of an instrument (and in fact Hipparchus is credited with making such observations with a 'four-cubit dioptra'), but can also be derived from eclipse observations, as can the diameter of the earth's shadow at a lunar eclipse, which is another element in this calculation. It can be shown that, given those two angles, in the configuration depicted, if the moon's distance EM is given, the sun's distance ES can be calculated, or vice versa. Hipparchus chose the latter possibility, estimating the sun's distance as 490 earth-radii, whence he derived a distance of $67\frac{1}{3}$ earth-radii for the moon. While this assumption of a very small value for the unknowable distance of the sun might appear perverse, it is justified by the observation that 490 earth-radii corresponds to a maximum parallax of 7': Hipparchus, being unable to find a measurable parallax for the sun, assumed that it must be less than 7', which means that the sun must be *at least* 490 earth-radii distant. It is an interesting consequence of the configuration that as the sun's distance increases, the moon's decreases: hence the amount $67\frac{1}{3}$ earth-radii represents the moon's *maximum* distance. Furthermore, as the sun's distance tends to infinity, the moon's tends not to zero, but to about 59 earth-radii. Hence Hipparchus was able to fix the moon's distance, when it has the same apparent diameter as the sun, as between about 59 and 67 earth-radii. This is close enough to the truth to produce parallax values resulting in good agreement with observation in the computation of solar eclipses.

Thus Hipparchus was able to produce a theory of the sun and moon which, even if it did not completely satisfy himself, could be (and was) used by his successors to calculate their positions and the resulting eclipses. For the planets, however, his work

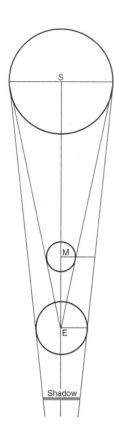

26 Hipparchus's determination of the distance of the sun and the moon from the apparent diameters of the sun, the moon and the shadow. In the situation depicted, the discs of the moon (centre M) and the sun (centre S) are seen under the same angle from the centre of the earth E. If this angle, and the angle under which the earth's shadow appears at a lunar eclipse for the same distance of the moon, are both given, then from the sun's distance ES (in terms of the earth's radius) can be computed the moon's distance EM, and vice versa.

was critical rather than constructive. He demonstrated that the theories that had been proposed hitherto to explain their motions were inadequate. His main objection must have been that the planetary models of his Greek predecessors, whether epicyclic or eccentric, had all assumed a single 'anomaly', or factor causing non-uniform motion, the period of which was the planet's return to the sun – hence it is known as the synodic anomaly. Hipparchus knew from his study of Babylonian planetary ephemerides that those who compiled them had recognised two anomalies, the synodic and the sidereal (the period of which was the planet's return to the same point in the ecliptic). He did not produce any alternative theory for the planets, but he did present period relations for the mean motions of all five planets (which were derived, like his lunar period relations, from Babylonian sources), and also compiled a list of all the planetary obser-vations he could extract from Babylonian and Greek sources, reduced to a common calendar, for the use of his successors.

Hipparchus must have spent many years observing and recording the positions of the fixed stars. The ostensible purpose of this was to delineate a star-globe, and some idea of his results can be obtained from his one surviving work, a commentary (highly critical) on the descriptions of the heavens by Eudoxus and Aratus. But in the course of comparing his observations with those of his Greek predecessors, he incidentally made the discovery for which he is most famous, of the 'precession of the equinoxes'. In Greek terms, the ecliptic longitudes of the fixed stars, measured from the intersection of ecliptic and celestial equator, very slowly increase. This motion is so slow that Hipparchus was uncertain both about the amount of it, and, at first, whether it affected all stars or only those in the vicinity of the ecliptic (his willingness to entertain the latter hypothesis is a remarkable example of his freedom from the trammels of ancient thinking about a 'sphere of the fixed stars'). He finally decided that it was common to all stars, and that the amount was at least 1° in 100 years.

Hipparchus was an innovator in other areas: in mathematical geography, in the design of instruments (there is reason to believe that he invented the plane astrolabe) and in astrology. But his real achievement is the revolution that he initiated in Greek astronomy. However, although he made enormous contributions to this revolution, it was far from complete when he ceased work, and it had to wait almost 300 years before a worthy successor arose to complete it. This was Ptolemy, whose *Almagest* makes frequent mention of Hipparchus, both admiring and critical, and remains the most important source for our knowledge of his great predecessor. Unfortunately Ptolemy says almost nothing about the development of astronomy in the intervening period, not because no astronomers were active, but because his contempt for their work was so great that he did not want to dignify it by even mentioning it. Since the success of the *Almagest* led to the disappearance of their work (as it did that of Hipparchus), we are very ill informed about this interesting period of Greek astronomy. Some information comes from finds of papyri in Egypt, but the best picture is provided by Indian astronomical works, for although the surviving Sanskrit texts were written at a later period, they are partly derived from a pre-Ptolemaic Greek tradition. The problem of discriminating between the native and Greek elements is a delicate one (see pp. 125–7).

Hipparchus had on occasion used a Babylonian lunar scheme, based on a 248-day cycle, to determine the moon's position in longitude, at least for short-term motion. This was adopted and developed by his successors, and indeed seems to have been the usual method of computing the moon's position, at least among the astrologers. For these the main gap left by Hipparchus was the lack of any computational planetary models. These were soon supplied, and indeed the enormous growth in horoscopic astrology in the first century BC is evidence for their availability, since the computation of the positions of the heavenly bodies at birth, conception or some other critical time was essential for this art. From the Indian texts we have some idea of the form these models took: they did indeed account for both synodic and sidereal anomalies, which were represented by epicycles, eccentrics or a combination, but in a manner which must have been thoroughly unsatisfactory to anyone who asked how the various parts of the model worked together. This explains Ptolemy's total silence about them apart from a couple of contemptuous asides. However, they did produce numerical results which satisfied the needs of the astrologers. Tables for computing astronomical positions were produced. In these the mean motions were represented by the number of integer revolutions the body made in some huge period of many thousands or millions of years. For computations involving the observer's position on the earth, in the absence of spherical trigonometry various ingenious methods akin to descriptive geometry were employed, which had first been developed in sundial theory in Greece (under the general name of *analemma*), and which had probably already been used by Hipparchus for other astronomical purposes. The basic theorem in spherics which was to provide the basis for the spherical trigonometry, which Ptolemy later substituted, was discovered towards the end of this period by Menelaus of Alexandria (*c.* AD 100).

Ptolemy

Ptolemy (Claudius Ptolemaeus) worked in Alexandria from about AD 130 to 175. His great astronomical treatise, usually known under its Arabic title of *Almagest*, in thirteen books, was published about AD 150, and although it is only his second publication, it is a work of maturity, on which he had spent many years. It is the first attempt by a Greek to provide a systematic treatment of the whole of mathematical astronomy (as the term was then understood). As a didactic work, it is a masterpiece of clarity and order, which deserved the authoritative status which it soon achieved and retained for many centuries. Proceeding from basic principles and taking his topics in logical order, Ptolemy explains, for each topic in turn, what the phenomena are, and how they may be explained by a mathematical model. Then, on the basis of a few well-chosen observational data, he determines the numerical parameters of the model, finally explaining the construction and use of the tables which he gives for each topic. At the appropriate place he explains how to construct each of the instruments which he used for the several kinds of observation. While superbly designed for the student, the work is less satisfactory for the historian of astronomy. Ptolemy does not always make it clear where he is incorporating traditional material and where he is proposing an innovation. He rarely mentions his predecessors, apart from Hipparchus (who was clearly the only one whom he considered his peer). He adduces only those observations which are to be used for his demonstrations, giving no details of the many more he certainly possessed, both by himself and by his predecessors, especially the Babylonian resources transmitted by Hipparchus. Yet in the nature of things some of these must have been in conflict with the results he produces. Hipparchus had been much more open in displaying discrepant results from different data, often in the same work. Ptolemy, who evidently found it difficult to conclude from these discussions what Hipparchus's real opinion was, preferred not to confuse his readers.

The *Almagest* begins with a description of the nature of the universe (in so far as it concerns the astronomer), and the phenomena which justify that description – for instance, the reasons for thinking that the earth is spherical, that it is motionless at the centre of the universe, and that it is negligibly small in comparison to the distance of the heavens. All of this was standard belief for educated Greeks since the time of Aristotle (the exceptions, such as the Epicureans, who argued for an infinite universe, were always a small minority). In this world view there was a fundamental difference between the region below the sphere of the moon, made up of the corruptible four elements of earth, water, fire and air (with a natural motion in a straight line towards or away from the centre), and the heavenly regions, which were composed of the incorruptible fifth element (*aether*), the natural motion of which was circular. For Ptolemy the heavenly bodies were not merely eternal but in some sense of the word 'divine', and in this he was a man of his time.

The rest of the first book is devoted to the mathematical apparatus necessary for much of what follows, trigonometry, both plane and spherical. Ptolemy explains in detail how to calculate his chord-table, and develops the fundamental theorem in spherical trigonometry which he uses to solve all problems involving the surface of the sphere. It is notable that he avoids all reference to analemma methods, the more so since he himself later wrote a work called *On the analemma* on sundial theory. It is clearly a methodological choice in the *Almagest*, which was designed to introduce mathematical rigour throughout the whole of astronomy, whereas, in Ptolemy's view, it had been sadly lacking in the work of his predecessors. Most of the applications of spherical trigonometry are found in Book 2, for the solution of problems connected with the annual variation in solar declination. The problem of 'rising-times' is treated at great length (the rising-time of an arc of the ecliptic is the time taken by that arc to cross the horizon at a given terrestrial latitude), and tables of rising-times are computed for many different latitudes. These are useful astronomically, especially for the conversion of the 'civil' hours in normal use in the ancient world (one civil hour is $\frac{1}{12}$ of the length of the actual day or night on a given date at a given place) to the unvarying 'equinoctial' hours used by the astronomers. However, their principal importance was in horoscopic astrology, and one may suspect that the space devoted to the topic by Ptolemy reflects that importance. The only observed numerical parameter involved in all this is the obliquity of the ecliptic, which Ptolemy claims to have found (by measurements with two different instruments which he describes – Fig. 27) as $23;51,20°$, in agreement with earlier results of Hipparchus and Eratosthenes (third century BC). The observation is inaccurate, for by Ptolemy's time the obliquity had become noticeably less than it was four centuries earlier: this is only one example of Ptolemy's tendency to 'confirm' accepted values if possible.

Book 3 treats the theory of the sun. Once again, Ptolemy confirmed an estimate of Hipparchus, that the length of the year is $365\frac{1}{4} - \frac{1}{300}$ days. He did this by comparing his own observations of equinoxes with those recorded by Hipparchus and of the solstice with that of Meton. This result is notoriously wrong (the last fraction should be about $\frac{1}{128}$), and the consequent error in Ptolemy's mean motion for the sun was serious, since

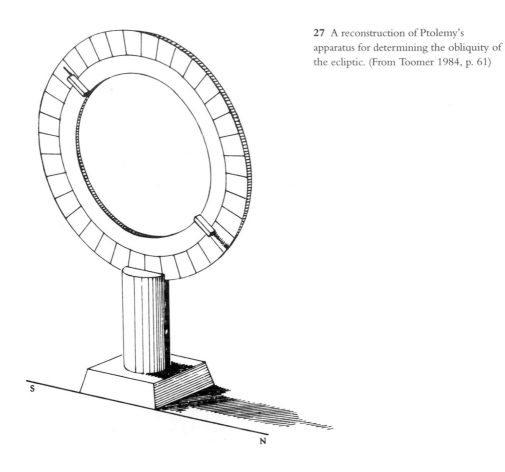

27 A reconstruction of Ptolemy's apparatus for determining the obliquity of the ecliptic. (From Toomer 1984, p. 61)

it was in turn transmitted to the mean motion of the moon and thence to the planets and fixed stars. Moreover, it meant that the mean position of the sun derived from his tables became progressively more incorrect over time, until by the ninth century AD it was so obviously discrepant with the facts that corrections were introduced by the Arab astronomers. Ptolemy's equinox and solstice observations, which he used to verify Hipparchus's year length, are in fact all about a day off, which is a gross error even by ancient standards. As a preliminary to demonstrating the explanation of the sun's anomaly (which accounts for the variation in the length of the seasons), Ptolemy introduces the epicyclic and eccentric models, and gives proofs of their geometric equivalence. He then derives the sun's eccentricity from the same season lengths as Hipparchus used, on the basis of an eccentric model, and arrives at the same solar eccentricity. This is slightly too large by modern computation, but a more serious fault is that he also necessarily finds that the sun's apogee is at the same position in the ecliptic as it was nearly three hundred years earlier, and concludes that the apogee is fixed with respect to the ecliptic. This is in contrast to the apogee of the planets, which, as he later finds, share in the motion of precession, i.e. are sidereally fixed and move through the ecliptic over the course of time. In fact the sun's apogee too undergoes the same motion: this defect in Ptolemy's solar theory also had to wait until the ninth century AD for correction.

In Book 4 Ptolemy treats the simple lunar theory, more or less as it had been left by Hipparchus, but correcting some of the details. He takes the Hipparchan mean motions (which we now know are in fact derived from Babylonian astronomy), and Hipparchus's method of finding the lunar eccentricity, to determine afresh the eccentricity or epicycle radius, which he does twice, once from three very early Babylonian eclipses, and again from a triple of eclipses observed by himself. In both cases he finds the same value for the eccentricity, $5\frac{1}{4}:60$, which was an improvement on the values found by Hipparchus, and at the same time showed that there was no variation in the eccentricity over the many centuries between the two sets of observations. This was important, since Hipparchus had found discrepant values from two sets of three eclipses, and Ptolemy goes to some pains to show that Hipparchus's results are due to errors in his calculations. A further benefit of solving the problem twice with two eclipse-triples far apart in time was that Ptolemy could use his results to verify the mean motions in longitude and anomaly, and was able to supply a small correction to the latter (amounting to only 17' in the 854 years between the observations, a tribute to the accuracy of the Babylonian period relations).

So far Ptolemy had merely been adapting and revising Hipparchus's procedures. But in Book 5 he proposed a profound modification to the lunar theory. Noticing that the simple model, as established so far, worked well at the syzygies (when the moon was in conjunction with or opposition to the sun), but often displayed considerable variation from the observed lunar position when the moon was at other elongations from the sun, and that this variation was greatest at quadrature (when it was 90° away on either side), he inferred that the moon's epicycle was not a constant distance from the earth. The model he constructed to effect this was as follows (Fig. 28): the moon's epicycle (centre C), while continuing to rotate uniformly with respect to the earth (O), is carried on a circle whose centre is not the earth but another point (M), which in turn rotates in the opposite direction about a circle with centre O. The uniform motion of M is that of the mean motion in elongation between moon and sun (denoted by η). This means that as the elongation increases from syzygy, the epicycle is pulled in towards the earth by the 'crank' OMC, 'jointed' at M, and of length OM + MC equal to the radius of the deferent in the simple model. The effect is greatest when the mean elongation η is 90°, but at mean conjunction and opposition the 'crank' is straightened out and the configuration is identical with that of the simple model. Thus the refined lunar model does not affect the situation at eclipses (from which Hipparchus's model had been derived and for which it worked well), but it does produce considerable differences (up to almost 3° according to the numerical parameters which Ptolemy established) at other elongations. Ptolemy's refined model does indeed represent the longitudes far better, accounting for most of what is known in modern times as the moon's 'evection', and

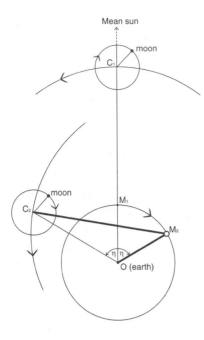

28 Ptolemy's revised lunar model. The epicycle is carried on a circle, the centre of which, M, itself moves on a circle about the earth O, in the opposite sense, with the motion of mean elongation between moon and sun. The epicycle (centre C) is depicted in two positions, at mean conjunction and near quadrature. In the first it is at maximum distance from the earth, while in the second the 'crank' has pulled it in close to minimum distance, hence it will appear much larger.

was a significant improvement. However, a great disadvantage of the 'crank' is that it greatly increases the amount by which the distance of the moon from the earth varies, so that its minimum distance is little more than half its maximum: this ought to be reflected by a similar variation in the apparent size of the moon's disc, whereas the observable variation is far smaller. Ptolemy simply ignores this obvious objection, and treats the resultant variation as a fact in his later computations of the moon's distance and parallax.

These occupy the last chapters of Book 5. Ptolemy adopts Hipparchus's method of computing the lunar and solar distances (with some corrections to the amounts assigned to the apparent diameters), but in reverse: that is, he first finds the lunar distance from a carefully chosen parallax observation. This turns out to be 59 earth-radii at mean distance, in approximate agreement with what Hipparchus had already found (and also close to the actual distance). There are so many imperfections in the observation and computation from which this number is allegedly derived that there can be little doubt that Ptolemy was aiming in advance at some such result and chose an observation which would produce it. From it he derives, by Hipparchus's method, a solar distance of 1,210 earth-radii. This is too small by a factor of twenty, which is of little consequence in the *Almagest*, since the resulting solar parallax is still negligible (we shall see that it had much greater consequences in Ptolemy's *Planetary Hypotheses*). Ptolemy can now construct a parallax table. This is the final preliminary step necessary for his exposition of eclipse theory, which occupies Book 6. This involves no new models, but does require a great deal of explanation of the computation of suitable tables for both solar and lunar eclipses, and rules for how to use them. For a solar eclipse Ptolemy's tables enable one to compute the circumstances (time, duration and magnitude) for any given place on earth, but not to determine its path (the latter is a comparatively modern idea). There is an interesting discussion of eclipse intervals: in order to construct his tables Ptolemy had already established 'eclipse limits', i.e. how far from the node the moon could be for an eclipse still to occur. From the same numbers he shows that both solar and lunar eclipses can occur at 6-month intervals, that solar eclipses can occur at both 5-month and 7-month intervals, and that lunar eclipses can occur at 5-month, but not 7-month intervals. He also discusses the question of solar eclipses at 1-month intervals. We know that Hipparchus had discussed the same topic: no doubt his interest in it was aroused by seeing examples of Babylonian eclipse-lists in which such intervals are apparent (see pp. 50, 51).

In Books 7 and 8 Ptolemy deals with the fixed stars. He begins by showing that they are indeed fixed in the sense that they do not change their relative positions (the effects of proper motions were too small to show up over the time interval for which he had observations). For this purpose he compares alignments recorded by Hipparchus with his own observations. He then discusses the phenomenon of precession, which he demonstrates in general by showing that there had been systematic changes in the declinations of certain stars from the time of Timocharis (early third century BC) through Hipparchus (mid second century BC) to his own epoch. The particular amount which he assigns to it, 1° in 100 years, is justified by comparing observations of prominent fixed stars with respect to the moon by Timocharis and later astronomers. There are many peculiarities about these observations, but undoubtedly the main reason for the error in this value is that Ptolemy's error in the solar mean motion has been transmitted via the computed lunar positions.

The main part of Books 7 and 8 is taken up by a star catalogue (a list to enable the reader to enter them on his own star globe, the construction of which is explained later) This consists of 1,022 stars arranged under 48 constellations (Col. Pl. IV). For each star Ptolemy gives a description of its position on the 'figure' of the constellation (supposed to be outlined on the globe), its longitude (for the epoch of the beginning of the reign of the Emperor Antoninus), latitude and magnitude. There has been much discussion about the extent to which this catalogue embodies Ptolemy's own work or whether it is simply derived from Hipparchus. While it seems probable that it is not simply the product of Ptolemy's own observations, but relies heavily on Hipparchus's results, it is not a mere updating of an existing catalogue. There seems no doubt, for instance, that Ptolemy was the first systematically to use the co-ordinates of latitude and longitude (the reason being, as he himself explains, that the catalogue can readily be adapted to another epoch by adding a constant for precession to the longitudes: this was in fact done by many of his successors down to Copernicus). Whether we regard it as the work of Ptolemy or the combined work of him and Hipparchus, it is, for all its individual faults, a remarkable achievement, both in size and comprehensiveness.

The remaining five books are devoted to the theory of the planets. The second chapter of Book 9 contains one of Ptolemy's rare discussions of the intellectual and historical difficulties associated with his task: in this case the fact that the planets have two anomalies, the effects of which are difficult to disentangle, and the inadequacy of most of the available observations, which are mostly of ill-defined events like stations or first and last visibilities (a clear reference to the characteristic Babylonian planetary observations). He excuses Hipparchus for failing to produce a planetary theory, and apologises in advance for any apparently unjustified assumptions which he may make in his own approach. He is undoubtedly thinking here of his introduction of the equant point, the most innovative and also the most controversial aspect of his planetary theory. Ptolemy decided to represent the synodic anomaly by means of an epicycle and the sidereal or zodiacal anomaly by an eccentric on which the epicycle rides. In this he may have had predecessors; but, unlike any of these, he was aware that such a model could not adequately represent the phenomena: in particular, any such arrangement which reproduced with sufficient accuracy the longitudes of Mars, for instance, at conjunction, would also result in arcs of retrogradation which were obviously wrong, being too small at apogee and too great at perigee. Ptolemy's solution (Fig. 29) was to make the epicycle revolve with uniform motion, not about the centre of the deferent, M, but about a point E removed towards the apogee of the deferent from M, by the same amount as the earth, O, is removed in the opposite direction. The point E is known by its medieval name of 'equant'. This device was severely criticised in later times on the grounds that it violates the principle of uniform circular motion, since the epicycle, although it moves

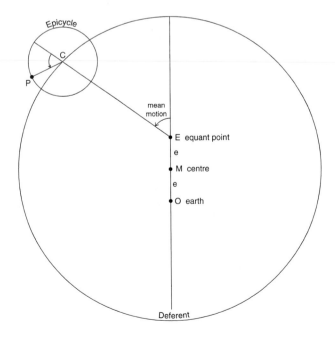

29 Ptolemy's equant model. The epicycle moves on the deferent (to which the earth O is eccentric), but its uniform motion is counted not about the centre M, but about the equant E, the distance of which, e, from M, is the same as the distance of O from M on the opposite side.

on a circle, moves non-uniformly with respect to that circle's centre. From a practical point of view, however, it turns out to be an extremely good approximation to the planet's true motion.

Unfortunately the introduction of the equant point greatly complicated the problem of determining the numerical parameters of the models for the planets. Without it, the calculation of the eccentricity of the deferent for the outer planets (Mars, Jupiter and Saturn) could be assimilated to Hipparchus's method for finding the moon's eccentricity, with the substitution of oppositions for eclipses. But with the equant point as the centre of uniform motion, the problem is no longer soluble by Euclidean geometry (in fact it could not be solved at all by the mathematics available in antiquity). Ptolemy's response is a tour de force: he solves it first for the simple, non-equant model, using three observed oppositions from which he derived the angles at the observer and the centre of mean motion: this gave him a 'preliminary' eccentricity and apogee, which he then transfers to the equant model, from which he computes 'corrected' angles, and from these in turn derived a corrected eccentricity and apogee, and so on. This iteration procedure does in fact converge on the true values, but Ptolemy's only way of demonstrating this is, after it does converge, to use them together with the equant model to produce the originally observed positions by computation.

Before he could embark on this procedure at all he had to have mean motions for the planets. These were easily provided: Ptolemy quotes from Hipparchus certain period relations, e.g. 'for Saturn, 57 returns in anomaly correspond to 59 solar years and to 2 revolutions in longitude'. We now know that these are all well-known Babylonian values. Ptolemy applied certain small corrections which he was to justify later, and on this basis computed mean-motion tables. More difficult was the problem of explaining the apparent motions of Mercury, which is complicated because the planet's orbit is extremely elliptical, and the observations available to Ptolemy were few and unreliable. He devised a very complicated model, most of the features of which are illusory and stem from inaccurate observations. (Yet it survived virtually unchanged down to the time of Copernicus.) Having demonstrated the theory for the longitudes of the planets in Books 9 to 11, Ptolemy devotes Book 12 to the resultant stations, retrogradations and other 'phases', and Book 13 to the planetary latitudes. It is here above all that the disadvantages of operating with a geocentric, rather than heliocentric system are apparent, for since the planes of the planets' orbits pass not through the earth but the sun, the transformation is difficult, and Ptolemy's solution is not successful in all respects (although here again no significant improvement was made before Kepler).

Ptolemy composed many works after the *Almagest*. Some of these are on non-astronomical topics (such as optics, music and geography), but others supplement the *Almagest* in particular ways (for instance, a work on the phases of the fixed stars). In general, however, they do not present improved results either in theory or in numerical parameters, except in the models for planetary latitudes, which Ptolemy simplified for his

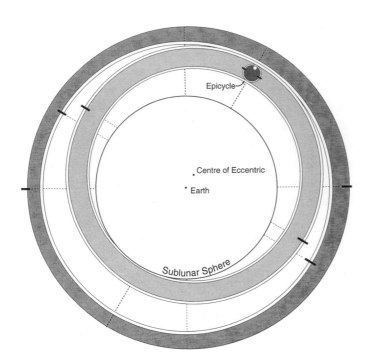

30 Cross-section of the physical model for the moon described in Ptolemy's *Planetary Hypotheses*. All parts are contiguous solid spheres or spherical shells. The 'pivots' and axes about which the latter rotate are indicated. Cf. Neugebauer 1975, p. 925.

Handy Tables (Fig. 33) and *Planetary Hypotheses*. The latter two works proved particularly influential. The *Handy Tables* are a complete set of astronomical tables, based on those already presented in the *Almagest*, but made more convenient for use in many ways. For instance, the mean motions are given at 25-year intervals instead of the 18-year groupings of the *Almagest* (which were chosen merely to fit the tables as closely as possible on the page), and Ptolemy changes the epoch from the first year of Nabonassar (747 BC) to the death of Alexander (324 BC, in fact in the year previous to Alexander's death because the epoch is taken as the beginning of the Egyptian year in which Alexander's successor began to rule). The format of these tables became the norm for the vast majority of astronomical tables throughout the whole of the medieval period both in Europe and the Islamic world, which persisted well into early modern times.

The ostensible purpose of the *Planetary Hypotheses* is to give instructions for constructing physical representations of the models for the heavenly bodies presented in the *Almagest*. However, Ptolemy implies that the physical structures he describes may in fact be real components of the universe: these are combinations of solid spheres, or rather spherical shells of varying thickness, arranged to completely fill the available space in a way which reproduces the eccentrics and epicycles of the *Almagest*. A cross-section of the moon's sphere is illustrated here (Fig. 30). Furthermore, Ptolemy was able to give precise absolute dimensions of all these elements (whereas in the *Almagest* he had done this only for the moon and the sun). For, under the assumption that there is no space wasted in the universe, the convex surface of each body's sphere must coincide with the concave surface of the sphere of the body immediately above, i.e. the greatest distance of Mars, for instance, is the same as the least distance of Jupiter. Under the assumption that the spheres of Mercury and Venus lie between those of the moon and the sun, and starting from the lunar mean distance of 59 earth-radii which he had already found, Ptolemy calculated by this method that the solar distance is almost exactly that which he had found by a completely different method in the *Almagest*. This must have appeared a striking confirmation of the correctness of his assumptions. He went on to calculate the distances (and also the sizes) of all the bodies out to the sphere of the fixed stars (which he computed as slightly less than 20,000 earth-radii, close to the modern estimate of the distance from the earth to the sun). This picture of a small finite universe, for which one knows precisely the dimensions of all the parts, was almost universally accepted right down to the end of the Middle Ages, in both Islamic and Christian lands (with minor modifications to accommodate the account in Genesis). It is recognisable, for instance, in Dante's *Divina Commedia*.

Ptolemy did not intend the *Almagest* to be the last word on astronomy, merely the best that could be done with the material available to him, and explicitly looked forward to seeing his work improved by his successors. This did not happen in antiquity, where nothing of significance in astronomy was done after him. Even in the medieval Islamic world, where real improvements were made to the solar theory, and other aspects of Ptolemy's work were subjected to intelligent criticism, the structure as a whole was not affected. Greek astronomy, as it culminates in the *Almagest*, is a remarkably successful blending of two approaches to the science: of theory based on geometry (the original Greek idea), and of numerical prediction based on observation (derived from Meso-potamia through Hipparchus). It is the outstanding example of the application of mathematical theory to a practical science in antiquity. As a system of explanation it cannot be regarded as successful in modern times, since it was always constrained by the inadequacy of Aristotelian physics. But in providing a means of accurately determining the positions of the heavenly bodies from mathematical theory, its place in the history of human endeavour is secure.

Bibliography

Neugebauer, O. 1975. *A History of Ancient Mathematical Astronomy*, 3 vols. Berlin, Heidelberg and New York: Springer Verlag.

Toomer, G. J. 1978. Hipparchus. *Dictionary of Scientific Biography* 15, 207–24. New York: Charles Scribner's Sons.

Toomer, G. J. 1984. *Ptolemy's Almagest*. London: Duckworth.

Toomer, G. J. 1988. Hipparchus and Babylonian astronomy. In *A Scientific Humanist: Studies in Memory of Abraham Sachs*, eds E. Leichty *et al*, pp. 353–62. Occasional Publications of the Samuel Noah Kramer Fund, 9, Philadelphia.

T.W. POTTER

Astronomy in Etruria and Rome

The Romans did not lack brilliant thinkers. Lucretius's *De Rerum Natura*, for example, is a poetic study of atomic theory, which has the power and the ideas to fire the imagination down to this day. Likewise Varro, who wrote on nearly every field of learning, achieved the reputation of a truly great scholar, whose influence has stretched far through the centuries. Yet in science in general, and astronomy in particular, the Romans were not innovative. This was recognised, as the words of the Elder Pliny, written shortly before his death in the aftermath of the volcanic eruption of Vesuvius in AD 79, which engulfed Pompeii and Herculaneum, make clear (*Natural History*, 2.45.117):

> Yet, now in these glad times of peace under an emperor [Vespasian] who so delights in productions of literature and science, no addition whatever is being made to knowledge by means of original research, and in fact even the discoveries of our predecessors are not being thoroughly studied.

Pliny himself was a prodigiously industrious scholar, whose working habits as recounted by his nephew are legendary. No moment, day or night, was spared in the assembly of facts. The surviving legacy is his monumental *Natural History*, an extraordinary encyclopedia culled altogether from the works of 146 Latin and 327 Greek writers (or so his index claims). The observations range from the acute to the bizarre, but overall reflect a world where manuals were perceived as practical and useful, and applied rather than pure science was what really counted. The Greeks might argue about theories concerning the motions of celestial bodies; but for most Romans it was pragmatic considerations, such as whether the aqueduct was functioning (and if not, why not?) that counted. There are no mathematical formulae amongst the numerous graffiti on the walls of Pompeii.

This is not to suggest that astronomical matters were of no importance to the peoples of Italy (and the Latin-speaking part of the Roman empire). Roman 'culture' drew heavily upon that of Etruscan (and Italic) predecessors, among whom the vault of heaven was believed to play a profound part in the affairs of humans. The sky was divided along the cardinal points, the *cardo* (N–S) and the *decumanus* (E–W) – the concept and terms later to become fundamental to Roman surveying and town planning – and was further subdivided into sixteen celestial regions, inhabited by different deities. Portents such as thunderbolts identified the will of a particular god, so that over time the Etruscan world in particular becomes dominated by the need to respond to such omens, combined with the results of divination. When in the fifth century AD, a North African called Martianus Capella wrote his account of the celestial marriage between Mercury and Philologia, the seven handmaids spoke of the seven liberal arts, astronomy amongst them, before gods called from the sixteen regions of the heavens. Here is surely a clear echo of Etruscan astrological practices.

This mystical view of heavenly bodies and phenomena would seem to stand in sharp contrast to the empirical approach of the classical Greeks. It is summed up by the philosopher Seneca, who was born in Spain at the end of the first century BC. 'The difference between us [i.e. Graeco-Romans] and the Etruscans is the following', he remarks (*Natural Questions*, 2.32.2):

> Whereas we believe lightning to be released as a result of the collision of clouds, they believe that clouds collide so as to release lightning: for, as they attribute all to the deity, they are led to believe not that things have a meaning in so far as they occur, but rather that they occur because they must have a meaning.

Seneca's *Natural Questions* included the subjects of physics, geography, meteorology and, of course, astronomy. As with Lucretius, Varro and Pliny, his sources were primarily Greek, not least the writings of Aristotle. Much the same is true of Vitruvius's *On Architecture*, published in the late first century BC. Astronomy is dealt with in Book IX, and reflects his belief that architects must have a working knowledge of the subject. It is, however, a very condensed summary, although it docs include the remark that 'the stars of Mercury and Venus make their retrograde motions and retardations about the rays of the sun, forming by their courses a wreath or crown about the sun itself as centre'. He is in fact quoting the conclusions of the Greek astronomer Heraclides of Pontos (fourth century BC, see p. 75).

So Greek-derived are the astronomical ideas of what has become known as the 'encyclopaedist' or 'handbook' approach of Roman writers on scientific matters that we need hardly dwell upon them. Indeed, they are sometimes dismissed as mere (and inaccurate) plagiarists; but this really will not do. The works of writers such as Pliny the Elder ensured that Greek thought upon matters of science was widely disseminated, both in their own day, and in subsequent centuries, and this did have a profound effect. An influential statesman and writer such as Cicero (born in 106 BC) may have thought the mathematical sciences were useful only in the sharpening of the minds of young men; but he did translate Aratus's astronomical poem, *Phaenomena* (see p. 73), and publish a commentary on planetary motions.

Perhaps what really matters is that astronomical knowledge was put to practical effect. Its importance in governing the agricultural cycle is a recurrent theme in the ancient writers, who are also at pains to spell out which stars to follow when navigating at night. Maritime commerce reached a peak in the first centuries BC and AD, with ships plying their way to all parts of the Mediterranean and beyond, and knowledge of the stars was undoubtedly a factor. Likewise, the collection of surveyors' manuals known as the *Corpus agrimensorum* shows that astronomy was an important part of the training of these men, charged with the task of shaping the landscape of a vast empire. To judge from finds from the house of a surveyor called Verus at Pompeii, their equipment was simple: a *groma* or cross-staff, a Greek-style portable sundial, and measuring devices. More elaborate instruments like the *dioptra*, which was another Greek invention, used for both surveying and astronomical observations, are not represented in either the archaeological or literary record of the Romans. Simple tools could do the job – and extraordinarily well.

Although no astronomical observations of the sort known at Alexandria and Rhodes (see pp. 73 and 76) are attested in Italy or the West, sundials are common (Fig. 31). At least thirty-five have been found at Pompeii, and there was one in the forum at Rome from the third century BC. Thereafter they proliferated in the city, in public places and in the houses of the rich. But all were outstripped by that built by Augustus to commemorate his victory in Egypt in 30 BC (see p. 121). Set in the Campus Martius, it measured 150 × 75 m, and the shadow was cast by an obelisk, brought from Heliopolis, which stood 22 m high and was surmounted by a bronze globe. Now in the Piazza di Montecitorio in Rome, it was a fitting symbol of the practical application of astronomical knowledge in the Roman world. Likewise, under Julius Caesar, Greek astronomical research was utilised to reform the calendar. The old lunar system had become so muddled by Caesar's day that the civic year (of 365 days) was about three months ahead of the solar, and all the seasons were out-of-step. From 45 BC onwards, a solar year of 365 days was adopted, with a leap year every four years, when an extra sixth day (the *bissextus*, hence bissextile year) was added before the kalends (first) of March. Although too long by about three days in every 400 years, the Julian calendar (Col. Pl. V) was to survive down to 1582, when 10 days disappeared by command of the Pope and the Gregorian calendar (which compensates for Caesar's too lengthy year) was adopted. It is that which, effectively, serves to this day.

It would be a mistake to imagine that Caesar's reformed calendar was instantly adopted in all parts of the Roman world. Many cities, especially in the Greek East, stuck resolutely to their old ways, often for centuries; but the long-term significance of the measure cannot be doubted, and it is another illustration of the Romans' extraordinary ability to make pragmatic exploitation of the skills of others.

There is, however, another strand of Roman thinking about astronomy which was less based on scientific fact, and yet of immense significance: namely the rise of astrology.

That the position and movement of celestial bodies might influence human destiny, and allow the prediction of terrestrial events, might seem alien to the Roman mentality. To many Romans indeed it was, and there are a number of instances where practitioners of astrology were expelled from Rome and Italy. Nevertheless, as we have seen, divination played a fundamental part in the world of the Etruscans (who for a time ruled Rome, and were eventually absorbed into the Roman state), so that the concept was already firmly planted in Italy. Indeed, the highly regarded politician and scholar of the first century BC, P. Nigidius Figulus (who was a good friend of Cicero), practised Etruscan

31 A fine Roman marble sundial, divided into 12 hours. Ht 53.6 cm. Unprovenanced. (British Museum, GR Sculpture 2545)

haruspicy. But astrology really evolved in Mesopotamia (often referred to by Classical authors as Chaldea), and was introduced into Rome as Hellenistic Greeks, themselves fascinated by the subject, increasingly permeated the new capital of the world in the second and first centuries BC (where astrologers were sometimes known as *Chaldaei*).

A significant figure in its spread was the Greek astronomer Posidonius, born about 135 BC, who taught amongst others Cicero, and whose writings strongly influenced people such as Lucretius and Pliny the Elder. He wrote some five works on astrology, a subject that was increasingly taken up by one of the most prevalent schools of philosophy in Roman Italy, that of the Stoics. Its spread to the West was also, however, much involved with the diffusion of the eastern so-called 'mystery' or 'personal' religions, such as those of Isis and Mithras. For initiates to the cult of Isis, the rites simulated a cosmic voyage and, although treated initially with suspicion by many Romans, in AD 38 a temple to the goddess rose in the heart of Rome itself. Thereafter these eastern religions, with their close cosmic links, spread rapidly in the Roman world. Even the sun became a focus of worship. Thus, when in AD 203, the emperor Septimius Severus built a huge three-tiered monument, the Septizodium, at the foot of the Palatine Hill in Rome, it was designed to hold seven great statues of the planetary deities, with at the centre the ruler himself in the guise of the sun. It was a true triumph for astronomy.

The ultimate victory in the Roman world for these oriental religions was, of course, for Christianity. The early Fathers of the Church were strongly opposed to astrology, since it seemed to represent a denial of free will. Tertullian, writing in the years around AD 200, could maintain that it was an invention of fallen angels (together with botany and metallurgy!); and for St Augustine (AD 354–430), astrologers could predict the future, but aided by evil spirits intent on deceiving man into believing 'this false and dangerous opinion of fate in the stars' (*City of God*, 5.7). Study of astronomy, on the other hand, was not excluded, as long as it was recognised that Christianity, not science, is wisdom. Thus Origen (born *c.* AD 185) could write that 'I would wish that you should take with you . . . so much of the geometry and astronomy as may be helpful for the interpretation of the Holy Scriptures' (*Epistle to Gregory*, 1).

There were indeed late-Roman writers in science and astronomy, men such as Macrobius, Chalcidius, Martianus Capella, Boethius and Cassiodorus. Although far from original, they helped to ensure the survival of at least some scientific learning into the early Middle Ages, especially through the preservation of manuscripts in monasteries. Whilst many fewer translations of Greek works were available than for Arab scientists of the ninth and tenth centuries AD, scholarship did not wholly wither with the onset of the Dark Ages in the West, and the Church was here to play a major role.

To sum up, there are doubtless many who would hold that the Romans have no place in a book of this sort, a view that we have sought to refute. While the main discoveries redound to the credit of Greek astronomers (not least Ptolemy during the Roman period), it was the eclectic attitude of the Romans that helped to disseminate, and take advantage of, this knowledge. A degree of awareness of astronomy must have been extraordinarily prevalent over the Roman empire, and it was not elitist: soldiers, sailors, farmers and craftsmen will have known something of the stars, as well as the *literati*. That is one of many achievements of a remarkable age.

Bibliography

Bickerman, E. J. 1968. *Chronology of the Ancient World*. London: Thames & Hudson.

Clagett, M. 1957. *Greek Science in Antiquity*. London: Abelard-Schuman.

Dilke, O. A. W. 1971. *The Roman Land Surveyors*. Newton Abbot: David & Charles.

Gibbs, S. L. 1976. *Greek and Roman Sundials*. New Haven and London: Yale University Press.

Iversen, E. 1968. *Obelisks in Exile, I: the Obelisks of Rome*. Copenhagen: G.E.C. Gad.

Le Boeuffle, A. 1989. *Le Ciel des Romains*. Paris: De Boccard.

Lindberg, D.C. 1992. *The Beginnings of Western Science*. University of Chicago Press.

Neugebauer, O. 1975. *A History of Ancient Mathematical Astronomy*, 3 vols. Berlin, Heidelberg and New York: Springer-Verlag.

Samuel, A.E. 1972. *Greek and Roman Chronology. Calendars and Years in Classical Antiquity*. Munich: Beck.

Scott, A.B. 1991. *Origen and the Life of the Stars*. Oxford University Press.

Stahl, W. H. 1962. *Roman Science*. University of Wisconsin Press.

Ulansey, D. 1989. *The Origins of the Mithraic Mysteries: Cosmology and Salvation in the Ancient World*. Oxford University Press.

ALEXANDER JONES

Later Greek and Byzantine Astronomy

During much of the thirteen centuries of Greek-speaking culture that intervened between the career of Claudius Ptolemaeus (Ptolemy) and the fall of Constantinople to the Turks in 1453, astronomy was a highly prized discipline. It became, however, a received science, one to be mastered, explicated, exploited, but scarcely to be tested or augmented. During the twilight of antiquity older writings were gradually lost and rival methods faded from use, until Ptolemy's models and tables became almost synonymous with astronomy. The Byzantine Greeks never quite forgot how to use the tables to predict the celestial phenomena, and their reconquest of the theoretical expanses of the *Almagest* after AD 1300 was among the intellectual highlights of the Middle Ages, preparing for the developments of the European Renaissance. From Byzantium the Islamic world too drew its knowledge of Ptolemaic astronomy; and in return Byzantine scholars studied and translated Arabic works. Their attempts to absorb this 'new astronomy' and confront it with their own Ptolemaic heritage make up one of the most interesting parts of the vast astronomical literature lying, largely unpublished and even unread, in numerous manuscripts.

Late antiquity

Ptolemy's principal astronomical works, the *Almagest*, *Handy Tables* and *Planetary Hypotheses*, appeared in succession during the decades following AD 150. What was the contemporary state of Greek astronomical practice, and what sort of reception did Ptolemy's writings obtain? Using contemporary documents recovered by archaeology, supplemented by clues from the ancient astrological literature, we can begin to answer these questions. And because of the special conditions of climate and culture that favoured the survival of numerous papyrus texts there, we are best informed about the astronomy of provincial Roman Egypt.

The more than 100 astronomical papyri currently known – most of them as yet unpublished – are predominantly numerical tables and texts concerned with the practical task of determining the positions of the heavenly bodies at specific dates. These were the papers of astrologers, whose activity is also witnessed by numerous papyrus horoscopes and fragments of astrological treatises. Temples belonging to the partially Hellenised Egyptian local cults provided one of the venues where horoscopes were computed and cast; but many astrologers were probably independent professionals. Although most of the papyri are written in Greek, Demotic Egyptian texts and tables occur as late as the second century of our era. Aside from the language and script, there is no important difference between the Demotic and Greek documents, so that the astronomy they record can best be described as 'Graeco-Egyptian'.

Unlike its Mesopotamian counterpart, Graeco-Egyptian astronomy had no observational component, except as reflected in the works of 'scientific' astronomers such as Ptolemy; and these theoretical works are almost wholly absent from the papyri. The conventions used in the papyri are fairly uniform: positions of heavenly bodies are given in sidereal ecliptic co-ordinates (longitudes are in degrees within zodiacal signs), and the fractional parts of numbers are expressed sexagesimally, a convention obviously adopted from Babylonian astronomy. The dates are sometimes in the civil ('Alexandrian') calendar established by Augustus; this had 12 months of 30 days each and 5 'epagomenal' days not belonging to any month at the end of the year, with a sixth epagomenal day every 4 years. The older Egyptian calendar, with the same 12 months but invariably five epagomenals (see p. 35), also survived in astronomical tables because of the convenience of computing with uniform years of 365 days. The Roman (Julian) calendar appears infrequently as early as the reign of Augustus, becoming more common in ephemerides from the fourth century AD and later.

The methods by which the Graeco-Egyptian astrologers computed the configuration of the heavens divide into two groups: versions of Ptolemy's tables, and representatives of an older astronomical tradition, largely arithmetical in character, that descends at no great remove from the mathematical astronomy of the Babylonian cuneiform texts (Fig. 32). In the non-Ptolemaic methods one began by tabulating the dates and longitudes of a succession of characteristic moments in the anomalistic cycles of a heavenly body, e.g. summer solstices or the first appearances of Mercury as morning star. The progress in time and longitude from one epoch to the next was found by simple arithmetical rules which, at least for the five planets, can sometimes be traced back half a millennium in Babylonian tablets. One then established how the body in question had moved between the preceding epoch and any given date either by linear interpolation between epoch positions or by looking up the progress in a table (or 'template') that set out a standard pattern of day-to-day motion during the anomalistic period. The templates, which seem to have been a specifically Greek innovation, were also usually computed according to straightforward arithmetical rules, e.g. keeping the speed constant or having it increase or decrease by constant differences.

Ptolemy's tables follow an essentially different structure reflecting the kinematic models compounded of circular motions that he deduces in the *Almagest*. The uniform,

32 A fragment of a late third-century AD papyrus almanac in codex format, excavated at Oxyrhynchus in Egypt. The planetary positions, pertaining to the reign of the Roman emperor Elagabalus (AD 218–22), were in part computed by methods known to us from Babylonian tablets five centuries older. (Egypt Exploration Society, P. Oxy. 3299)

or 'mean', motions are tabulated as a linear function of time since a single epoch date in the remote past; complex trigonometrical tables ('equation tables') are then used to find the longitude and latitude of the body as a function of the mean motions. It was not the original tables embedded in the theoretical exposition of the *Almagest* that achieved wide circulation, but rather the *Handy Tables*, which Ptolemy had adapted for more convenient use. Fragments of several manuscripts of the *Handy Tables* survive on papyrus, some as old as the early third century AD; and almanacs from about this date were often computed using the *Handy Tables*. We also have examples of tables that were modified in various ways from Ptolemy's. On the other hand, Egypt has so far yielded no tables independent of Ptolemy that use the analysis of planetary motion into mean motion and equation tables.

Ptolemy's tables came into common use quickly, but did not immediately drive the Babylonian-style arithmetical methods into oblivion. As late as the middle of the fourth century AD both approaches to astronomical computation are found side by side. This is hardly surprising, since the *Handy Tables* were not only more accurate and theoretically sophisticated than the older techniques, but also much more difficult to use. Vestiges of the Mesopotamian heritage lived on into the Middle Ages in the perpetual almanacs, tables that presented the longitudes of the planets at intervals of 5 or 10 days for a Babylonian Goal Year period ranging from 8 years (for Venus) to 83 years (Jupiter), after which the cycle could be repeated.

The astronomical literature of late antiquity that has come down to us through medieval manuscript copies is dominated by commentaries of one sort or another on Ptolemy's works. On the one hand, a great demand existed for instruction manuals for the *Handy Tables*, which Ptolemy's own curt preface did not satisfy. On the other, the *Almagest* found its way into the curriculum of academic mathematical education, as a sequel to the reading of Euclid's *Elements* and the treatises on spherical astronomy by Autolycus and Theodosius. The earliest known commentary on Ptolemy's tables is an anonymous fragment dating from about AD 203, barely a generation after Ptolemy himself, that rather ineptly discusses the lunar tables of the *Handy Tables* and their relation to the *Almagest*. The third century, a period of political and social instability throughout the Roman empire, has left us little else of an astronomical nature.

During the fourth century AD, by contrast, mathematical and astronomical education flourished in Alexandria at the hands of the pedagogues Pappus and Theon of Alexandria. Pappus, who is deservedly better known for his mathematical *Collection*, also composed a long commentary on the *Almagest* about AD 320, of which parts concerning Ptolemy's lunar and eclipse theory survive. Pappus still had access to some of the older literature that is lost to us – he was well informed about Hipparchus's measurements of the distances of the sun and moon, for example – but for the most part he was content to clarify and fill out Ptolemy's mathematical expositions, and there is no allusion to developments after Ptolemy. The better preserved commentary on the *Almagest* by Theon, dating from the 360s, exceeds in bulk the work it explicates. It is more thorough and carefully written than Pappus's, but even poorer in material of historical interest. Among Theon's other works are a lucid manual to the *Handy Tables* (Fig. 33), and a larger treatise that attempts to show how Ptolemy derived the *Handy Tables* from the tables of the *Almagest* – a difficult undertaking in which Theon acquits himself surprisingly well. The manual is noteworthy for giving the earliest account of the supposed 'trepidation' or oscillation in longitude of the equinoctial points, a concept that survived in Islamic and European solar theory as late as Copernicus (see pp. 149, 153 and 185). Of other commentaries by Serapion, Arcadius and Theon's fabled daughter Hypatia, we know next to nothing.

On a higher plane than the voluminous commentaries of Theon and Pappus is the *Hypotyposis* (or *Outline of the Astronomical Models*) by the Neoplatonist philosopher Proclus, who was active during the mid fifth century AD at Athens. Proclus wished to describe briefly and non-technically the problems, hypotheses and methods of astronomy, which he equated with the contents of the *Almagest*, and his book is one of the best introductions to that work. His descriptions of some of Ptolemy's observational instruments go into more detail than the *Almagest* does, and he knew something about the theory (first set out in a part of Ptolemy's *Planetary Hypotheses* that now exists only in Arabic translation) of models composed of nested spheres for the sun, moon and planets. Proclus rejects Ptolemy's theory of precession, ostensibly on observational grounds, but perhaps also influenced by the lingering preference of the astrologers for

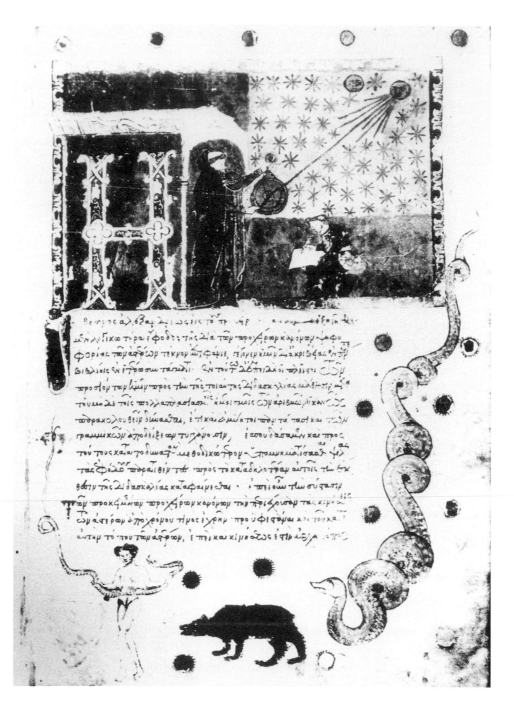

33 An illustrated manuscript of the *Handy Tables* with Theon's manual, copied in 1358. The picture depicts an astronomer holding what is presumably meant to be an astrolabe, with a seated assistant to his right. The figures surrounding the text represent (from left to right) Ophiuchus, one of the Bears, and Draco. (Biblioteca Ambrosiana, Milan, MS H57 Sup., f. 1r)

using sidereal longitudes. He also chastises the astronomers for believing in the physical reality of their models, and maintains a strictly 'instrumentalist' interpretation of Ptolemy's astronomical models as computational devices that have no physical validity.

Given the seemingly total neglect of observational and theoretical astronomy in the 300 years following Ptolemy, it comes as a surprise to encounter in the prefatory matter to some medieval copies of the *Almagest* a list of seven observations made between AD 475 and 510 at Athens and Alexandria. The observers were the brothers Heliodorus and Ammonius and a third man who may have been their uncle Georgius, all Neoplatonists intellectually allied to Proclus. The observations were naked-eye sightings in which the moon or a planet either occulted or passed close to another planet or a fixed star. The only instrument mentioned is the plane astrolabe, which was used to convert equinoctial to seasonal hours. What was the purpose of these observations? Some are compared with positions computed from the *Handy Tables*, but it is difficult to discern in them any systematic effort to check the tables' accuracy.

The plane astrolabe, whose history up to this point is mired in obscurity, emerges about the time of Theon in its normal medieval form combining a stereographic projection of the celestial sphere on one face with a sighting instrument on the other. No ancient specimens survive, but we do possess technical treatises on the astrolabe and its applications by John Philoponus (a pupil of Ammonius, early sixth century AD) and, in Syriac, Severus Sebokht (seventh century); both appear to be reworkings of a lost book by Theon.

Byzantine astronomy

The transition between the periods labelled 'antiquity' and 'Byzantium' may be very crudely characterised as astronomy's passage from one intellectual centre in Alexandria to another in Constantinople, and from a pagan to a Christian milieu. More detailed inspection of course tends to obscure these neat distinctions. Hypatia's murder at the hands of a Christian mob in AD 415 was perhaps no more typical of intellectual conditions in Alexandria than the fact that her Christian pupil Synesius of Cyrene was able to reconcile a dilettantish interest in the pagan sciences with political and ecclesiastical activities, finally becoming bishop of Ptolemais in AD 410. And a century and a half later, the pagan Neoplatonist philosopher Olympiodorus could deliver lectures on astrology, the text of which survives, to an Alexandrian audience that must have been overwhelmingly Christian.

In the meantime Constantinople, although the imperial seat since AD 330 and the site of a 'university' (that is, a school with several endowed teaching positions) since AD 425, was not notable for its astronomers until the time of Emperor Heraclius (AD 610–41), when Olympiodorus's pupil Stephanus of Alexandria came to the city, supposedly at the emperor's summons. Very little is actually known about Stephanus's activity as astronomer and astrologer, except that he is alleged to have cast Heraclius's horoscope.

Heraclius himself appears as the author of a prolix manual of instructions for the *Handy Tables*, modelled on Theon's shorter commentary and adapted for use at the latitude of Constantinople; modern scholars, doubting an emperor's competence in the subject, have presumed that the book must have been ghost-written by Stephanus.

This first establishment of a tradition of technical astronomy in the capital was short-lived. The century and a half between the reign of Heraclius and the beginning of the ninth century has left us negligible traces of astronomical writings, a silence that may be partly the effect of the iconoclastic religious movement of the eighth century and its hostility to scientific and intellectual institutions. Elsewhere in the Greek-speaking empire, however, one finds scattered traces of manuscripts and of practitioners who maintained the ability to read and use them. Not only was the survival in outlying districts of astronomy, even if it was at a modest level, crucial for the revivification of studies in Greek during the ninth century and after, but its role in transmitting Greek astronomy into other cultures should not be forgotten. Thus parts of the *Handy Tables*, with instructions compiled from Theon and other sources, were translated into Latin in the sixth century at Rome; and at the other geographical extremity the astrologers Stephanus and Theophilus of Edessa, rather shadowy figures of the late eighth century, served as points of contact between Greek and Islamic astrology.

In the course of the next century, the intense interest in Greek scientific writings on the part of Arabic scholars must have depleted the stock of old manuscripts still in Byzantine hands; and of such copies as survived from antiquity, often in the form of papyrus rolls, many were damaged or fragile. Fortunately the ninth century also inaugurated a period during which the remains of ancient literature were sought out and recopied in durable parchment codices. The beginnings of this process are often associated by modern historians with Leo the Mathematician, a scholar whose scientific attainments are alleged to have inspired Caliph al-Ma'mūn (see p. 161) to invite Leo to Baghdad about 830. According to the tale, Leo refused, but managed to obtain a teaching position at the church of the Forty Saints in Constantinople, and in time appointment to the archbishopric of Thessalonica. The overthrow of iconoclasm brought an end to Leo's tenure, but he eventually re-emerged as head of another school under imperial patronage in the Magnaura Palace, living at least until 869. Leo's personal library is known to have included manuscripts of mathematical authors (Archimedes, Apollonius), at least one astrological collection, and perhaps the *Almagest*; and although none of his copies survive, their importance for the survival of these texts cannot be doubted. More questionable is Leo's ability to interpret these highly technical works, for his own surviving writings are meagre and unimpressive.

Among the surviving Greek manuscripts from the time of Leo and shortly after, copies of scientific works are remarkably prominent – an emphatic contrast to the obscure place these writings have held in modern Classical studies since the eighteenth century. The astronomical manuscripts include four of the *Almagest* (some incorporating minor writings of Ptolemy), another four of the *Handy Tables*, two containing commentaries by Theon and Pappus, and one with the writers on spherical astronomy. These codices were written by skilled professional calligraphers, usually employing the new Greek minuscule script, and the cost of both the parchment and the scribe was high. The versions of the texts that they contain are often important for modern editors because they are closer to the authors than other manuscript copies and exhibit few attempts to correct or improve upon the text as received. But this same paucity of corrections by scribe or owner also suggests that, for all their splendour, these manuscripts were more for display than for study. Original writings from the ninth and tenth centuries, whether in the margins of the extant contemporary codices or in later copies, are pitiful and scarce. One concludes that practical understanding of astronomy was sustained by few besides the astrologers, whose working copies of the old texts were presumably more perishable than the bibliophiles' treasures that have come down to us, but whose existence is revealed by the odd horoscope or anecdote.

The eleventh century brings a reversal of this situation: almost no surviving contemporary manuscripts, but renewed activity in scholarship, attested by later and, alas! very unsatisfactory copies. The earliest of these texts, a long marginal note to the *Handy Tables* originally written about 1032, already shows that Byzantine astronomy was now embarked on a new course. For the unknown author mentions various parameters of the sun's motion that differ from Ptolemy's values, parameters that were in fact measured by the astronomers of al-Ma'mūn at Damascus in the first half of the ninth century, and he also discusses astronomical tables of a certain 'Alim' who can be identified as the tenth-century astronomer Ibn al-A'lam (see pp. 153 and 163). In the latter case, at least, we are dealing with an actual translation of Arabic tables into Greek, which, though not now extant, was still available in the twelfth century; we are told moreover that the Greek version of Ibn al A'lam's tables utilised the Byzantine (Julian) calendar.

Further acquaintance with the works of the ninth-century Arabic astronomers is revealed by an anonymous manual, mostly assembled in the 1060s, that gives instructions and examples of astronomical computations such as the components of a horoscope and the characteristics of eclipses, by methods that are wholly non-Ptolemaic. The text refers to the tables of Ḥabash al-Ḥāsib (see pp. 151 and 155), whereas the long section concerning solar eclipses proves to be a competent Greek translation of the original instructions of al-Khwārizmī's *zīj* (see pp. 148 and 151), followed by a worked example using the eclipse of 20 May 1072. The author actually observed this eclipse, using an astrolabe to sight the solar altitudes (cf. Fig. 34). A more chaotic collection of astronomical chapters and tables, dating from the twelfth century, mentions further Arabic astronomical authorities. It contains a brief handbook on the astrolabe, 'compiled from various methods taken from a Saracen book', employing transliterations of Arabic technical vocabulary. These texts are manifestly the fragmentary remains of a larger

34 The only surviving Byzantine astrolabe, this instrument is signed by one Sergius, a 'Persian', and dated July 1062. It is thus contemporary with the earliest Arabic influences on Byzantine astronomy. IC no. 2 (Civici Musei d'Arte e di Storia, Brescia)

transmission of Islamic astronomy into the Byzantine world, although it is not necessary to assume that Greek translations existed of every Arabic author named; some of the information was doubtless second hand, and some tables may have been consulted directly in the Arabic. Of the scholars participating in this transmission of the eleventh and twelfth centuries, only one, the polymath Symeon Seth (late eleventh century) is known by name. The chronicler Anna Comnena, who had much experience of astrologers at her father Alexius I Comnenus's court, was satisfied of Seth's competence by his predicting the death of the Norman Robert Guiscard in 1085. His knowledge of Arabic astronomy may have been at a less technical level than that of the anonymous texts, and might perhaps be connected with a sojourn in Egypt, where he had observed a solar eclipse in 1058.

The Islamic tables that were now becoming accessible in Byzantium did not differ profoundly in arrangement, notation, or purpose from Ptolemy's tables, on which, after all, the Arabic *zījes* had to a great degree been modelled. We have already seen that the Greek writers were aware of differences between the old and new tables in some of the numerical parameters, e.g. the mean motions of the sun, moon and planets and the rate of precession. Other tables, such as the star catalogues and the lists of co-ordinates of principal cities, were presumably more up to date in the Arabic works; it was from such a source, rather than observation, that the eleventh-century astronomers first corrected Ptolemy's erroneous latitude for Constantinople. And the indirect influence of Indian astronomy was felt in the introduction of sine tables and in features of al-Khwārizmī's methods of eclipse prediction that had little resemblance to Ptolemy.

To judge from the career of Seth, and from the largely astrological contents of the later manuscripts that preserve the anonymous manuals, the astronomical writings of the eleventh and twelfth centuries were still the work of practising astrologers. The great men of letters of this time, such as Psellus and Eustathius, at best betray knowledge of the rudiments of astronomy; nor do we find prominent figures of church and state writing on the science, as their counterparts were to do during the Palaeologan period. That fine manuscripts of astronomical classics were nevertheless esteemed – if not actually read – by the mighty is suggested by the emperor Manuel I Comnenus's choice of a tenth-century copy of the *Almagest* as a gift to the Sicilian king William I about 1158.

The sack of Constantinople by the Venetians and their allies in the Fourth Crusade (1204) was a cultural as well as a political disaster for the Byzantine Empire, and astronomy seems to have experienced a hiatus that lasted through the 57-year interval during which the seat of the empire was displaced to Nicaea, and indeed until the reign of Andronicus II Palaeologus (1282–1328). The recovery, when it came, was fast and vigorous. Manuscripts were produced in great numbers, and the many surviving copies of astronomical works, old and new, that date from the late thirteenth and fourteenth centuries prove to have been scholars' books, written in often crabbed hands on paper rather than parchment, but bristling with corrections and annotations.

One of the new lines of astronomical scholarship begins with Theodorus Metochites (1270–1332), minister of Emperor Andronicus II and a scholar of wide interests. In 1314, when he was forty-three, Metochites began to study astronomy with the help of Manuel Bryennius, and he eventually produced a vast treatise, entitled *Elements of Astronomy*, that constituted the first significant Greek attempt to master the theoretical and mathematical content of Ptolemaic astronomy since the time of Theon. Metochites has nothing to say about Arabic contributions, and elsewhere asserts that Ptolemy had left nothing for his posterity to discover in this field.

Metochites's pupil Nicophorus Gregoras inherited his master's hard-won grasp of Ptolemy, and exploited it by predicting the solar eclipse of 16 July 1330. He undertook this most difficult of computations not merely as a scientific exercise, but as a sally in his bitter polemic with the Calabrian monk Barlaam of Seminara. Barlaam, however, rose to the challenge by using Ptolemy's tables to predict the solar eclipses of 14 May 1333 and 3 March 1337. It is not known whether either Gregoras or his adversary, for whom mathematical astronomy was only a minor issue in their wide-ranging dispute on religious and other questions, bothered to observe the eclipses. Gregoras's other works include two handbooks on the astrolabe; and he suggested a revision of the calendar based on a value for the length of the tropical year more accurate than Ptolemy's. Although Gregoras claims to have established this parameter by observation, it is actually equivalent to a value for the rate of precession of 1° in 66 years which the Byzantine astronomers of the eleventh century had already learned of from Arabic sources.

But at the same time as Metochites was reviving the Ptolemaic tradition, more recent astronomical texts were passing into Greek from Persia. In contrast to the Arabo-Byzantine contacts of the eleventh and twelfth centuries, the new transmission took the form of translations or adaptations of complete *zījes* (see pp. 150–55), including the *Zīj al-Sanjarī* of al-Khāzinī (*c.* 1120) and the *Zīj al-ʿAlaʾī* of al-Fahhad (*c.* 1150). Some or all of these versions are almost certainly to be attributed to one Chioniades. According to the account of George Chrysococces a generation later, Chioniades had studied the mathematical sciences and medicine at Constantinople but, in order to master astronomy, found it necessary to travel to Trebizond and beyond to Persia, whence he brought back various treatises that he translated into Greek.

The translations that have come down to us date from about the years 1295–1302, and display a progressing grasp of Persian and Arabic and of the technical vocabulary of astronomy. We have two manuscripts containing these works that are nearly contemporary with the translations. For reasons that are not yet clear, however, the Persian tables seem to have attracted little attention in Constantinople until a generation later, when several copies were made. It was in the 1340s as well that Chrysococces, who had learned astronomy from a pupil of Chioniades, produced the *Persian Syntaxis*, a set of tables originally translated by Chioniades, with new instructions by Chrysococces himself. Nearly twenty copies of this work survive. The central tables for computing the longitudes of the heavenly bodies in the *Persian Syntaxis* derive from the *Zīj-i Īlkhānī* of Naṣīr al-Dīn al-Ṭusī (*c.* 1270). This is not the only evidence of Byzantine familiarity with the quite recent work of al-Ṭusī and the Maragha astronomers (see pp. 150, 151 and 164), for among the ʿChioniadesʾ materials dating to the beginning of the century is a short illustrated text that sets out al-Ṭusīʾs innovative model for the moon.

But documents of such theoretical interest were not typical productions of the astronomers of the fourteenth and fifteenth centuries, who remained preoccupied with the adaptation and manipulation of a bewildering profusion of tables. Chrysococcesʾs tables reappear in slightly modified form as the topic of the last part of the *Astronomical Tribiblos* (*c.* 1352) of Theodorus Meliteniotes, who, among other lofty ecclesiastical functions, directed the patriarchal school at Constantinople; the first two parts of this unusual ʿsynthesisʾ of astronomical traditions comprise a manual for the astrolabe and a commentary on Ptolemyʾs tables. Unusually, Meliteniotes vehemently deprecates the astrological concerns that had been prominent in Chrysococcesʾs work, as in most of the sets of tables of this period. About the same date, Cyprus emerged as a centre for up-to-date methods, producing another revision of the tables of the *Persian Syntaxis* and a translation from Latin of the *Toledan Tables*. Byzantium was by now drawing as much from the West as from the East in its appetite for tables; and it is not surprising to find Greek adaptations of Hebrew works (by way of Latin) in the fifteenth century.

The Turkish capture of Constantinople in 1453 seems to have been accompanied by less violence and destruction than the Venetian sack two and a half centuries earlier.

But the blow it inflicted on Byzantine astronomy proved fatal, because the flow of Greek scholars and Greek manuscripts to Italy and other parts of western Europe, both before and after the final defeat, drained the Greek world of the resources necessary for yet another scientific revival. Nearly every astronomical manuscript that we now possess was in the West by the sixteenth century, brought over by refugees such as Cardinal Bessarion and Isidore of Kiev, or by Western humanists visiting the East in search of literary treasures. European scholars made uneven use of this inheritance, neglecting almost entirely the Persian tables and the other Islamic materials (although it has been suggested that Copernicus somehow learned of the Maragha models for the moon and planets, which recur in his works, from the Byzantine manuscripts). The Renaissance's greatest debt to Byzantium in astronomy was ironically for its most conservative aspect, the preservation of good texts of the writings of antiquity, especially the *Almagest* and its commentators, which helped to prepare the ground for Regiomontanus and Copernicus.

The course of medieval Greek astronomy was shaped by several forces. The retrospective, antiquarian character of Byzantine secular culture was only imperfectly counteracted by intercourse with the Islamic world and the western Mediterranean until the very latest period. Moreover, astronomical tables were the indispensible ancillary to astronomy, and mastery of the most difficult varieties of prediction (eclipses) and of the geometrical principles underlying the tables proved a scholar's intellectual prowess; but the reputed perfection of Ptolemy's works left little role for observation or new theoretical speculation. And finally, the vicissitudes of politics and religion periodically subjected learning to long interruptions, which were particularly disastrous for a highly technical science dependent for its continuity on competent instruction. One is astonished, not so much that later Byzantine astronomy was an astronomy of manuscripts and computations, but that, being such, it could flourish so vigorously.

Bibliography

Neugebauer, O. 1975. *A History of Ancient Mathematical Astronomy*, 3 vols. Berlin, Heidelberg and New York: Springer-Verlag.

Pingree, D. 1964. Gregory Chioniades and Palaeologan Astronomy. *Dumbarton Oaks Papers* 18, 133–60.

Tihon, A. 1981. L'astronomie byzantine (du ve au xve siècle). *Byzantion* 51, 603–24.

Tihon, A. 1983–. *Corpus des Astronomes Byzantins*. (Editions and translations of Byzantine astronomical writings, in progress.) Amsterdam.

Tihon, A. 1990. Tables islamiques à Byzance. *Byzantion* 60, 401–25.

Tihon, A. 1993. L'astronomie a Byzance à l'époque iconoclaste (viiie–ixe siècles). In *Science in Western and Eastern Civilization in Carolingian Times*. Basel.

Wilson, N.G. 1983. *Scholars of Byzantium*. London: Duckworth.

J.V. FIELD

European Astronomy in the First Millennium: the Archaeological Record

Other chapters in this book consider the astronomical instruments described in surviving texts. On the whole, such instruments were intended for astronomical research. They would have been used by people with a serious interest in astronomy for its own sake, or perhaps in connection with some learned application of it, say in an attempted reform of the calendar. The archaeological record tells a very different story. First, it is extremely thin: possibly due to instruments largely being made of perishable materials, such as wood, or obviously recyclable ones such as bronze or brass. Second, what we find overlaps only to a very small degree with the evidence from written sources. Sometimes it shows up the inadequacy of written sources, most particularly in regard to relatively sophisticated devices for use by the layman, but it also leads to elegantly scholarly disputes about such matters as exactly what Vitruvius (fl. *c.* 40 BC) may have meant by some particular Greek term. Such disputes can probably never be completely resolved.

Most of what survives in museum collections has to do with time-keeping or with the calendar. This is hardly surprising, since these were the two most obvious uses of astronomy in everyday life, for citizens and rulers alike. We shall begin, however, with less visibly practical objects, namely celestial globes.

Celestial globes

One celestial globe survives from the ancient world, the marble sculpture known as the Atlante Farnese, shown in Col. Pl. VII, which seems to date from about 200 BC. At least, we must presumably consider it as a celestial globe, though the giant Atlas is clearly at least as important a component of the composition as the globe is. Moreover, the constellations, shown in relief, are notably inaccurate. A fairly large globe such as this – its diameter is 65 cm – can be made more accurate than a small one, but this consideration does not seem to have weighed overmuch with the maker of the present example. What we have is a decorative symbolic sphere of the heavens. Its intellectual connection is not with technical astronomy but rather with the astronomical poem of Aratus (*c.* 310–240/239 BC), *Phaenomena* (Φαινομενα), which describes the heavens as patterned with figures embodying the cultural history and traditions of Aratus's world. Whereas to the Psalmist the heavens were, rather vaguely, telling the Glory of God, to Aratus they are telling the story of the civilisation to which he belongs. Given this symbolic meaning, which is characteristically classical in its intermingling of the human and the divine, it is not surprising that we also find images of the celestial sphere used on coins, from the first century BC onwards. The terrestrial globe also appears. Images on coins have the advantage of generally being among the most precisely and securely datable of our records, but they have the more than compensating disadvantage of being very small. The celestial globe is identifiable, for instance, by the presence of marks that seem to denote stars, or two circles crossing at an angle that seems intended to be the $23\frac{1}{2}°$ between the equinoctial circle (the celestial equator) and the ecliptic.

Similarly rudimentary celestial globes appear as a symbol to identify Urania, the Muse of Astronomy, say on mosaic pavements. The globe is made apple-sized, so as to fit comfortably into her hand. As for the images on coins, viewers are presumably expected to recognise the object from rather slender clues. A globe also sometimes appears to identify Aratus himself. He seems to be the favourite astronomer for decorative schemes. A particularly clear example occurs on a mosaic floor from Mérida, Spain (Emerita Augusta, founded in AD 40 as a model city to house retired legionaries). A photograph of the relevant part of the elaborate floor mosaic is shown in Fig. 35. The mosaic was first uncovered in 1834, and a drawing made at that time shows details which

35 Detail from a mosaic pavement showing Aratus with a celestial globe. Ht of human figure about 80 cm. (Museo Nacional de Arte Romano, Mérida)

36 The 'Philosophers' Plate'. Silver, 43 × 28.5 cm. About AD 300–700? (Collection of the J. Paul Getty Museum, Malibu, California, 83.AM.342)

have since been lost. Among these details are the lower parts of the four legs of the stand which holds Aratus's globe. The stand is such a realistic touch that it seems likely that globes like this were really made. It would appear that the sphere is merely resting in the stand, since its orientation, with one equinoctial point clearly visible, does not seem to relate in any way to a possible axis designed to allow the globe to be turned. The circle supporting the globe might perhaps have been used to represent the observer's horizon. Aratus, as an astronomer, has a pointer (like the one held by Urania elsewhere in the pavement) and also, as a poet, a wreath of laurel round his head.

A considerably smaller globe, on a rather similar stand (it may have only three legs, but probably has four), appears in a fragment of a silver plate acquired by the Getty Museum (Malibu, California, USA) in 1983 – see Fig. 36. The plate has imprecise provenance and cannot be dated exactly, but its classicising style suggests a date earlier than about AD 700. There have, however, also been claims that the piece dates from the sixteenth century, or from the twentieth. Greek inscriptions identify the man on the left, writing in a book, as Ptolemy, and the female personification behind him as Skepsis (Doubt). On the other side we have Hermes, presumably Hermes Trismegistus, since he too appears as an author, and a female personification pointing upwards to a figure who is probably Christ in Majesty (Christ Pantocrator, that is Christ Ruler of All). The purpose of the globe is no doubt to identify the subject of discussion as being astronomy. There are no literary references to a debate between Ptolemy and Hermes, in any period. Absence of evidence is notoriously not evidence of absence. However, given the extent of surviving records from the periods concerned, this sensible rule seems more rigorously applicable to the first few centuries AD than to the sixteenth century, and therefore seems to favour the earlier dating. The globe in fact fits in rather well with our other examples. There are marks on it which appear to represent stars, though in no identifiable pattern, and there is also a circle, directly facing us, centred on the knob at the top of the nearest leg and about the right size for the $23\frac{1}{2}°$ circle round a pole. So the globe might be designed to rotate about an axis through its poles.

This small amount of evidence for the existence of simple celestial globes is supported by an important piece of literary evidence, namely Plutarch's statement that Archimedes (c. 287–212 BC) made a model to show the motions of the heavens, and that this device pleased him so much that, contrary to his habit of not writing about his mechanical inventions, he wrote a book on it, called *Sphere-making* ($\Sigma\phi\alpha\iota\rho\sigma\pi\sigma\iota\eta\sigma\iota\varsigma$). This title sounds as if it referred to something resembling a celestial globe, and until the middle of the twentieth century it was usual for historians to suggest that the object involved may have been something like an armillary sphere (see pp. 165, 232 and 259– 62). This judgment now seems hopelessly condescending. Strong evidence has emerged that at least by the first century BC the Greeks were capable of making complicated trains of gearing, and there is even some evidence to suggest that the invention of the toothed wheel may go back to the third century BC. Perhaps the inventor concerned was Archimedes? In any case, we now have every reason to suppose that his 'sphere' was something a great deal more complicated than a fixed armillary.

Mathematical gearing

It had long been known that the Romans used heavy gearing to transmit power in water mills, and perhaps also in other machines. The gearing of water mills is described in some detail by Vitruvius. It is not known when it first came into use. In this kind of machinery, the precise ratios between the numbers of teeth on the wheels concerned are not crucial to the functioning of the machine, though some care must be taken to ensure that components move at reasonable speeds.

Here we are concerned with a rather different kind of gearing, for which there is no literary evidence, namely what may be called 'mathematical gearing', in which the

37 The Antikythera mechanism, fragment A, photograph and X-radiograph. Bronze, first century BC. (National Archaeological Museum, Athens, X.15087)

exact ratios between different motions is crucial to the functioning of the device – as it is, for instance, in a modern geared clock. The evidence for the existence of such mathematical gearing in ancient times emerged in 1900, though it was not properly understood until about 50 years later.

The evidence is archaeological, and it in fact 'emerged' in the most literal sense, since the object in question was recovered from a wreck. The discovery was accidental. Some sponge divers were driven off course by high winds and took refuge close to the small island of Antikythera (so called because it lies off the larger island of Kythera), off the south coast of Greece. While diving in these enclosed waters, they came across the wreck of an ancient ship. It was the first such wreck to be investigated by the methods of underwater archaeology. The objects recovered included some high-quality bronze sculpture and some equally high-quality items made of glass, which are beautifully well preserved. There was also a heavily corroded metal object shaped like a small version of a shoe-box. After some months, it cracked apart. The reason for this turned out to be that part of it was made of wood which had now dried out and crumbled away. The fracture revealed substantial traces of a large number of gear wheels. Later, the pieces broke again, so that the object is now in a number of fragments (Fig. 37).

The other objects found in the Antikythera wreck dated the ship to the first century BC, but scholars were at first very reluctant to face up to the fact that what had now become known as 'the Antikythera mechanism' was also of this date. Ideas about ancient technology had to be changed. Until that happened, there always seemed to be the possibility that the Antikythera mechanism might be what archaeologists call an 'inter-loper', that is, an object of more recent date that had somehow come to rest in the wreck. There is no equivalent for metals to the carbon-14 dating that can be carried out on organic materials.

Chemical tests showed that the Antikythera mechanism was made of bronze, though nearly two millennia under the sea had ensured that what remained consisted almost entirely of corrosion products. As we have seen, it was the consequent fragility of the piece that led to the discovery that it contained gear wheels. Fragmentation also revealed that the gears were arranged in depth, and in the 1950s X-radiographs showed some additional wheels. Enough detail had now been revealed to make it possible to attempt a reconstruction of the gear trains.

Fortunately, enough inscriptions had been preserved to make it certain that the output of the gearing (if we may be permitted the modern hi-tech term) included the positions of something on the ecliptic. For example, one piece of scale has the inscription XHΛAI (*chelai*), meaning 'claws', which was the older name for the constellation now known as Libra. (The name refers to the claws of the adjacent constellation of the Scorpion.) However, when it came to the task of counting teeth, it was found that there were no wheels that were complete and completely visible. The reconstruction of the trains thus necessarily involved a fair amount of highly educated guesswork. This relied partly on what was known of the astronomical numbers current in the first century BC, and partly on ideas about how the gear wheels might have been made. As can be seen in Fig. 37, the teeth have the shape of an equilateral triangle, which suggests that they have been cut with a 60° file of the kind used to sharpen saws. The more interesting part of the problem is to decide what procedure has been employed to divide the rim of the wheel in order to find the positions of the teeth. At certain points in the reconstruction it had to be decided, for instance, whether a partially visible wheel was more likely to have, say, sixty teeth or fifty-nine. In such a case, the supposed method of laying out positions of teeth is clearly relevant.

Notwithstanding the inevitable uncertainties, it proved possible to show beyond reasonable doubt that the Antikythera mechanism had originally contained upwards of thirty gear wheels, including a differential arrangement, and that its outputs had included the positions of the sun and the moon on the ecliptic. The mechanism was apparently turned by hand. This reconstruction was published in 1974.

The context in which the piece had been found suggested it was an expensive item, but that followed in any case from its being made of metal. Metal was an expensive material. However, the gear wheels are thin and small, and appear to be more or less at the limit of size for a mechanism of this complexity, which does suggest that technical skill was being displayed to the full. The sophistication of the mechanism also shows that we are certainly not looking at a tradition that is in its stumbling infancy. Perhaps we have a descendant of Archimedes' mechanical sphere?

It is now extremely unfashionable to doubt the antiquity of the Antikythera mechanism, because in 1983 another piece of Greek gearing appeared (Fig. 38). These gears seem to date from the early sixth century AD and, since no-one is eager to claim gearing as a Byzantine invention, historians of technology accept this later gearing as a product of an ancient tradition that was still active in the Early Byzantine period.

The four fragments of the device are made of brass. Dating was from the style of the heads of the planetary gods used to identify the days of the week, the style of the script, and the place names in the list of latitudes supplied for the sundial to which the gearing was attached. We shall turn to the sundial, and calendrical matters, below. Conveniently, the sundial is of a type known in a number of other examples. The gear wheels look like those of the Antikythera mechanism, but are of much stouter

38 Early Byzantine sundial-calendar. Brass, diam. of sundial plate 13.5 cm. About AD 520. (Science Museum, London, 1983–1393)

construction, and are very well preserved. Small traces of silicon, found by the micro-probe of a scanning electron microscope, suggest that the sundial-calendar had been buried in sand. The absence of substantial corrosion indicates that the sand was dry.

The gear trains of the Byzantine device can be reconstructed much more easily than those of the Antikythera mechanism, for two reasons. First, there is only one simple way in which the surviving pieces can be assembled, and the resulting partial trains are convincing both mechanically and astronomically. Second, this reconstruction gives part of a gear train for which there is literary evidence, albeit in Arabic and dating from the eleventh century AD. The text concerned describes something that the author, al-Bīrūnī, calls 'the Box of the Moon'. Other devices mentioned in his book are ascribed to particular inventors, but this one is not. It is thus possible to take the Box of the Moon as a traditional and well-known mechanical device. The discovery of a Byzantine example accordingly suggests that Islamic interest in Greek culture, which became active from the ninth century AD onwards, extended not only to mathematics and natural philosophy but also to ingenious mechanical devices.

The outputs of the Byzantine gearing are in principle astronomical, namely an indication of the shape of the moon, and the positions of the sun and the moon in the ecliptic, but their significance is almost certainly essentially calendrical, since we have the days of the week and the device has been provided with a ratchet which allows it to be clicked forward one day at a time. There is no ratchet in al-Bīrūnī's Box of the Moon. The ratchet prevents the mechanism from being turned backwards, which makes re-setting awkward and suggests that the instrument was not intended for calculation.

Because the Byzantine gears are so well-preserved, it is possible to see how their teeth were laid out. On the moon wheel, which has 59 teeth, numbered 1 to 29 and then 1 to 30, there is a radial line down each tooth, indicating the position for filing. Interestingly, some of the teeth are canted over, suggesting that the workman was not carrying out his task with anything more than what he regarded as an adequate degree of precision. Although the object is made of metal, and must therefore have been expensive, it was clearly a workaday device to the man who made it. There is no exact geometrical way of dividing a wheel blank into fifty-nine equal parts, but there are any number of approximate ways that were used by clockmakers from the thirteenth century onwards. Most of them could have been used by a craftsman in the early sixth century. What is of interest is not exactly how he made the division, but that he did make it. If the Byzantines could make a wheel of fifty-nine, then it seems highly likely that a Greek of the first century BC could also have done so. The discovery of the Byzantine device accordingly changed some of the self-imposed rules for educated guesswork that needed to be applied in reconstructing the Antikythera mechanism.

Since 1983, further investigations of the Antikythera fragments have been carried out, employing some new mechanical insights that are partly the result of using images obtained by X-ray tomography. Further attempts are also being made to read more of the fragmentary inscriptions. Some of the new information obtained proves that the reconstruction of 1974 cannot be correct in all its details, though there is no doubt about the general astronomical nature of the device. Unfortunately, a new reconstruction has not yet become possible. It is, however, certain that the device must indeed have been at least as sophisticated as the older reconstruction proposes.

Calendrical considerations

The complexity of the Antikythera mechanism is such that one cannot suppose that instruments like it were ever common. Insofar as the piece appears to have a 'use' one may say that it seems to have been capable of demonstrating the motions of the sun and the moon, of showing something of changing astronomical appearances (such as the phases of the moon) and perhaps also indicating the dates of eclipses. That is, its uses must have been essentially the same as those of the instrument Plutarch says was made by Archimedes. One must of course bear in mind that before the introduction of the Julian calendar (46 BC) the position of the sun in the ecliptic did not have a simple correspondence to the date. So the possible uses of the Antikythera mechanism are not all quite as rarefied as they may at first seem.

The gear trains of the Byzantine piece are not (as far as we know) merely parts of the trains in the Antikythera mechanism, and the essentially practical purpose of the

39 Reconstruction of the Early Byzantine sundial-calendar (Fig. 38). Brass, made by M.T. Wright. (Science Museum, London, 1985–222)

40 Roman portable sundial. Bronze, diam. 6 cm. About AD 250 (Museum of the History of Science, Oxford, inv. LEI 1)

instrument is fairly clear. The portable sundial to which the gearing is attached is, moreover, of a type that is known from more than a dozen examples, some the product of recent and well-recorded archaeological excavation. Like all sundials, this type requires the user to have some additional astronomical information, in this case his latitude and the position of the sun in the ecliptic. A list of latitudes is provided on all known examples of this kind of sundial (a different list on each dial). But once we begin to think about the correspondence between the date and the position of the sun in the ecliptic it is clear that the Byzantine sundial-calendar contains some internal contradictions. To understand these we shall first need to examine the working of the sundial.

The sundial markings are in two parts. Round one quadrant of the outer rim there is a scale, in degrees, which is used to adjust the position of the suspension arm according to the latitude at which observations are to be made. Running across the centre of the disc is a double fan scale, marked with dates, using the months of the Julian calendar. This is used to adjust the position of the combined gnomon and scale which is mounted on an axis passing through the hole in the centre of the disc. This piece is missing in the Byzantine example. The reconstruction shown in Fig. 39 is based on the corresponding part of the only dial of this type to survive in a complete form. This complete example, which is inscribed in Latin, is earlier than the Byzantine sundial-calendar and its method of adjustment is slightly different, but the principle of the calendrical scale for adjusting the position of the gnomon-hour scale is the same.

The Roman dial (Fig. 40) has dates for the solstices (25 December, 25 June) which give the conventional date for the spring equinox of 25 March. Radial lines in principle show divisions between the zodiac signs, so one would need a calendar to find the corresponding dates. This dial presumably came with a scroll of instructions. The Byzantine sundial-calendar is much larger than the Roman dial, which allows its scales to show more detail. The scale to adjust the gnomon is marked with the months of the Julian calendar, but each of the signs on either side of the equinoxes is divided into three – as if they were zodiac signs divided into decans in the Egyptian manner. This does not fit very well with the fact that the calendrical gearing makes use of the 7-day week (a Judaeo-Christian unit). Worse still are the dates given for the equinoxes. The line that marks them runs between March and April and between September and October. (A much better fit with astronomical reality could have been obtained by running the equinox line through the 20/21 March line between the second and third decans of the month.) All the remaining eight dials of this type with inscriptions in Greek also have equinoxes of 31 March and 30 September. The explanation would seem to be that the dial was designed using the Egyptian or Alexandrian calendar, in which the spring equinox does fall very close to the division between two months, and that the names of the months were then changed to those of the nearest months in the Julian calendar. In fact, among the recorded examples of dials of this type there is one, found in Egypt in the mid nineteenth century, which uses the Egyptian months. Unfortunately, this dial now appears to survive only in the form of a drawing. However, the evidence, such as it is, does tend to suggest that even by the sixth century the Julian calendar reform could sometimes be treated rather casually.

The calendar scale on this type of dial is in any case strongly non-linear, making correct adjustment rather awkward. In fact, the hour scale attached to the gnomon is also strongly non-linear, making reading difficult, but this is probably not important. Classical texts never refer to divisions of an hour, so it was presumably enough for the dial to indicate that the time was, say, between the third and fourth hours. To use this kind of dial, one did, however, need to know whether the hour was before or after midday, since the scale only shows how far the time is from noon (using the standard system whereby the period from dawn to sunset was divided into 12 hours).

Sundials, portable and fixed

The kind of portable sundial just described is the commonest one in the archaeological record. Quite a lot of astronomical knowledge has gone into its design, together with some use of approximation. Modern investigation shows that the approximations used could introduce errors of up to a quarter of an hour. Incorrect adjustment for date or latitude would also introduce errors. However, errors may well have passed unnoticed. The expected outcome of an attempt to check against the time given by another instrument may be guessed from the remark Seneca (d. AD 65) puts into the mouth of one of his characters: 'I cannot tell you the hour exactly: it is easier to get agreement among philosophers than among clocks.' ('Horam non possum certam tibi dicere:

facilius inter philosophos quam inter horologia conveniet,' *Apocolocyntosis*, 2.2.) Considering how simple the motion of the sun appears to be, rising roughly east and setting roughly west each day, sundials are surprisingly complicated. There seems to have been incomprehension at the highest political level when a sundial captured in the sack of Syracuse (latitude about 37°) in 212 BC failed to show the right time when transported to Rome (latitude about 42°).

Earlier forms of the portable dial we have discussed probably had flat scales, but the introduction of the curved hour scale is not a gratuitous complication, since it allows noon to be shown, and brings this kind of dial closer to the commonest kind of fixed sundial, in which the scale was on the inside of a sphere (Fig. 31). In fact we have a single example of another kind of portable sundial that also has a curved scale, a rather elegant ring dial found at Philippi and apparently dating from the fourth century AD. The advantage of drawing the sundial scale on the inside of a sphere is that the path of the shadow of the tip of the gnomon will then be a circle, since the sphere of the sundial surface mimics the sphere of the heavens, around which the sun is moving. As the sun moves, the line joining it to the tip of the gnomon sweeps out a cone: one of the standard ancient definitions of a cone is that it is the surface made by a line with a fixed point (here the tip of the gnomon) and another that moves round a circle (the sun). So the path that the line will trace on a flat surface is a conic section. Accordingly, if you want a flat sundial, you need to draw the conic swept out by the tip of the shadow at each solstice, and mark the hour points along each. Reasonable hour lines are then obtained by joining corresponding points with straight lines. This is only fairly easily said, and not at all easily done. Elaborate, but incomplete and unexplained, instructions are given by Vitruvius. With some help from the less practical but more mathematically coherent account given by Ptolemy (in a short book called *On the analemma*), Renaissance commentators made a thorough job of explaining matters properly. Unlike today's writers on sundials, they were not merely commenting on a classical text, but also contributing to the understanding of a problem that was still of current interest. Even as late as the eighteenth century, an expensive mechanical clock was liable to need restarting, or at least resetting, by a sundial every now and then.

Most of the sundials known from the first millennium are quite small. However, the connection with the calendar, which is a genuine astronomical connection, also seems to have allowed sundials to be given a larger cultural significance, rather like that apparently attached to representations of the celestial sphere. One of the most spectacular monuments of Western astronomy was the huge sundial set up by Augustus using one of the obelisks from Heliopolis as a gnomon and incorporating the *Ara Pacis* into its calendrical scheme (see p. 94). Careful archaeology, partly carried out in the cellars of a fashionable part of modern Rome, has led to the discovery of parts of the dial, for instance the Greek letters pi, alpha, rho, theta, in bronze inlaid in travertine marble, giving the beginning of the name of the constellation of Virgo (Parthenos, ΠΑΡΘΕΝΟΣ in Greek). However, this grandiose scientific instrument seems destined to remain largely hidden from view for the foreseeable future. It is nonetheless an excellent reminder that though the archaeological record teaches us that the literary record is not complete, the visible archaeological record has its deficiencies also. All the same, in so far as she deals with the history of astronomy, a modern representation of Clio should certainly show her not only with the traditional basket of scrolls (as she is shown in the mosaic pavement from Mérida), but also with a spade.

Acknowledgement

I am grateful to M.T. Wright for his helpful comments on an earlier draft of this chapter.

Bibliography

Buchner, E. 1982. *Die Sonnenuhr des Augustus.* Reprint from *Römische Mitteilungen* 1976 and 1980, and *Nachtrag über die Ausgrabung 1980/1981.* Mainz: Philipp von Zabern.

Field, J.V. 1990. Some Roman and Byzantine portable sundials and the London sundial-calendar. *History of Technology* 12, 103–35.

Gibbs, S.L. 1976. *Greek and Roman Sundials*, New Haven and London: Yale University Press.

Price, D.J. de S. 1974. Gears from the Greeks: the Antikythera mechanism – a calendar computer from *ca.* 80 BC. *Transactions of the American Philosophical Society,* new ser. 64 (7).

Wright, M.T. 1990. Rational and irrational reconstruction: the London sundial-calendar and the early history of geared mechanisms. *History of Technology* 12, 65–102.

DAVID PINGREE

Astronomy in India

The science of astronomy in India is part of a broad intercultural development of knowledge, theory and practice that criss crossed the major civilisations of Eurasia between the late second millennium BC and the triumph (at least on an international basis) of the modern Western variant in the nineteenth century. The primary source from which the various astronomies of this great tradition descended was Mesopotamia, but each culture shaped its borrowings from others so that its astronomy conformed to its cultural needs and its intellectual traditions. Seen in this context, Indian astronomy was an off-shoot of external sciences, a creative force in its own right, and a contributor to other sciences. This chapter will attempt to sketch the main outlines of this historical movement.

The Babylonian Period

The oldest traces in India of an awareness of the order and periodicity of celestial phenomena are found in some late hymns of the *Ṛgveda*, in passages in Yajurvedic and Atharvavedic texts, and in a few Brāhmaṇas and Āraṇyakas. Except for the Ṛgvedic hymns, these texts are liturgical, giving instructions and advice for the practice of the great śrauta rituals; the dates of their compositions lie roughly in the first half of the last millennium BC. The astronomical knowledge that these texts evince was of importance for the performance of the śrauta rituals because many of those sacrifices were to be performed at given times within the solar year or within a system of synodic months.

The details of this knowledge are clearly related to the content of a late second-millennium Mesopotamian text called *MUL.APIN* (see pp. 46–9 and Col. Pl. III). The cuneiform text includes a catalogue of sixty constellations in the order of their risings and of seventeen constellations in the path of the moon, beginning with MUL.MUL (the Pleiades); corresponding to this are the late Vedic lists of twenty-seven or twenty-eight nakṣatras, also beginning with the Pleiades (Kṛttikā in Sanskrit) and associated with the moon. As one means to determine the need for intercalation, *MUL.APIN* gives the dates of the heliacal risings of the constellations in terms of an 'ideal calendar' in which one

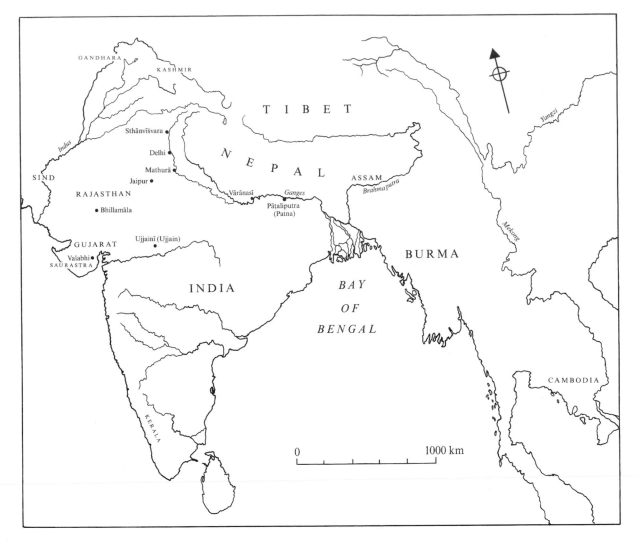

41 The Indian subcontinent.

year contains 12 months, each of which has 30 days, and consequently exactly 360 days; a late hymn of the *Ṛgveda* refers to the same 'ideal calendar'. And *MUL.APIN* describes the oscillation of the rising-point of the sun along the eastern horizon between its extremities when it is at the solstices; the same oscillation is described in the *Aitareyabrāhmaṇa*. These are just some examples of the parallels between the Mesopotamian and the Sanskrit texts.

The Indians developed an intercalation cycle only in the fifth century BC, during the Achaemenid occupation of Gandhāra and the Indus Valley. This cycle is expounded in the *Jyotiṣavedāṅga* associated with Lagadha. Its contents again reflect Mesopotamian astronomy, as it employs an outflowing water-clock whose operation is governed by a linear zigzag function with the ratio of the longest to the shortest day of the year taken to be three to two. All of the elements of this instrument and its use are Babylonian. So also is the very idea of an intercalation cycle, though Lagadha's is influenced by the already existing Vedic 5-year period, and he adopts an Egyptian year of exactly 365 civil days – a year-length whose use in Achaemenid Iran is probable. The result is a cycle in which 5 years contain 1830 days and 2 intercalary months; this is quite inaccurate, as it is off by 1 synodic month in just 25 years.

Contemporaneously with Lagadha – that is, in about 400 BC – a large corpus of omen texts, both celestial and terrestrial, were introduced into India from Iran; these omens are related to those in the Mesopotamian series called *Enūma Anu Enlil* (see pp. 42 and 44) and *Šumma ālu*. Embedded in the celestial omens are theories of planetary motion meant to be used in making rough predictions of the dates of the occurrences of the ominous planetary phenomena – first and last visibilities, retrogressions and conjunctions with constellations. First attempts at constructing mathematical models of planetary motions in Mesopotamia are found precisely in the Achaemenid period; and they characteristically divide the ecliptic for each planet into various arcs in each of which it has a given velocity, and employ standard intervals of time and longitude between phenomena. Precisely the same pattern appears in the Indian omen collections, the saṃhitās, with the arcs identified by sequences of nakṣatras, and with very similar intervals of time and longitude between the phenomena.

The Graeco-Babylonian period

A more advanced form of Babylonian planetary theory, that which was fully developed by about 300 BC, is reflected in later Indian texts. Again, this material was not transmitted directly to India from Mesopotamia, but this time passed first through Greek inter-mediaries, and appears in texts based on Sanskrit translations of Greek astrological and astronomical treatises made between the second and the fourth centuries AD, when trade between Western India and the Roman Empire was at its height. The earliest of these texts, on astrology, was translated near Ujjayinī, which came to be regarded as lying on the Indian prime meridian, by a man bearing the title Yavaneśvara (Lord of the Greeks) under the Śaka Mahākṣatrapa, Rudradāman I; Yavaneśvara was a leader of the Greek colonies in Western India, as was also Sphujidhvaja, who versified Yavaneśvara's lost translation into the *Yavanajātaka* – the basic text from which Indian genethlialogy descends – in AD 269/270. The astronomy of the *Yavanajātaka* is a mixture of both Babylonian ideas, coming from Lagadha, the saṃhitās and the Greeks, and some elements of Hellenistic astronomy. The original work of Yavaneśvara introduced into India its characteristic calendar, whose first attested occurrence is in the inscriptions of Rud-radāman I. This calendar uses the Babylonian division of the synodic month into thirty, originally equal parts, called *tithis* in Sanskrit (see p. 64). Other Sanskrit translations of

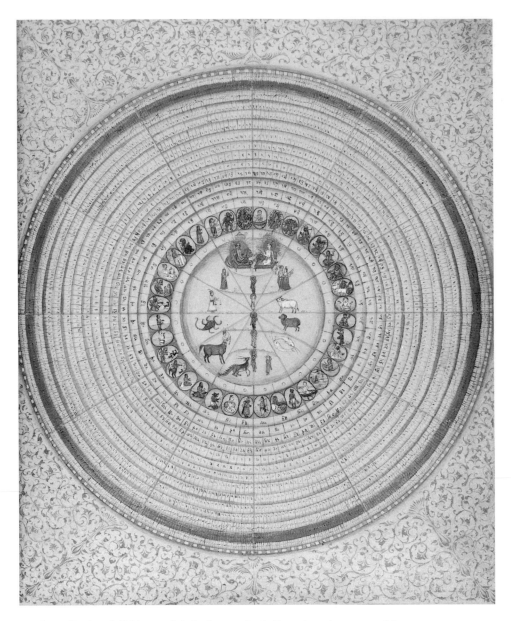

42 The zodiac, its subdivisions and their planetary lords. From the unique copy of the *Sarvasiddhāntatattvacūḍamaṇi*, composed by Durgāśaṅkara Pāṭhaka in the 1830s. (British Library, MS Or. 5259, f. 75)

Graeco-Babylonian texts in this period include the *Vasiṣṭha*, *Romaka*, and *Pauliśa siddhāntas*, which we know of only through the somewhat incompetent summaries provided by Varāhamihira in the *Pañcasiddhāntika* that he wrote in about 550. Included with the Greek adaptations of Babylonian astronomy of the Seleucid period are many elements of Hellenistic astronomy, especially some derived from Hipparchus, including the Indian transformation of his table of chords into a table of sines, and the first application of analemmata to problems in spherical trigonometry, a field in which Indian astronomers were brilliantly innovative.

One of the most astonishing statements in the *Pañcasiddhāntikā* is that the time differences between Yavanapura (Alexandria in Egypt) and Ujjayinī and Vārāṇasī are 7; 20 and 9 nāḍīs respectively, corresponding to 44° and 54° of longitude. The correct longitudinal differences are 45; 50° and 53; 7°. The only way that the level of accuracy displayed in the *Pañcasiddhāntikā* could have been achieved was by determining the local time for simultaneous observations of a lunar eclipse in Alexandria and in India. This required both international organisation and highly accurate time measurements.

While the *Romaka* and the *Pauliśa* also contain such Hellenistic features as a table of solar equations based on either an epicyclic or an eccentric model and the computation of lunar and solar eclipses, including, for the latter, the computation of lunar parallax, the first Sanskrit text representing fully developed geometric models of the terrestrial and celestial spheres and of the mechanisms accounting for planetary motions was the *Paitāmahasiddhānta*, which is the basic text of the Brāhmapakṣa. It was composed in about AD 425 and preserved because of its incorporation into the gigantic *Viṣṇudharmottarapurāṇa*, which was apparently compiled in Kashmir in the early eighth century. The author of this text takes the traditional cosmology of the *Purāṇa*s, in which the earth is a flat circle in the middle of which is a lofty mountain, Meru, causing day and night, encircled by seven ring-continents alternating with seven ring-oceans; above the earth, circling about Mt Meru as their axis, are giant wheels on which the sun, the moon, the planets and the nakṣatras revolve (a theory perhaps related to those of Anaximander and Anaximenes), and converts it into a spherical universe in the centre of which is the sphere of the earth with Mt Meru at its North Pole, through which extends the axis around which the spheres of all the planets rotate.

One of the conflicts in Greek planetary theory had arisen from the Aristotelian theory that the spheres of the planets move naturally with uniform circular motion concentric with the centre of the earth, which is the centre of the universe, and the realisation by later Greek astronomers – particularly Apollonius and Hipparchus – that the two anomalies of the planets' motions could be quantitatively explained by eccentric circles and by epicycles, or by some combination of them. The choice that astronomers faced, then, was between saving concentricity while sacrificing uniformity, or saving uniformity while sacrificing concentricity (both the eccentric and the epicycle cause a planet's distance from the centre of the universe to vary). We know from Ptolemy that some of his predecessors adopted a simple model with an eccentric deferent on which moves the centre of the epicycle that bears the planet; all the motions are uniform and circular, but the planet does not remain at a fixed distance from the earth. Ptolemy himself improved this model by introducing the equant, which destroys the uniformity of the motion of the epicycle centre on the deferent.

The *Paitāmahasiddhānta* and the Brāhmapakṣa school

The theory of the *Paitāmahasiddhānta* represents another choice, to preserve concentricity at the expense of uniformity. For the planet (Fig. 43) travels on a concentric deferent, while around it as their centre revolve two epicycles, one for each anomaly (the manda corresponding to the equation of the centre (μ), due to the ellipticity of the planet's orbit around the sun, while the śīghra corresponds to the equation of the anomaly (σ), due to the fact that the planet is observed from the earth (O), which also is orbiting around the sun). Each of these two epicycles exerts a pull on the planet, which dislodges it from its mean longitude ($\bar{\lambda}$); the combination of the two pulls at any moment with the mean motion produces the instantaneous true longitude. The resulting motion is clearly discontinuous rather than uniform.

Since the geometrical device (epicycles) and the problem they are used to solve (that of preserving concentricity) are both Greek in origin, so in basic form is also the whole model. However, the Indians, not being committed to Aristotle, felt free to manipulate it, and did so with some success, inventing various approximative algorithms as substitutes for the geometrical solutions of the triangles formed by the radii of the

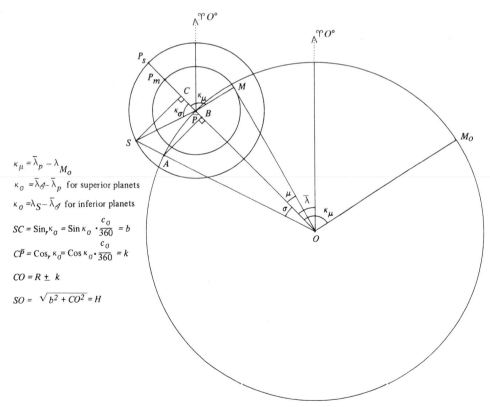

$$\kappa_\mu = \bar{\lambda}_p - \lambda_{M_o}$$

$$\kappa_\sigma = \bar{\lambda}_\sigma - \bar{\lambda}_p \quad \text{for superior planets}$$

$$\kappa_\sigma = \lambda_S - \bar{\lambda}_\sigma \quad \text{for inferior planets}$$

$$SC = \mathrm{Sin}_r \kappa_\sigma = \mathrm{Sin}\,\kappa_\sigma \cdot \frac{c_\sigma}{360} = b$$

$$C\bar{P} = \mathrm{Cos}_r \kappa_\sigma = \mathrm{Cos}\,\kappa_\sigma \cdot \frac{c_\sigma}{360} = k$$

$$CO = R \pm k$$

$$SO = \sqrt{b^2 + CO^2} = H$$

43 The planetary theory of the *Paitāmahasiddhānta*. (From D. Pingree, History of mathematical astronomy in India, *Dictionary of Scientific Biography* 15, Charles Scribner's Sons, New York, 1978, p. 558)

Plate I Aerial view of Stonehenge, showing the alignment along the Heel Stone axis, with the Stonehenge avenue stretching away into the distance.

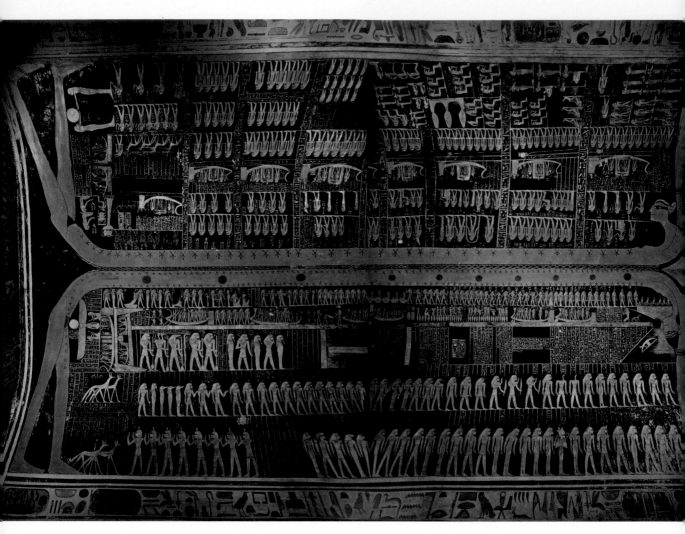

Plate II The astronomical ceiling above the sarcophagus of Ramesses VI (1143–1136 BC) in the Valley of the Kings, Luxor, showing the sky goddess Nut in Siamese-twin form, representing the northern (upper) and southern (lower) skies. The lower twin with sun discs indicates that the sun courses the southern sky in Egypt. Ten discs within her body plus one each at the mouth and birth canal represent the 12 months it takes Ra to return to his birthplace each year. The river is that along which the solar barque with Ra must sail through the underworld each night. Its emergence on the left from the Egyptian hieroglyph ⏢ for heaven shows that it represents the nightly equivalent of Ra's passage across the sky during the day. Arrayed along the river and in some of the other panels are stations of the solar barque poised in front of the Gates of the Duat. Also shown are numerous deities, demigods and demons with which Ra must contend to pass through these portals, which represent the hours of night.

Plate III A Late Babylonian copy of the astronomical compendium *MUL.APIN*. The tablet, only 8.4 cm high, is a masterpiece of miniature cuneiform writing. About 500 BC. (British Museum, WA 86378)

Plate IV (*above*) A Latin version of the star catalogue from Ptolemy's *Almagest*. The manuscript was prepared *c.* 1490 for King Henry VII. To allow for precession the copyist has added the longitudes of the stars at the time of Adam, set at 3496 BC, and in the mid-fifteenth century AD. The somewhat crude medieval pictures of Libra and Scorpio contrast with the precision of the astronomical data. (British Library, MS Arundel 66, f. 41)

Plate V (*left*) Drawing from the seventeenth-century Paper Museum of Cassiano dal Pozzo, showing a Julian calendar, probably found in Rome. The months are listed with their astrological signs, hours of daylight and darkness, and festivals. (British Museum)

Plate VI (*right*) A fifteenth-century manuscript of the *Phaenomena* of Aratus, written for one of the Medici family, whose arms appear in the lower border. The manuscript includes a brief biography of Aratus and comments on the *Phaenomena* by Theon (a grammarian of the first century AD), Hipparchus, Eratosthenes and Achilles Tatius. (British Library, Add. MS 11866, f. 1)

ΑΡΑΤΟΥ
ΦΑΙΝΟΜΕΝΑ

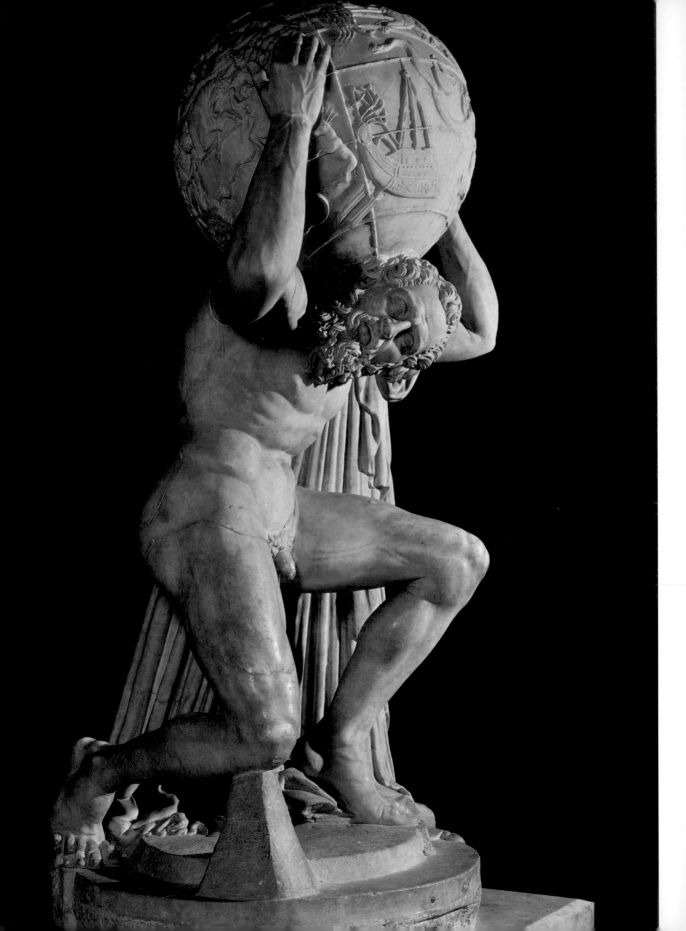

Plate VII (*left*) The Atlante Farnese. Marble,
diam. of globe 65 cm. About 200 BC.
(Museo Nazionale Archeologico, Naples)

Plate VIII (*above*) Indian astronomers determining
the co-ordinates of a star. From the
Suryasiddhantatattvacūḍamaṇi composed by
Durgāśaṅkara Pāṭhaka in the 1830s (British
Library, MS Or. 5259, f. 29)

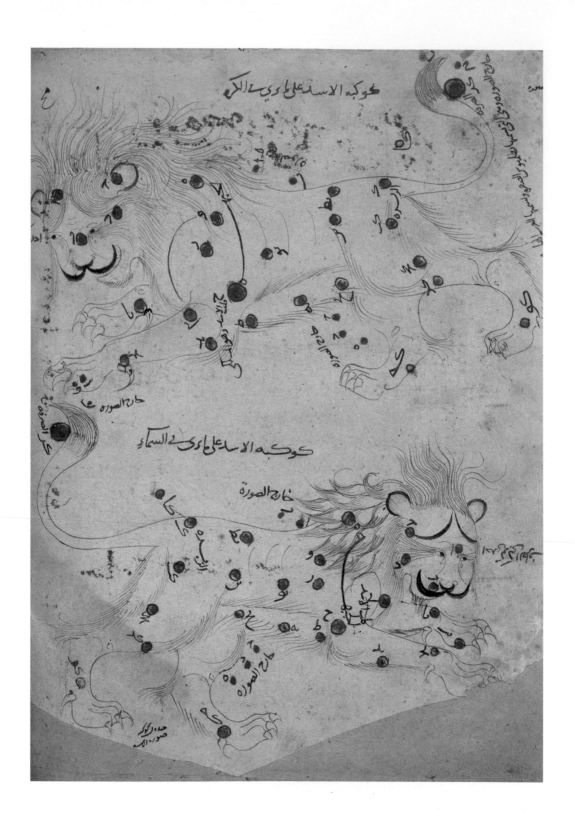

Plate IX (*left*) The constellation Leo as depicted in a
thirteenth-century manuscript of al-Sūfī's *Ṣuwar
al-kawākib* (*Book of Constellation Figures*). The
constellation is shown in double aspect, as seen in the
sky and as seen on a globe. (British Library, MS Or.
5323, f. 45v)

Plate X (*right*) A celestial globe made and inscribed by
'Muḥammad ibn Hilāl the astronomer of Mosul', in
674 (= AD 1274–5). Brass inlaid with silver, diam.
24 cm. Forty-eight constellation figures are drawn
and about 1,000 stars indicated by inlaid silver dots:
their names are given in Kufic characters. The
constellation figures follow for the most part the
tradition of al-Sūfī's *Ṣuwar al-kawākib* (see Pl. IX).
The stand is later than the globe. (British Museum,
OA 1871,3-1,1)

Plate XI Astrolabe with seven latitude plates,
made in Isfahan by 'Abd al-'Ali and
Muḥammad Bāqir in 1124 (= AD 1712) for the
Safavid Shāh Ḥusayn. Brass, ht 53 cm. This
spectacular scientific work of art well illustrates
the quality of workmanship and scientific
accuracy that could still be achieved after the
creative period of Islamic science during the
ninth to fifteenth centuries. IC no. 33. (British
Museum, OA + 369)

mnient occidente. Constat ergo qd n a statu suo sed ali
tu nostro partes celi sub tali stent uocabulo. Mouet itaq;
firmamentum ab oriente ascendendo ad meridiem. inde p
occidente iterum ad oriente. planete u motu gmnario ab
occidente in oriente naturali mouent. sed i eode cursu
impetu firmamenti retracti. ab oriente i occidentem cum
ipo ferunt. semp tam in cursu sua n no sine officia
eor quilibet uenitur firmamento. Que aute uis quis
impetus cursu tam celeri firmametum trahat i plane
tas contra firmametu tam potenter ferat no est qui
dicere ualeat. sed rerum indagatores subtilium
probabilib; hoc argumentis astruit
qd nisi uiolentus planetarum
cursus quo contra firma
mentum intuitur in
nua firmamenti
celeritate re
morando
opeleceret

mundi machina deperiret. Cursus ergo planetaru
uersus suo motari nitutur firmamentu i ido de ari
ete in pisces. de piscibus in aquarii. i sic ulterius
sed de ariete in tauru. de tauro in geminos. i ita uen
ceps. eorum est cursus. Planeta quoq; hec eade ratione.
quanto a firmamento est remotior tanto est uelocior.
quanto uicinior firmamento graditur. tanto plus
in suo naturali cursu ab impetu firmamenti tardat.

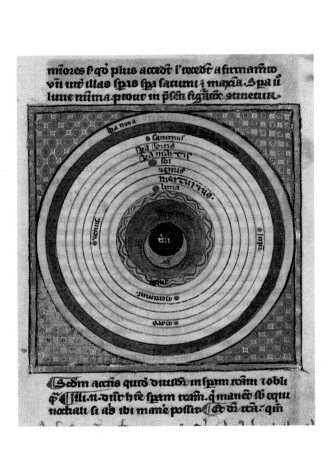

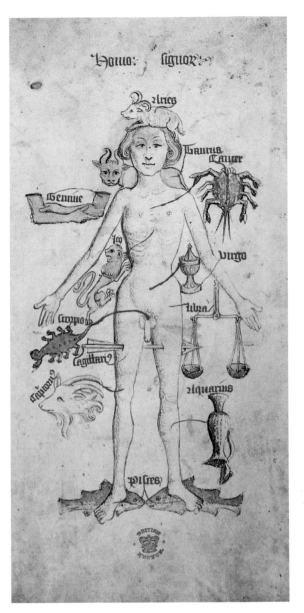

Plate XII (*left*) The geo-heliocentric planetary system of the early Middle Ages. From an anonymous treatise. (Royal Library of Copenhagen, MS GKS 277 2°, f. 49r)

XIII (*above*) The usual system of spheres in medieval cosmology. Outside the eighth sphere (or the firmament) containing the fixed stars is placed a ninth sphere that was unknown to Aristotle. It was empty and served only as a background for the slow motion of precession of the stars. From an early manuscript (AD 1240) of the *Tractatus de Sphera* of John of Sacrobosco. (Royal Library of Copenhagen, MS NKS 275a 4°, f. 11v)

Plate XIV (*above right*) Astrology was based on the conception of man as a microcosm with limbs and organs correlated to the twelve signs of the zodiac. From the fifteenth-century Guild-Book of the Barber-Surgeons of York. (British Library, MS Egerton 2572, f. 50v)

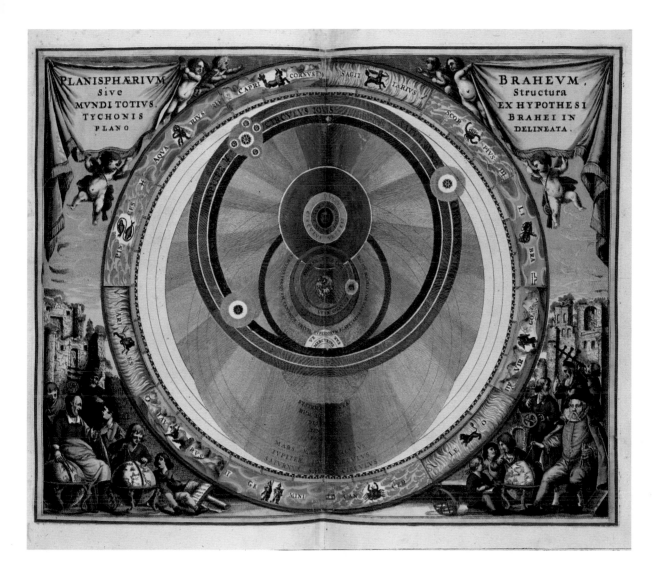

Plate XV (*left*) Tycho Brahe, *Astronomiae instauratae mechanica* (1598), f. A4v: Tycho, his great mural quadrant and his dog. Note the scale of the quadrant, divided by transversal dots, and the sight with parallel slits near *F*. In the foreground is an elaborate but insufficiently accurate clock, and in the background a cross-section of Uraniborg with various instruments, the library and the alchemical laboratory. The pictures at *Y* and *Z* are of King Frederick and Queen Sophie. (British Library)

Plate XVI (*above*) Tycho's planetary system (see Fig. 71), as represented in Andreas Cellarius, *Atlas Coelestis*, Amsterdam, 1660, with the addition of the satellites of Jupiter first discovered by Galileo in 1609. (British Library)

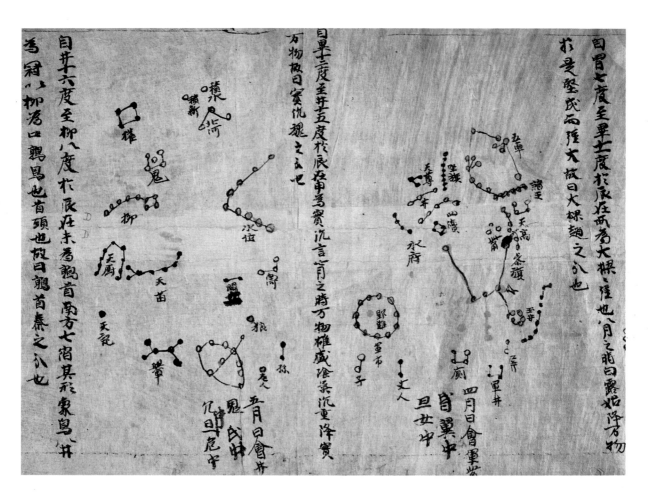

Plate XVII (*left*) Astronomical compendium made in Heilbronn, Germany, in 1596 by Johann Anton Linden. It is constructed in gilded silver. The main features include an astrolabe with a rete for 34 stars, a table with the longitude and latitude of 29 stars, a lunar aspectarium, a table with the longitude and latitude of 70 towns, and a calendar for the years 1596–1625. (British Museum, MLA 57,11-16,1)

Plate XVIII (*above*) A section of the Dunhuang star chart; the constellation Orion may be seen on the right (British Library, MS Stein 3326)

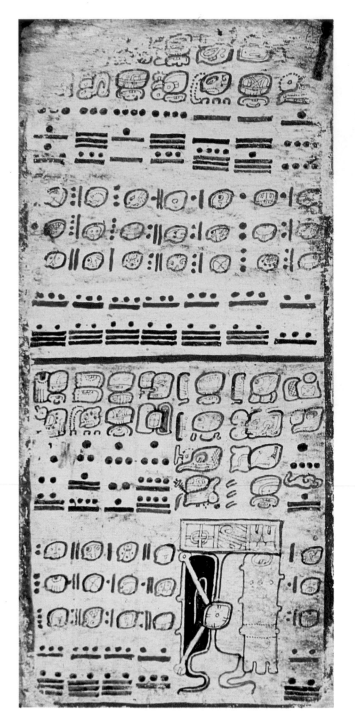
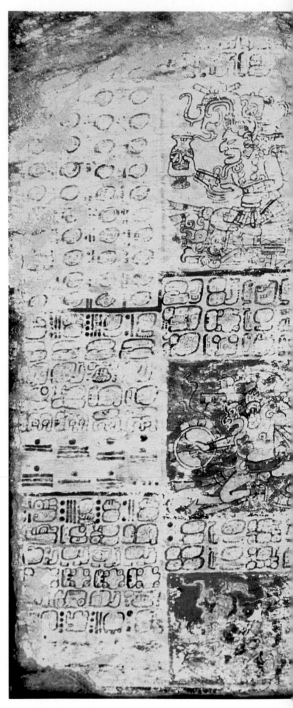

Plates XIX, XX The Mayan Dresden Codex (see Fig. 96). (Sächsische Landesbibliothek, Dresden, MS L310, ff. XLVI, LIV)

deferents and the circles and the lines joining the mean and the true longitudes of the planet to the centre of the universe.

The method of determining the mean longitudes of the planets in the *Paitāmahasiddhānta* was a combination of Babylonian, Greek, and Indian ideas. The earliest yuga in the sense of a period of time containing an integer number of rotations of two 'planets' was the Metonic cycle, in which 19 solar years contain 235 conjunctions of the sun and the moon (see pp. 46 and 52); this was known to Babylonian astronomers, perhaps as early as the sixth or seventh century BC, and Lagadha had introduced such a cycle into India in about 400 BC, though it was not very accurate. For each star-planet the Babylonians had devised by 300 BC several periods in which an integer number of occurrences of a phase or transits of a fixed star occurred in an integer number of years, that is, of rotations of the sun. Plato had already spoken of a 'perfect year' in which all the planets will return to the positions(s) from which they started. The *Paitāmahasiddhānta* combines this idea with an ancient Indian chronological scheme whose basic unit is a Babylonian period of 432,000 years (sexagesimally 20, 0, 0, 0). This length of time, in a text of perhaps the second century AD that was copied by the *Mahābhārata* and the *Manusmṛti*, is called a Kaliyuga. Ten Kaliyugas – 4,320,000 years – constitute a Mahāyuga, also called a Caturyuga; and 1000 Mahāyugas – 4,320,000,000 years – make a Kalpa. Kalpas are alternately daytimes and night-times of Brahmā, so that 720 Kalpas constitute a year of Brahmā, who exists in this universe for 100 such years: 72,000 Kalpas equalling 311,640,000,000,000 years.

A Mahāyuga consists of four yugas, which are in the ratios of $4:3:2:1$ to each other:

Kṛtayuga	$4 \times 432,000 =$	1,728,000
Tretāyuga	$3 \times 432,000 -$	1,296,000
Dvāparayuga	$2 \times 432,000 =$	864,000
Kaliyuga	$1 \times 432,000 =$	432,000

And the Kalpa of 10,000 Kaliyugas is divided into 14 Manvantaras; each Manvantara, then, contains 710 Kaliyugas. The residue of 60 Kaliyugas is divided into 15 Sandhis of 4 Kaliyugas, or 1 Kṛtayuga, each; a Sandhi precedes the first Manvantara, and one follows each of the 14 Manvantaras.

The *Paitāmahasiddhānta* uses a Kalpa as the equivalent of Plato's 'perfect year'. To each planet and to each node and apogee is assigned an integer number of rotations in a Kalpa (*B* in Table 1 overleaf) – for the planets these numbers can be arrived at initially from the Babylonian period relations – and they were all at sidereal Aries 0° at the beginning of the Kalpa, and will be there again at its end.

In order to produce correct mean longitudes in the present time, the author of this system determined a time at which there had been a configuration of all the planets near sidereal Aries 0°; the moment chosen corresponds to sunrise at Laṅkā (on the

Table 1

Body	R	$\dfrac{R \times 6,0}{Y}$	$\dfrac{R \times 6,0}{C}$
Saturn	146,567,298	12; 12, 50, 11, 21, 50, 24°	0; 2, 0, 22, 51, 43, 54, . . .°
Manda	41	0; 0, 0, 0, 44, 16, 48°	
Node	− 584	− 0; 0, 0, 10, 30, 43, 12°	
Jupiter	364,226,455	30; 21, 7, 56, 11, 24°	0; 4, 59, 9, 8, 37, 23, . . .°
Manda	855	0; 0, 0, 15, 23, 24°	
Node	− 63	− 0; 0, 0, 1, 8, 2, 24°	
Mars	2,296,828,522	3, 11; 24, 8, 33, 23, 45, 36°	0; 31, 26, 28, 6, 47, 12, . . .°
Manda	292	0; 0, 0, 5, 15, 21, 36°	
Node	− 267	− 0; 0, 0, 4, 48, 21, 36°	
Sun	4,320,000,000	6, 0°	0; 59, 8, 10, 21, 33, 30, . . .°
Manda	480	0; 0, 0, 8, 38, 24°	
Venus's			
śīghra	7,022,389,492	3, 45; 11, 56, 50, 51, 21, 36°	1; 36, 7, 44, 35, 16, 45, . . .°
Manda	653	0; 0, 0, 11, 45, 14, 24°	
Node	− 893	− 0; 0, 0, 16, 4, 26, 24°	
Mercury's			
śīghra	17,936,998,984	54; 44, 59, 41, 42, 43, 12°	4; 5, 32, 18, 27, 45, 33, . . .°
Manda	332	0; 0, 0, 5, 58, 33, 36°	
Node	− 521	− 0; 0, 0, 9, 22, 40, 48°	
Moon	57,753,300,000	2, 12; 46, 30°	13; 10, 34, 52, 46, 30, 13, . . .°
Manda	488,105,858	40; 40, 31, 45, 26, 38, 24°	0; 6, 40, 53, 56, 32, 54, . . .°
Node	− 232,311,168	− 19; 21, 33, 21, 1, 26, 24°	− 0; 3, 10, 48, 20, 6, 41, . . .°

Indian prime meridian on the equator) of 18 February −3101 Julian. This moment was the beginning of the current Kaliyuga. Its distance from the beginning of the Kalpa had to be a number of Kaliyugas close but not equal to half of the Kaliyugas in a Kalpa; the multiple chosen was 4,567, so that the Kalpa began 1,972,944,000 years before the beginning of the present Kaliyuga. Now, the initial numbers of rotations of the planets in a Kalpa could be adjusted so that their mean longitudes were close to sidereal Aries 0° on 18 February −3101, and the planets' apogees and nodes could be given small numbers of rotations in a Kalpa such that they would be in their correct positions in our time.

The mean positions of the planets at the beginning of any sidereal year then could be found by simple proportion. But one normally wants the planets' mean longitudes for particular days within the year. For this one must use a proportion involving not the ratio of lapsed years to the years in a Kalpa (Y), but that of lapsed days (the ahargaṇa) to the days in a Kalpa (C). The latter number was defined as 1,977,916,450,000, so that each sidereal year contains 6, 15; 15, 30, 22, 30 days. The ahargaṇa had to be computed from the calendar date, which was expressed as the tithi-number of the current sunrise day of a named synodic month in a given sidereal year lapsed from the epoch of the Śaka era, which began at the sun's entry into sidereal Aries 0° 3,179 years after the beginning of the present Kaliyuga, i.e. in AD 78. For the purpose of determining the tithi-number of the day, the first tithi in a synodic month begins at true conjunction of the sun and the moon and lasts until the true elongation increases by 12°, when a new tithi begins. This means that the lengths of the tithis are constantly changing. A day receives the tithi-number of the tithi which was current at local sunrise of that day; local terrestrial latitude and local terrestrial longitude determine when local sunrise occurs, so that each locality must have its own calendar (pañcāṅga) for which tithi-endings, sunrises and other astronomical phenomena are computed. This requires many astronomers.

The actual computation of the ahargaṇa is further complicated by the later rules for the intercalation and suppression of months in the calendar. A synodic month-name is associated with each sidereal zodiacal sign, beginning with Caitra and Aries. Then that month is called Caitra which is current when the sun enters Aries, and so on; but if there are two conjunctions after the sun enters Pisces, the month beginning with the first conjunction but ending before the sun enters Aries is an additional Caitra, an intercalary month (adhimāsa). However, there are periods during which the sun passes through a zodiacal sign without conjoining with the moon; then the month-name associated with the following zodiacal sign is omitted (this is called a malamāsa). Furthermore, in some calendars, the months begin with opposition (pūrṇimānta), in others with conjunction (amānta). There is no space here to explain the complexity of the computations both of the calendar and of the ahargaṇa that result from these rules and their local variants.

Another idea that had been adumbrated by Plato is that each planet travels the same distance in the same time, so that the distances of their spheres from the earth are inversely proportional to their angular velocities. The *Paitāmaha* employes this idea to compute the dimensions of the universe. The distance of the moon is determined from the fact that its horizontal parallax is the radius of the earth seen at the moon's distance. Since horizontal parallax is the moon's motion in $\frac{1}{15}$ days or about 0;52° and the circumference of the earth is taken to be 5,000 yojanas, so that its radius is approximately 800 yojanas, each minute in the moon's orbit is about 15 yojanas long, and its circumference is 324,000 yojanas. The circumferences of the orbits of the remaining planets and their radii in yojanas follow. The circumference of the sphere of the fixed stars is sixty times that of the sun; and the circumference of the sphere of heaven is the product of the circumferences of the orbits of the seven planets (O in Table 2 overleaf).

The *Paitāmaha* employs one epicycle each for the sun and the moon, two each for the five star-planets. The latitude of the moon is computed on the assumption that it travels on its deferent at a fixed angle of 4;30° to the ecliptic; the latitudes of the planets are computed on the assumption, contradicting the double-epicycle model, that they are located on their śīghra epicycles. For the superior planets, their deferent orbits are inclined to the plane of the ecliptic while their śīghra epicycles are parallel to that plane, while for the inferior planets, their deferent orbits fall in the plane of the ecliptic while their śīghra epicycles are inclined to that plane. These must be Hellenistic models also since the latitudinal models in Ptolemy's *Almagest* are modifications of them.

Table 2

	O/R
Saturn	$127{,}668{,}787 \, \dfrac{8412079}{24427883}$ yojanas
Jupiter	$51{,}374{,}821 \, \dfrac{54182089}{72845291}$ yojanas
Mars	$8{,}146{,}916 \, \dfrac{82430924}{1148414261}$ yojanas
Sun	$4{,}331{,}497 \, 1/2$ yojanas
Venus's śīghra	$2{,}664{,}629 \, \dfrac{1627580383}{1755597373}$ yojanas
Mercury's śīghra	$1{,}043{,}210 \, \dfrac{1561237670}{2242124873}$ yojanas
Moon	$324{,}000$ yojanas

Other planetary data considered in the *Paitāmaha* are the arcus visionis of each expressed in terms of the right ascensional difference between the planet and the sun necessary for the planet to be visible. Similarly, arcs of elongation between the sun and each planet are given at which first and last visibility and first and second station occur. For conjunctions of the planets with each other and with the fixed stars, polar longitudes and latitudes are used; these are undoubtedly based on the system employed by Hipparchus. This means that two planets are in conjunction when they both lie on the same circle of declination, and the same holds true for the conjunctions of a planet with a fixed star. To facilitate the computation of the latter type of conjunction, the co-ordinates in longitude and latitude of the yogatārās of the nakṣatras are polar. To make possible the computation of conjunctions between planets, an algorithm is presented for transforming ecliptic into polar longitudes and latitudes.

The computations of the times and longitudes of the five phases of lunar and solar eclipses are similar to those already found in the *Romaka* and *Pauliśa* siddhāntas. The most disturbing feature is that the orbit of the moon is assumed to be parallel to the plane of the ecliptic throughout the eclipse. Projections are also to be drawn of the different phases of eclipses because of their usefulness in divination.

In the *Paitāmaha*, as in all later Indian siddhāntas, considerable ingenuity is devoted to the elaboration of analemmata that permit the transformation of solar declination (δ in Fig. 44), local time since sunrise, local terrestrial latitude (ϕ) and the length of the shadow of a gnomon into each other, and the derivation from them of the rising amplitude of the sun (η), the half-equation of daylight, the cardinal directions and the solar longitude and altitude. Later texts apply the same sort of analemmata to the employment of the moon as a time-keeper at night. The *Paitāmaha* starts a tradition of computing the elevation of the horns of the moon and the width of its sickle.

Many later Sanskrit texts, especially in Central, Western, and Northern India follow the tradition of the *Paitāmaha*; the school thus formed was called the Brāhmapakṣa. It

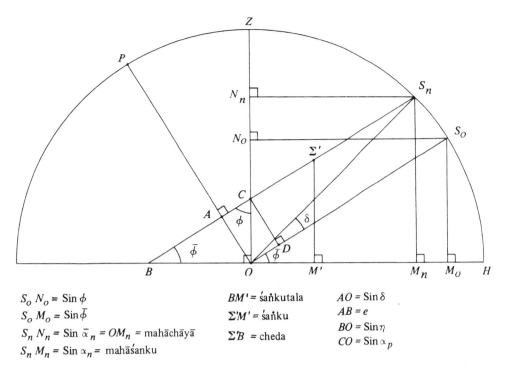

$S_O N_O = \mathrm{Sin}\, \phi$

$S_O M_O = \mathrm{Sin}\, \bar{\phi}$

$S_n N_n = \mathrm{Sin}\, \bar{\alpha}_n = OM_n = $ mahāchāyā

$S_n M_n = \mathrm{Sin}\, \alpha_n = $ mahāśanku

$BM' = $ śankutala

$\Sigma'M' = $ śanku

$\Sigma'B = $ cheda

$AO = \mathrm{Sin}\, \delta$

$AB = e$

$BO = \mathrm{Sin}\, \eta$

$CO = \mathrm{Sin}\, \alpha_p$

(projection in the plane of the meridian)

44 The analemmata of the *Pañcasiddhāntikā*. (From D. Pingree, History of mathematical astronomy in India, *Dictionary of Scientific Biography* 15, Charles Scribner's Sons, New York, 1978, p. 551)

had a wide influence outside India also, beginning with the adaptation of at least some of its parameters by the authors of the Pahlavī *Zīk-i Shahriyārān* in Sasanian Iran in about 450, and in the *Sindhind* tradition in Arabic, Samaritan, Greek, Hebrew and Latin (see pp. 137 and 148).

The Āryapakṣa school

At the end of the fifth century, at Kusumapura (traditionally identified with Pāṭaliputra, the modern Patna), Āryabhaṭa (b. 473) started a new pakṣa, strongly influenced by the Brāhmapakṣa; named after its author, it is called the Āryapakṣa. As Āryabhaṭa desired to simplify the computation of the longitudes of the planets, he decided to shorten the yuga in which the planets should have integer numbers of revolutions from a Kalpa to a Caturyuga a thousandth of its length, 4,320,000 years. This did not offer enough time to have a true conjunction of all the planets at its beginning and end since the apogees would have to move too fast to be at Aries 0° at the beginning and end of the yuga. Āryabhaṭa, therefore, fixed the longitudes of the apogees in accordance with astrological and numerological principles, and those of the nodes in a geometrical progression, but he assumed mean conjunctions of the planets at the beginning and the end of the Caturyuga and of each of the four yugas that constitute it. However, if the traditional system were retained, this would entail a mean conjunction of all the planets at the beginning and end of a Kaliyuga of just 432,000 years. In order to avoid this consequence, Āryabhaṭa asserted that all four yugas were of equal length, 1,080,000 years. This meant that the number of rotations of each planet within a Caturyuga had to be divisible by four, as in Table 3.

Table 3

Planet	R	$\dfrac{R \times 6,0}{Y}$	$\dfrac{R \times 6,0}{C}$
Saturn	146,564	$12; 12, 49, 12°$	$0; 2, 0, 22, 41, 41, 32, \ldots°$
Jupiter	364,224	$30; 21, 7, 12°$	$0; 4, 59, 9, 0, 38, 51, \ldots°$
Mars	2,296,824	$3, 11; 24, 7, 12°$	$0; 31, 26, 27, 48, 54, 22, \ldots°$
Sun	4,320,000	$6, 0°$	$0; 59, 8, 10, 13, 3, 31, \ldots°$
Venus's śīghra	7,022,388	$3, 45; 11, 56, 24°$	$1; 36, 7, 44, 17, 4, 45, \ldots°$
Mercury's śīghra	17,937,020	$54; 45, 6°$	$4; 5, 32, 18, 54, 36, 24, \ldots°$
Moon	57,753,336	$2, 12; 46, 40, 48°$	$13; 10, 34, 52, 39, 18, 56, \ldots°$
Manda	488,219	$40; 41, 5, 42°$	$0; 6, 40, 59, 30, 7, 38, \ldots°$
Node	$-232,226$	$-19; 21, 8, 48°$	$-0; 3, 10, 44, 7, 49, 44, \ldots°$

The actual numbers were computed, it seems likely, in the following way. Initial estimates of these numbers of rotations could have been derived from either taking a thousandth of the Brāhmapakṣa's rotations in a Kalpa, or from the period-relations of Babylonian origin that had long been known in India. These numbers were refined by computing, from a table of mean motions of the planets derived from a Greek source, the positions of the planets exactly 3,600 (sexagesimally 1, 0, 0) years from the beginning of the Kaliyuga at 6 am (dawn) at Laṅkā on 18 February −3101. According to the year-length of the Brāhmapakṣa 1, 0, 0 years ended at $0; 7, 30$ days after noon of 20 March 499. Āryabhaṭa's table was computed for noon epoch; so he added to the 1, 0, 0 years $0; 52, 30$ days, and computed the mean longitudes of the planets for noon of 21 March 499. To add $0; 52, 30$ days in 1, 0, 0 years is equivalent to adding 1050 days in the 20, 0, 0, 0 years of a Caturyuga. This is why he changed the number of days in a Caturyuga according to the Brāhmapakṣa, 1,577,916,450, to the Āryapakṣa's 1,577,917,500. This changes the length of a year from $6, 5; 15, 30, 22, 30$ to $6, 5; 15, 31, 15$ days.

The *Āryabhaṭīya* in which Āryabhaṭa set forth the results of his computations consists of four pādas: the daśagītikā containing the parameters, the gaṇita on mathematics, the kālakriyā on the computation of time, and the gola on the heavenly spheres. Like the *Paitāmaha* the *Āryabhaṭīya* presents a double-epicycle theory, though it is aware of the existence of planetary models employing an epicycle whose centre rotates on an eccentric deferent, a model to which Ptolemy refers. The description of this model in the *Āryabhaṭīya* shows that Āryabhaṭa had access to Greek material in addition to that which was incorporated into the *Paitāmaha*. Āryabhaṭa's epicycles, however, pulsate between limits attained when the anomaly (manda or śīghra) reaches 0°/180° and 90°/270°; the motivation for such a strange device is not at all apparent.

However, despite its curious features, the Āryapakṣa was also long-lasting in India, flourishing especially in Southern India and, to a lesser extant, in Western India; and it played a minor role in the development of early Islamic astronomy.

The Ārdharātrikapakṣa school

Āryabhaṭa later devised yet another system, the midnight or Ārdharātrika. He kept much of his Āryapaksa, changing the numbers of rotations of only Jupiter and Mercury's śīghra (see Table 4), stabilising the epicycles, and computing the longitudes of the apogees according to a new numerological pattern. But he changed the day epoch from dawn, which varies with terrestrial latitude, to midnight. To achieve this he shifted the beginning of the Kaliyuga to midnight of 17/18 February −3101. To add 0;15 days to 1, 0, 0 years is equivalent to adding 300 days to 20, 0, 0, 0 years; so the number of days in a Caturyuga grew to 1,577,917,800, and the length of a year to 6, 5; 15, 31, 30 days.

Table 4

Planet	R	$\dfrac{R \times 6,0}{Y}$	$\dfrac{R \times 6,0}{C}$
Saturn	146,564	12; 12, 49, 12°	0; 2, 0, 22, 41, 36, 36, . . .°
Jupiter	364,220	30; 21, 6°	0; 4, 59, 8, 48, 36, 56, . . .°
Mars	2,296,824	3, 11; 24, 7, 12°	0; 31, 26, 27, 47, 36, 55, . . .°
Sun	4,320,000	6, 0°	0; 59, 8, 10, 10, 37, 48, . . .°
Venus's śīghra	7,022,388	3, 45; 11, 56, 24°	1; 36, 7, 44, 13, 7, 53, . . .°
Mercury's śīghra	17,937,000	54; 45°	4; 5, 32, 17, 45, 23, 13, . . .°
Moon	57,753,336	2, 12; 46, 40, 48°	13; 10, 34, 52, 6, 50, 56, . . .°
Manda	488,219	40; 41, 5, 42°	0; 6, 40, 59, 29, 51, 10, . . .°
Node	− 232,226	− 19; 21, 7, 48°	− 0; 3, 10, 44, 7, 41, 54, . . .°

The Ārdharatrikapakṣa flourished principally on the borders of India – in Kashmir, Nepal, Tibet and Assam. In the seventh century a version whose epoch was 638 was transmitted to Burma; from there it was adopted in Cambodia and elsewhere in South-east Asia. In the middle of the sixth century an Ārdharātrika text was translated into Pahlavī as the *Zīk-i Arkand*, which was the basis of the *Zīk-i Shahriyārān* of Khusrō Anūshirvān in 556 and of all succeeding *Royal Canons* in Sasanian Iran up till and including the last, that of Yazdigird III, through which it exercised considerable influence on early Islamic astronomy.

The next significant astronomers after Āryabhaṭa and Varāhamihira both flourished in the seventh century. The first, Bhāskara, was a follower of the Āryapakṣa; besides a commentary on the *Āryabhaṭīya* written at Valabhi in Saurāṣṭra in 629, he composed a *Laghubhāskarīya* summarising and supplementing the astronomy contained in that work; and a *Mahābhāskarīya*, which provides a more detailed treatment of the astronomy of the school of Āryabhaṭa as well as a summary of that of the Ārdharātrikapakṣa. Among his 'innovations' is the adaptation of a model employing a concentric deferent with an equant in place of an eccentric deferent or a manda epicycle; the concentric with equant was probably derived from a Greek source.

While Bhāskara adheres to Āryabhaṭa's parameters, his mathematics is frequently more similar to that of his contemporary, Brahmagupta, than it is to that of the founder of his school. Brahmagupta, who was born in 598, published his lengthy and highly influential exposition of the Brāhmapakṣa, the *Brāhmasphuṭasiddhānta*, at Bhillamāla in southern Rājasthāna in 628. In the first ten adhyāyas (chapters), which are collectively known as the daśādhyāyī, he recapitulates the material contained in the *Paitāmahasiddhānta*, but in an order that became standard for Sanskrit treatises on astronomy, and interspersed the *Paitāmaha's* rules with many of his own, some of which are mathematically ingenious while others are simply wrong. The order of topics followed in the daśādhyāyī is: mean motions, true longitudes, the 'three questions' (local latitude, local time and directions), lunar eclipses, solar eclipses, risings and settings, the elevation of the moon's horns, the lunar 'shadow', conjunctions of planets with each other, and conjunctions of planets with fixed stars. Adhyāyas 13–17 contain supplementary rules to the topics of the daśādhyāyī. In adhyāya 11 Brahmagupta criticises his predecessors, especially Āryabhaṭa; adhyāya 12 is devoted to mathematics, including a brilliant series of rules for dealing with circumscribed quadrilaterals, and adhyāya 18 with the solution of indeterminate equations of the first degree and, for some cases, of the second degree as well. Adhyāyas 19–23 deal respectively with gnomons, metrics, cosmology, instruments and metrology. The last adhyāya gives the names of the preceding chapters and the number of verses in each; the grand total of verses is 1,008, a mocking allusion to the number of Mahāyugas in one of Āryabhaṭa's non-canonical Kalpas.

Among the new concepts introduced by Brahmagupta are: that component of the equation of time that is due to the solar anomaly, numerous elaborations of analemmata for solving problems in spherical trigonometry, and elaborate rules for computing the times of conjunctions of planets with each other or with fixed stars, in which a conjunction is assumed to occur when two bodies fall on the same circle of declination so that their polar longitudes are equal. In this last assumption Brahmagupta, following the *Paitāmaha*, falls into a tradition that emanates from Hipparchus.

A large part of the *Brāhmasphuṭasiddhānta*, in a derivative apparently entitled *Mahā-siddhānta*, was translated into Arabic at Baghdad in the early 770s, and became the basis of various zījes by al-Fazārī, by Ya'qūb ibn Ṭāriq and, in about 830, by al-Khwārizmī (see pp. 148 and 151). Al-Khwārizmī's *Zīj al-Sindhind*, as corrected by al-Majrīṭī, was translated into Latin by Adelard of Bath in 1126. This translation, along with others made from Arabic of associated texts, provided the basis for an Indo-Arab stage of Western European astronomy.

Late in the seventh century – its epoch is 23 March 665 – Brahmagupta composed his *Khaṇḍakhādyaka*. The first half of this text summarises the astronomy of the Ārdharātrikapakṣa of the author whom Brahmagupta previously despised; but the second part contains numerous corrections and revisions of the first, some of which suffice to bring parts of it into conformity with the Brāhmapakṣa.

Toward the middle of the eighth century there was composed at Sthānvīśvara a *Pauliśasiddhānta* that also belongs to the Ārdharātrikapakṣa; the text is now lost, but considerable knowledge of its contents can be gained from citations in both Sanskrit (e.g. by Bhaṭṭotpala) and Arabic (e.g. by al-Bīrūnī) sources. The main innovation of this *Pauliśa* was to reintroduce the unequal divisions of the traditional Mahāyuga into the chronological framework of one of Āryabhaṭa's systems while keeping the same numbers of revolutions of the planets in that period. This is achieved by beginning the planets' motions with a mean conjunction at Aries 0° 648,000 years after the beginning of the current Mahāyuga, that is, just 1,080,000 × 3 years before the beginning of the current Kaliyuga, at which time another mean conjunction of the planets at Aries 0° occurred.

The Saurapakṣa school

In about AD 800 a modification of the Ārdharātrikapakṣa was produced and entitled the *Sūryasiddhānta*. In this text the Brāhmapakṣa's description of a Kalpa is adapted, but the numbers of rotations of the planets in that period are kept multiples of four by the introduction of a period of creation lasting for the first 17,064,000 years of the Kalpa, at the end of which period the planets began their rotations. The parameters of the *Sūryasiddhānta* differ somewhat from those of the Ārdharātrikapakṣa (see Table 5), but

Table 5

Planet	R	$\dfrac{R \times 6,0}{Y}$	$\dfrac{R \times 6,0}{C}$
Saturn	146,568	12; 12, 50, 24°	0; 2, 0, 22, 53, 25, 46, ...°
Jupiter	364,220	30; 21, 6°	0; 4, 59, 8, 48, 35, 47, ...°
Mars	2,296,832	3, 11; 24, 9, 36°	0; 31, 26, 28, 11, 8, 56, ...°
Sun	4,320,000	6, 0°	0; 59, 8, 10, 10, 24, 12, ...°
Venus's śīghra	7,022,376	3, 45; 11, 52, 48°	1; 36, 7, 43, 37, 16, 52, ...°
Mercury's śīghra	17,937,060	54; 45, 18°	4; 5, 32, 20, 41, 51, 16, ...°
Moon	57,753,336	2, 12; 46, 40, 48°	13; 10, 34, 52, 3, 49, 4, ...°
Manda	488,203	40; 41, 0, 54°	0; 6, 40, 58, 42, 31, 5, ...°
Node	− 232,238	− 19; 21, 8, 24°	− 0; 3, 10, 44, 13, 35, 59, ...°

Table 6

Planet	R (Ārya) (yr = 6, 5; 15, 31, 15)	R (Ārdharātrika) (yr = 6, 5; 15, 31, 30)	Bīja	R (Saura) (yr = 6, 5; 15, 31, 31, 24)	Bīja
Saturn	146,564	146,564	0	146,568	+4
Jupiter	364,224	364,220	−4	364,220	−4
Mars	2,296,824	2,296,824	0	2,296,832	+8
Venus's śīghra	7,022,388	7,022,388	0	7,022,376	−12
Mercury's śīghra	17,937,020	17,937,000	−20	17,937,060	+40
Moon	57,753,336	57,753,336	0	57,753,336	0
Manda	488,219	488,219	0	488,203	−16
Node	−232,226	−232,226	0	232,238	−12

the motivations for the changes remain obscure (see Table 6); there is no reason to believe that they included new data derived from observations.

Apparently contemporaneously with the composition of the *Sūryasiddhānta* and the founding of the Saurapakṣa, Lalla composed his *Śiṣyadhīvṛddhidatantra*. In this he follows the Āryapakṣa with regard to parameters, but the Brāhmapakṣa with regard to the compositions of the Kalpa and the Mahāyuga.

An eclectic work mingling Āryapakṣa parameters with some from the Ārdharātrikapakṣa and others of unknown origin is the *Laghumānasa* of Muñjāla, who was alive in 932. But his most intriguing innovation is a formula for finding the evection of the moon. This brings about a maximum value of the evection of 2; 29°; this is probably an Indian approximation to Ptolemy's maximum value, 2; 39°. Muñjāla could have learned of Ptolemy's lunar model through some Islamic intermediary.

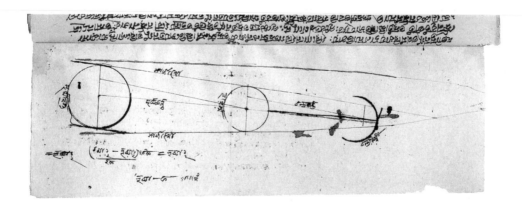

45 Diagram illustrating the computation of the diameter of the earth's shadow at the moon's distance, written in Bengali script. This is one of seventeen folios appended to the main part of a manuscript; that part contains the grahagaṇitādhyāya from Bhāskara's *Siddhāntaśiromaṇi*, but the extra seventeen folios were not originally part of that manuscript. (Indian Institute, Oxford, MS Chandra Shum Shere C.403, f. 159v)

Another author who gives a rule for finding the moon's evection (his maximum value is 2; 40°) was Śrīpati, who wrote his *Siddhāntaśekhara* in imitation of and dependent on the *Brāhmasphuṭasiddhānta* in about 1050. Besides the rule for evection, in which he had been anticipated by Muñjāla, Śrīpati introduced a method for computing the second component of the equation of time, that dependent on the changing declination of the sun.

In the Brāhmapakṣa the next is the most popular treatise, the *Siddhāntaśiromaṇi* that a second Bhāskara composed in 1150 (Fig. 45). His main contributions fall in trigonometry, in which he may have been influenced by Islamic mathematics. Bhāskara later wrote a *Karaṇakutūhala* in which he presented several ingenious approximations to functions based on modifications of trigonometric functions; the epoch of this karaṇa is 24 February 1183.

Southern India: Vākyas and Mādhava's school

In Southern India a set of easily memorised numbers, called vākyas, that gave the longitude of the moon according to Āryapakṣa parameters for each of the 248 days in 9 anomalistic months had been in use for several centuries when, in about 1300, the system was extended in a text entitled *Vākyakaraṇa* to the five planets as well as the moon. For each celestial body several cycles were devised in which, as the cycles become longer, the body returns closer to its epoch longitude. These cycles range up to 500,000 or 600,000 days of mean motion for each planet, and allow the astronomer to eliminate from his computations large periods of lapsed time since epoch while preserving an accurate record of the deviation of the planet's mean motion from an integer number of complete rotations. The basic ideas of this method are found in Mesopotamian texts and in Greek adaptations of them, and were transmitted to India in the third or fourth century AD; the Indian vākyas, however, are based on the Indians' own parameters, except for the two smallest cycles for the moon.

The culmination of Southern Indian astronomy, in which both the Āryapakṣa and the Saurapakṣa were represented, was in the tradition begun by Mādhava in Kerala shortly before 1400. Mādhava himself was most famous for his derivation of the infinite series for π and the power series for trigonometric functions. His pupil, Parameśvara, attempted to correct solar and lunar parameters by making a long series of eclipse observations between 1393 and 1432; in these he clearly utilised an astrolabe to determine the angle of altitude of the eclipsed body, and possibly also the time of the phases of the eclipses. This is particularly remarkable since the astrolabe had been introduced into Sanskrit in a translation (or adaptation) of a Persian text by Mahendra Sūri only in 1370, and Mahendra worked in Northern India, under Fīrūz Shāh. An attempt to revise planetary parameters on the basis of observations was made by Nīlakaṇṭha, a pupil of Parameśvara's son, Dāmodara; Nīlakaṇṭha, while not successful in this effort, also wrote an impressive but uninfluential exposition of the ideal role of observation in astronomy,

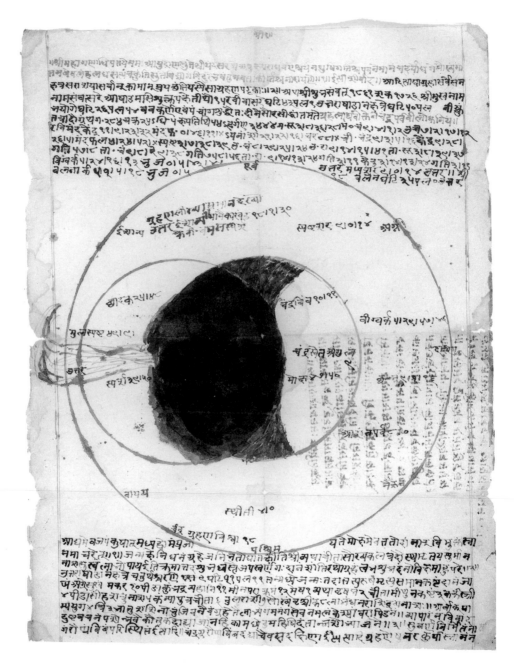

46 Drawing of the lunar eclipse of 22 July 1804. (Indian Institute, Oxford, collection of Robert Fraser)

the *Jyotirmīmāṃsā*. He did not succeed in persuading his readers in Southern India to continue to make and to utilise observations for determining correct parameters.

Attempts to employ observations not to generate new parameters, but to provide a basis for a choice among the parameters of the various pakṣas, were made by two contemporaries of Nīlakaṇṭha living in Gujarāt, Keśava and his son, Gaṇeśa, after whom the resulting Gaṇeśapakṣa was named.

Islamic influence

Some appreciation of what could be accomplished by careful comparisons between observation and models came only with the translations of various Arabic and Persian texts into Sanskrit in the seventeenth and eighteenth centuries. These translations include one of al-Qūshjī's *Risālat dar hay'a*, known in Sanskrit as the *Hayatagrantha*; the *Siddhāntasindhu* in which Nityānanda rendered Farīd al-Dīn's *Zīj-i Shāh Jahān*, with which must be mentioned the *Siddhāntarāja* in which Nityānanda unsuccessfully attempted to persuade orthodox Brāhmaṇas that Persian astronomy falls within the traditions of the Indian deities and sages from whose revelation the siddhāntas and karaṇas were claimed to be derived; and the series of texts that were translated at the court of Jayasiṃha (Jai Singh) in the 1720s and 1730s: Ptolemy's *Almagest*, Euclid's *Elements*, Theodosius's *Spherics*, and chapter 11 of the second book of the *Tadhkira* in which Naṣīr al-Dīn al-Ṭūsī (see pp. 148 and 150) had explained his new planetary and lunar models, together with the commentary on that chapter authored by al-Birjandī.

Jayasiṃha also constructed the mammoth stone instruments, in imitation of those built by Ulugh Beg at Samarqand, whose substantial remains can still be studied at Delhi, Jaipur and Ujjain, and which once could also be found at Mathurā and Benares. With little input from the observations made at these observatories Jayasiṃha's assistants composed in Persian a *Zīj-i Muḥammad Shāh*. The translations of Ptolemy, Euclid and Theodosius enjoyed some circulation in Northern India in the late eighteenth and early nineteenth centuries, and Jayasiṃha's instruments were described both by himself and his paṇḍitas in the *Siddhāntasamrāṭ* and the *Yantraprakāra*, and by Nandarāma Miśra in the *Yantrasāra* that he wrote in 1771. But no one attempted again to construct such massive instruments.

The attempts by a few scholars in Benares in the seventeenth century – in particular Muniśvara and Kamalākara – to combine elements of the Islamic interpretation of Ptolemaic astronomy with the tradition of the Saurapakṣa did not meet with widespread acceptance. While Mathurānātha Śukla described the contents of a Persian text on 'ilm al-hay'a in Sanskrit in his *Jyotiḥsiddhāntasāra* of 1782, most Indian astronomers continued to use one of the five pakṣas well into the nineteenth century and, if they wrote at all, to compose works that fell into one of those traditional schools. By the end of the nineteenth century, however, while some calendar-makers and horoscope-casters continued to utilise the old astronomical tables, the tradition was no longer a living one.

Those young Indians interested in astronomy as a field of research began universally to turn to the West for their training, and have indeed contributed significantly in the work of men such as Chandrasekhar to the development of modern astrophysics.

Epilogue

Indian astronomy is unusual in that observation played a very minor role – that of checking the validity of accepted parameters rather than that of discovering new parameters and/or new models. The tradition of Ptolemy is simply not applicable; and this evidently remained true until observational instruments were introduced into India from the Islamic Ptolemaic tradition in the late fourteenth century. This new Indo-Muslim tradition of observational astronomy culminated in the massive masonry instruments erected by Jayasiṃha in the 1730s, when they were already anachronistic.

Instead, Indian astronomers received both models and parameters from external sources, and adapted them to their own traditions and needs. Those needs were particularly the computation of their complex calendars (pañcāṅgas), time-keeping, the casting of horoscopes, and the prediction of solar and lunar eclipses and of conjunctions of planets with fixed stars or with each other. To these ends a sophisticated mathematics of approximation was employed and elaborate arrangements of tables developed. Bījas (epoch and periodic corrections) were introduced to convert the mean longitudes of one pakṣa into those of another. But astronomy was advanced by mathematics rather than by deductions from 'facts' in nature.

Some of these mathematical innovations had a profound effect on neighboring cultures, as, for instance, trigonometry and analemmata had on Islamic astronomy and its dependents in medieval Western Europe; others were never transmitted out of India. In its reception of external ideas and its influence on others India played a pivotal role in the development of astronomical sciences in Eurasia, but, of course, preserved its unique characteristics which mark it as an integral part of one of the world's great cultures.

Bibliography

Extensive bibliographies will be found in the several volumes of D. Pingree, *Census of the Exact Sciences in Sanskrit*, Series A, vols 1–5, American Philosophical Society, Philadelphia, 1970–94; these volumes also contain information both biographical and bibliographical on individual Indian astronomers. See also D. Pingree, *Jyotiḥśāstra*, Otto Harrassowitz, Wiesbaden, 1981. The technical aspects of Indian astronomy are described in D. Pingree, History of mathematical astronomy in India, in *Dictionary of Scientific Biography* 15, Charles Scribner's Sons, New York, 1978, pp. 533–633. For an opposing interpretation of Indian astronomy see R. Billard, *L'astronomie indienne*, Ecole française d'Extrême-Orient, Paris, 1971.

DAVID A. KING

Islamic Astronomy

The nature of Islamic astronomy

From the ninth century to the fifteenth, Muslim scholars excelled in every branch of scientific knowledge. In particular their contributions to astronomy and mathematics are impressive. There are an estimated 10,000 Islamic astronomical manuscripts and close to 1,000 Islamic astronomical instruments preserved in libraries and museums in the Near East, Europe and North America, but it is clear that even if all of them were properly catalogued and indexed – and we are still very far from this state of affairs – the picture that we could reconstruct of Islamic astronomy, especially for the eighth, ninth and tenth centuries, would be quite deficient. Most of the available manuscripts and instruments date from the later period of Islamic astronomy, that is, from the fifteenth to the nineteenth century, and although some of these are based or modelled on earlier works many of the early works are extant in unique copies and others have been lost almost without trace, that is, we sometimes know only of their titles. The thirteenth-century Syrian scientific biographer Ibn al-Qifṭī relates that the eleventh-century Egyptian astronomer Ibn al-Sanbadī heard that the manuscripts in the library in Cairo were being catalogued and so he went to have a look at the works relating to his field. He found 6,500 manuscripts relating to astronomy, mathematics and philosophy. Not one of these survives amongst the 2,500 scientific manuscripts preserved in Cairo today.

The surviving manuscripts thus constitute but a small fraction of those that were actually copied; nevertheless they preserve for us a substantial part of the Islamic scientific heritage, certainly enough of it for us to judge its level of sophistication. Only in the past few decades has the scope of the activity and achievements of Muslim scientists become apparent, and the days are long past when they were regarded merely as transmitters of superior ancient knowledge to ignorant but eager Europeans. Islamic astronomy is to be viewed on its own terms. The fact that only a small part of the available material, mainly Greek and Indian material in Arabic garb, was indeed transmitted to Europe is to be viewed as an accident of Islamic history. There is no need to apologise for using the expression 'Islamic astronomy'. Within a few decades of the death of the

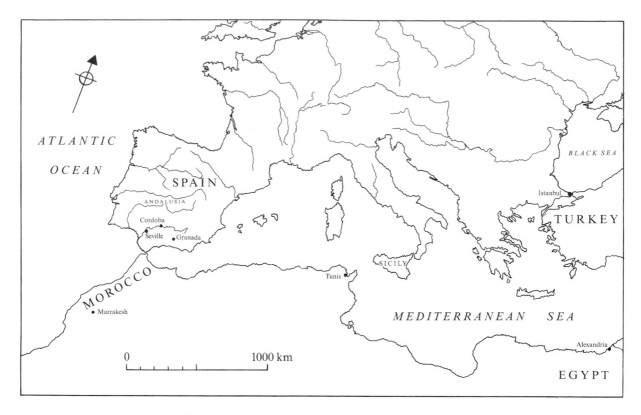

47 (*above and opposite*) The Islamic world.

Prophet Muḥammad in 632 the Muslims had established a commonwealth stretching from Spain to Central Asia and India. They brought with them their own folk astronomy, which was then mingled with local traditions, and they discovered the mathematical traditions of the Indians, Persians and Greeks, which they mastered and adapted to their needs. Early Islamic astronomy was thus a pot-pourri of pre-Islamic Arabian star-lore and Indian, Persian and Hellenistic astronomy, but by the tenth century Islamic astronomy had acquired very distinctive characteristics of its own. A. I. Sabra labels this process 'appropriation and naturalisation'.

We should point out at the outset that astronomy flourished in Islamic society on two different levels: folk astronomy, devoid of theory and based solely on what one can see in the sky, and mathematical astronomy, involving systematic observations and mathematical calculations and predictions. Folk astronomy was favoured by the scholars of the sacred law (*fuqahā'*), not least because of various religious obligations that demanded a basic knowledge of the subject; these legal scholars generally had no time (or need) for mathematical astronomy. That discipline was fostered by a select group of scholars, most of whose activities and pronouncements were, except in the case of astrological predictions, of little interest to society at large.

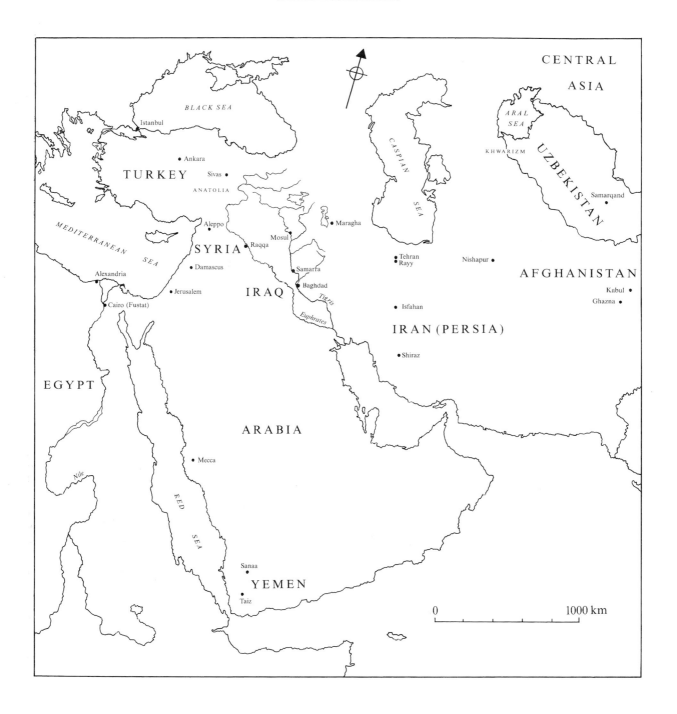

As we shall see, the astronomers also played their part in applying their discipline to certain aspects of Islamic religious practice. It was not Islam that encouraged the development of astronomy but the richness of Islamic society, a multiracial, highly-literate, tolerant society with one predominant cultural language, Arabic. But neither did Islam, the religion, stand in the way of scientific progress. The Prophet had said: 'Seek knowledge, even as far as China'. To be sure, over-zealous orthodox rulers occasionally pursued, killed or otherwise attacked 'scientists' or destroyed or burnt their libraries, but these were exceptions. The scholars of the religious law, who saw themselves as the representatives of Islam, generally ignored the pronouncements of the scientists, even on matters relating to religious practice. Astronomy was the most important of the Islamic sciences, as we can judge by the volume of the associated textual tradition, but a discussion of it in the broader context of the various branches of knowledge, which has been attempted several times elsewhere, is beyond the scope of this chapter.

Arab star-lore

The Arabs of the Arabian peninsula before Islam possessed a simple yet developed astronomical folklore of a practical nature. This involved a knowledge of the risings and settings of the stars, associated in particular with the cosmical settings of groups of stars and simultaneous heliacal risings of others, which marked the beginning of periods called *naw'*, plural *anwā'*. These *anwā'* eventually became associated with the twenty-eight lunar mansions (Fig. 48), a concept apparently of Indian origin. A knowledge of the passage of the sun through the twelve signs of the zodiac, associated meteorological and agricultural phenomena, the phases of the moon, as well as simple time-reckoning using shadows by day and the lunar mansions by night, formed the basis of later Islamic folk astronomy, which flourished separately from mathematical astronomy in Islamic society.

More than twenty compilations on the pre-Islamic Arabian knowledge of celestial and meteorological phenomena as found in the earliest Arabic sources of folklore, poetry and literature, are known to have been compiled during the first four centuries of Islam. The best known is that of Ibn Qutayba, written in Baghdad about the year 860. Almanacs enumerating agricultural, meteorological and astronomical events of significance to local farmers were also compiled: several examples of these survive from the medieval Islamic period, one such being for Córdoba from the year 961. The Yemen possessed a particularly rich tradition of folk astronomy, and numerous almanacs were compiled there.

Since the sun, moon and stars are mentioned in the Quran, an extensive literature dealing with what may well be labelled Islamic folk cosmology arose. This was inevitably unrelated to the more 'scientific' Islamic tradition based first on Indian sources and then predominantly on Greek ones. Since it is also stated in the Quran that man should use these celestial bodies to guide him, the scholars of the religious law occupied themselves with folk astronomy. We shall mention below various treatises dealing

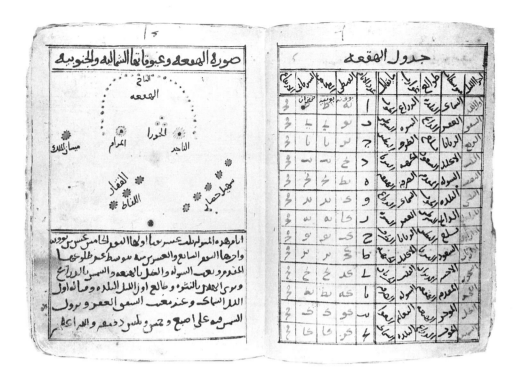

48 An illustration of the stars of one of the lunar mansions (al-haq'a) in an Egyptian treatise on folk astronomy. The table on the right identifies the mansions culminating, rising and setting, and the mansions opposite them (at 180°) at different times of the night when the sun is in that mansion; it also gives the date in the Coptic, Western and Syrian calendars and the midday solar altitude when the sun is in each of the 13 degrees of the mansion (360 ÷ 28 ≈ 13). The associated text repeats some of this information and adds the midday shadow length. (Chester Beatty Library, Dublin, MS 4538)

with simple time-keeping and the determination of the direction of Mecca by non-mathematical means.

Persian and Indian sources

The earliest astronomical texts in Arabic seem to have been written in Sind and Afghanistan, areas conquered by the Muslims already in the seventh century. Our knowledge of these early works is based entirely on citations from them in later works. They consisted of text and tables and were labelled *zīj* after a Persian word meaning 'cord' or 'thread' and by extension 'the warp of a fabric', which the tables vaguely resemble. The Sasanian *Shahriyārān Zīj* in the version of Yazdigird III (see p. 135) was translated from Pahlavi into Arabic as the *Shāh Zīj*, and the astronomers of the Caliph al-Manṣūr chose an auspicious moment to found his new capital Baghdad using probably an earlier Pahlavi version of this *zīj*. The various horoscopes computed by Māshā'allāh (Baghdad, *c.* 800) in his astrological world history are based on it.

Significant for the subsequent influence of Indian astronomy in the Islamic tradition was the arrival of an embassy sent to the court of al-Manṣūr from Sind *c.* 772. This embassy included an Indian well versed in astronomy and bearing a Sanskrit astronomical text apparently entitled the *Mahāsiddhānta* and based partly on the *Brāhmasphuṭasiddhānta* (see pp. 136–7). The Caliph ordered al-Fazārī to translate this text into Arabic with the help of the Indian. The resulting *Zīj al-Sindhind al-kabīr* was the basis of a series of *zījes* by such astronomers as al-Fazārī, Yaʿqūb ibn Ṭāriq, al-Khwārizmī, Ḥabash, Ibn Amājūr, al-Nayrīzī, and Ibn al-Ādamī, all prepared in Iraq before the end of the tenth century. The *Sindhind* tradition (see pp. 133 and 137) flourished in Andalusia, mainly through the influence there of the *Zīj* of al-Khwārizmī (see below). As a result, the influence of Indian astronomy is attested from Morocco to England in the late Middle Ages.

Greek sources

Ptolemy's *Almagest* was translated at least five times in the late eighth and ninth centuries. The first was a translation into Syriac and the others were into Arabic, the first two under the Caliph al-Maʾmūn in the middle of the first half of the ninth century, and the other two (the second an improvement of the first) towards the end of that century. All of these were still available in the twelfth century, when they were used by Ibn al-Ṣalāḥ for his critique of Ptolemy's star catalogue. The translations gave rise to a series of commentaries on the whole text or parts of it, many of them critical and one, by Ibn al-Haytham (*c.* 1025), actually entitled 'Doubts about Ptolemy' (*al-Shukūk*). The most commonly used version of the *Almagest* in the later period was the recension of the late ninth-century version by the polymath Naṣīr al-Dīn al-Ṭūsī in the mid thirteenth century. Various other works by Ptolemy, notably the *Planetary Hypotheses* and the *Planisphaerium*, and other Greek works, including the short treatises by Autolycus, Aristarchus, Hypsicles and Theodosius, and works on the construction known as the analemma for reducing problems in three dimensions to a plane, were also translated into Arabic; most of these too were later edited by al-Ṭūsī. In this way Greek planetary models, uranometry and mathematical methods came to the attention of the Muslims. Their redactions of the *Almagest* not only contained reformulations and paraphrases of its contents but they also 'corrected, completed, criticized and brought [the contents] up to date both theoretically and practically' (Saliba); most of this material has not been studied in modern times.

Developments in astronomy

Theoretical astronomy

The geometrical structure of the universe conceived by Muslim astronomers of the early Islamic period (*c.* 800–1050) is more or less that expounded in Ptolemy's *Almagest*, with the system of eight spheres being regarded essentially as mathematical models. However, already in Ptolemy's *Planetary Hypotheses* these models are taken as representing physical

49 A non-Ptolemaic planetary model for the moon, found in a copy of the treatise on planetary astronomy (*al-Tuḥfa al-shāhiyya*) written by Quṭb al-Dīn al-Shīrazī in Sivas (Anatolia) in AD 1285. Only since the 1950s have these models been investigated by modern scholars; the discovery that a series of Muslim astronomers concerned themselves with such models from the eleventh to the sixteenth centuries and developed models without the problems inherent in the Ptolemaic ones has promoted considerable interest in medieval Islamic planetary theory. (Egyptian National Library, Cairo, MS K3758)

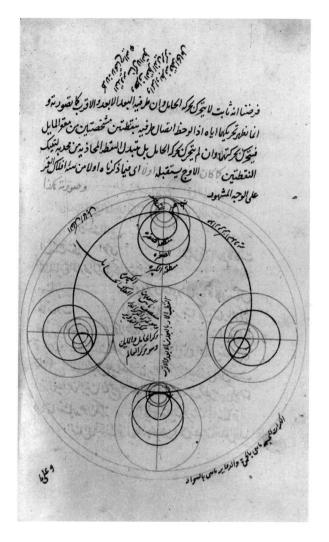

reality; this text also became available in Arabic. Several early Muslim scholars wrote on the sizes and relative distances of the planets, and one who proposed a physical model for the universe was Ibn al-Haytham (fl. *c.* 1025). In order to separate the two motions of the eighth sphere, the motion of the fixed stars due to the precession of the equinoxes and the motion of the fixed stars due to the apparent daily rotation, he proposed a ninth sphere to impress the apparent daily rotation on the others.

Of considerable historical interest are various Arabic treatises on the notion of the trepidation of the equinoxes. This theory, developed from Greek sources, found followers who believed that it corresponded better to the observed phenomena than a simple theory of uniform precession. The mathematical models proposed were complex and have only recently been studied properly (notably those of Pseudo-Thābit (date unknown) and Ibn al-Zarqāllu (Andalusia, *c.* 1070), who seems to have relied on his predecessor Ṣāʿid al-Andalusī). The theory of trepidation continued to occupy certain Muslim scholars (in the late period mainly in the Maghrib), as it did European scholars well into the Renaissance. The history of this notion has yet to be written.

Other significant Islamic modifications to Ptolemaic planetary models, devised to overcome the philosophical objections to the notion of an equant and the problem of the variation in lunar distance inherent in Ptolemy's lunar model, belong to the later period of Islamic astronomy (Fig. 49). There were two main schools, one of which reached its fullest expression in Maragha in North-west Iran in the thirteenth century (notably with al-Ṭūsī and his colleagues) and Damascus in the fourteenth (with Ibn al-Shāṭir), and the other of which developed in Andalusia in the late twelfth century (notably with al-Biṭrūjī). The latter tradition was doomed from the outset by a slavish adherence to (false) Aristotelian tenets and mathematical incompetence. The former was based on sophisticated modifications to Ptolemy's models, partly inspired by new observations; Ptolemy himself would have been impressed by it, as have been modern investigators, for the tradition has been rediscovered and studied only in the latter half of this century. In the 1950s E. S. Kennedy discovered that the solar, lunar and planetary models proposed by Ibn al-Shāṭir in his book *The Final Quest Concerning the Rectification of Principles* (*Nihāyat al-suʾl*) were different from those of Ptolemy, indeed that they were mathematically identical to those of Copernicus some 150 years later. In this work Ibn al-Shāṭir 'laid down the details of what he considered to be a true theoretical formulation of a set of planetary models describing planetary motions, and actually intended as alternatives to the Ptolemaic models' (Saliba). He maintained the geocentric system, whereas Copernicus proposed a hypothesis, which he was unable to prove, that the sun was at the centre of things. Nevertheless this important discovery raised the interesting question whether Copernicus might have known of the works of the Damascene astronomer (see p. 202). Since the 1950s we have progressed to a new stage of inquiry: we now know that there was a succession of Muslim astronomers from the eleventh century to the sixteenth who concerned themselves with models different from those of Ptolemy, all designed to overcome what were seen as flaws in them. The question we may now ask is: was Copernicus influenced by any of these Muslim works? The answer is unsatisfactory, namely, that he must have been; definitive proof is, however, still lacking.

Mathematical astronomy – the tradition of the zījes

The Islamic *zīj*es constitute an important category of astronomical literature for the historian of science, by virtue of the diversity of the topics dealt with, and the information that can be obtained from the tables. In 1956 E.S. Kennedy published a survey of about 125 Islamic *zīj*es. We now know of close to 200, and material is available for a revised version of the *zīj* survey. To be sure, many of these works are lost, and many of the extant ones are derived from other *zījes* by modification, borrowing, or outright plagiarism. Nevertheless, there are enough *zījes* available in manuscript form to reconstruct a reasonably accurate picture of the Islamic activity in this field.

Most *zījes* consist of several hundred pages of text and tables; the treatment of the material presented may vary considerably from one *zīj* to another. The following aspects of mathematical astronomy are handled in a typical *zīj*:

1) chronology
2) trigonometry
3) spherical astronomy
4) solar, lunar and planetary mean motions
5) solar, lunar and planetary equations
6) lunar and planetary latitudes
7) planetary stations
8) parallax
9) solar and lunar eclipses
10) lunar and planetary visibility
11) mathematical geography (lists of cities with geographical co-ordinates), determination of the direction of Mecca
12) uranometry (tables of fixed stars with co-ordinates)
13) mathematical astrology.

As noted above, already in the eighth century in India and Afghanistan there were compiled a number of Arabic *zījes*. These earliest examples, based on Indian and Sasanian works, are lost, as are the earliest examples compiled at Baghdad in the eighth century. With the *zījes* compiled in Baghdad and Damascus in the early ninth century under the patronage of the Caliph al-Ma'mūn we are on somewhat firmer ground. These follow either the tradition of the *Almagest* and *Handy Tables* or the Indian tradition. Manuscripts exist of the *Mumtaḥan Zīj* of Yaḥyā ibn Abī Manṣūr and the Damascus *Zīj* of Ḥabash, each of which was based on essentially Ptolemaic theory rather than Indian. The *Zīj* of al-Khwārizmī, based mainly on the Persian and Indian traditions, has survived only in a Latin translation of an Andalusian recension. The *Ṣābi' Zīj* of al-Battānī of Raqqa *c.* 910; the *Ḥākimī Zīj* of Ibn Yūnus, compiled in Cairo at the end of the tenth century; the *zīj* called *al-Qānūn al-Mas'ūdī* by al-Bīrūnī, compiled in Ghazna about 1025; the *Zīj* of Ibn Isḥāq, compiled in Tunis *c.* 1195; the *Īlkhānī Zīj* of Naṣīr al-Dīn al-Ṭūsī, prepared in Maragha in the mid thirteenth century; and the *Sulṭānī Zīj* of Ulugh Beg from early fifteenth-century Samarqand: these are amongst the most important later works of this genre, and also the most influential.

The only *zījes* from the early period of Islamic astronomy that have been published with translation and commentary are those of al-Khwārizmī (in the much modified later recension) and al-Battānī. The Arabic text of the *Zīj* of al-Bīrūnī has been published and a Russian translation and commentary are available. The observation accounts in the introduction of the *Ḥākimī Zīj* of Ibn Yūnus and the texts (but not the tables) of the *Zīj*es of Ibn al-Bannā' (Marrakesh, *c.* 1300) and of Ulugh Beg have been published and translated. Also a Byzantine translation of one of the *zīj*es of al-Fahhād (Iran, *c.* 1150; see p. 108) has been published. No other *zīj*es have received such attention.

Although the *zīj*es are amongst the most important sources for our knowledge of Islamic mathematical *astronomy*, it is important to observe that they generally contain extensive tables and explanatory text relating to mathematical *astrology* as well. Islamic astrological texts form an independent corpus of literature, mainly untouched by modern scholarship. Often highly sophisticated mathematical procedures are involved. It should also be pointed out that in spite of the fact that astrology was anathema to Muslim orthodoxy, it has always been (and still is) widely practised in Islamic society.

We shall now consider various aspects of the *zīj*es and the related literature.

NUMERICAL NOTATION AND BASIC MATHEMATICAL AUXILIARY TABLES   All early Islamic astronomical tables have entries written in Arabic alphanumerical notation and expressed sexagesimally, that is, to base 60. A number written in letters equivalent to '23 30 17 seconds' (Ulugh Beg's value for the obliquity) stands for $23 + \frac{30}{60} + \frac{17}{3600}$ degrees, that is, 23°30'17". In sexagesimal arithmetic, more so than in decimal arithmetic, it is useful to have a multiplication table at hand, and such tables, with 3,600 or even 216,000 entries, were available.

Already in the early ninth century Muslim astronomers had restyled the cumbersome Indian sine function using the Greek base 60 (which the Greeks had used for their even more cumbersome chord function). Likewise the Indian shadow functions, unknown in Greek astronomy, were adopted with different bases (12, 6, $6\frac{1}{2}$ and 7, and also 60 and occasionally 1). Most *zīj*es contain tables of the sine and (co)tangent function for each whole, or half, or quarter degree of arc. Entries are generally given to three sexagesimal digits, corresponding roughly to five decimal digits. But certain Muslim scholars compiled more extensive sets of trigonometric tables that were not included in *zīj*es. Already in the early tenth century al-Samarqandī prepared a set of tables of the tangent function with entries to three sexagesimal digits for each minute of arc. Later in the same century Ibn Yūnus tabulated the sine function to five sexagesimal digits, equivalent to about nine decimal digits, for each minute of arc, also giving the differences for each second. He also tabulated the tangent function for each minute of arc, and the solar declination for each minute of solar longitude. His trigonometric tables were not sufficiently accurate to warrant this number of significant figures, and indeed over four centuries were to elapse before the compilation in Samarqand of the magnificent trigonometric tables in the *Sulṭānī Zīj* of Ulugh Beg, which display the values of the sine and tangent to five sexagesimal digits for each minute of argument and are generally accurate in the last digit.

PLANETARY TABLES AND EPHEMERIDES   Given the Ptolemaic models and tables of
the mean motion and equations of the sun, moon and planets such as were available to
Muslim astronomers in the *Almagest* and *Handy Tables*, or the corresponding tables based
on Indian models that exemplify the *Sindhind* tradition, Muslim astronomers from the
ninth to the sixteenth century sought to improve the numerical parameters on which
these tables were based. Most of the leading Muslim astronomers of the early period
made solar observations and computed new solar equation tables. Ibn Yūnus is the only
astronomer from the first four centuries of Islam known to have compiled a new set of
lunar equation tables. The majority of Islamic planetary equation tables are Ptolemaic,
and where exceptions do occur, such as in the tables of Ibn al-Aʿlam and Ibn Yūnus for
Mercury, we find that they are based on a Sasanian parameter rather than on any new
observations. Pending a new edition of Kennedy's survey of Islamic *zījes* to include
all available parameters as well as bio-bibliographical data and a reclassification of the
200-odd known *zījes* into clearly defined families, it is better not to say more at this
time.

Ptolemy used the same data as Hipparchus for his determination of the solar apogee
and hence obtained the same result. The Muslims thus inherited the notion that the
solar apogee is fixed with respect to the fixed stars (although the planetary apogees move
with the motion of precession), and it is to their credit that their earliest observations
established that the solar apogee had moved about 15° since the time of Hipparchus.
Most early Muslim astronomers accepted the *Mumtaḥan* value of 1° in $66\frac{2}{3}$ Persian years
(actually a parameter attested in earlier Persian sources) for both precession and the
motion of the apogees. Ibn Yūnus possessed all the necessary data that could be used to
demonstrate that the motion of the solar apogee is not the same as the motion due to
precession, but he chose to use the same value for both, 1° in $70\frac{1}{4}$ Persian years, which
happens to be remarkably close to the actual rate of precession. Al-Bīrūnī (Central Asia,
c. 1025) seems to have been the first to distinguish the proper motion of the solar apogee
from the motion of precession (this discovery is sometimes erroneously attributed to al-
Battānī). It was Ibn al-Zarqāllu (Andalusia, *c.* 1070) who was the first to assign a numerical
value to both motions, although he also subscribed to the theory of trepidation.

All Islamic *zījes* contained tables of mean motions and equations for computing
solar, lunar and planetary positions for a given time (Fig. 50). Some of the equation
tables are arranged in a form more convenient for the user (so that one simply has to
enter the mean motions, and calculations are avoided). Auxiliary tables were sometimes
available for generating ephemerides without the tedious computation of daily positions
from mean-motion and equation tables. From the ninth to the nineteenth century
Muslim astronomers compiled ephemerides displaying solar, lunar and planetary positions of each day of the year, as well as information on the new moons and astrological
predictions resulting from the position of the moon relative to the planets. Al-Bīrūnī
described in detail how to compile ephemerides in his astronomical and astrological

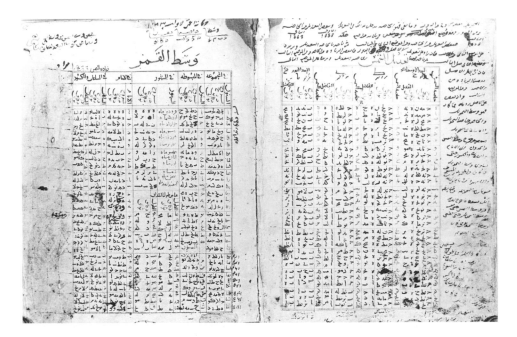

50 Astronomical tables typical of the type found in Islamic *zījes*. These serve the solar equation (right) and the lunar mean motion (left). They are found in a Yemeni manuscript of the *Zīj* of the Persian astronomer Kūshyār ibn Labbān, compiled *c.* AD 1000 and copied here *c.* AD 1250. The manuscript has various marginalia giving modifications to the tables for the longitude of Sanaa. The Yemen was an important centre of astronomy in the Middle Ages. (Egyptian National Library, Cairo, MS DM 400)

handbook *Instruction in the Art of Astrology* (*Tafhīm*). Manuscripts of ephemerides had a high rate of attrition since the tables could be dispensed with at the end of the year: the earliest complete extant examples are from fourteenth-century Yemen, discovered in Cairo in the 1970s and still unpublished; on the other hand, literally hundreds of ephemerides survive from the late Ottoman period.

STELLAR CO-ORDINATES AND URANOGRAPHY   Most *zījes* contain lists of stellar co-ordinates in either the ecliptic or the equatorial system, or occasionally in both systems. A survey of the stellar co-ordinates in Islamic *zījes*, which has not yet been conducted, would be a valuable contribution to the history of Islamic astronomy, and could help determine the extent to which original observations were made by Muslim astronomers. An impressive amount of research on Arabic star names and their later influence in Europe has been conducted in the last few years by P. Kunitzsch.

In his *Book of Constellation Figures* (*Ṣuwar al-kawākib*, Col. Pl. IX) the tenth-century Shiraz astronomer al-Ṣūfī presented lists of stellar co-ordinates as well as illustrations of the constellation figures from the Hellenistic tradition and also information on the lunar mansions following the Arab tradition. Later Islamic works on uranography are mostly restricted to Persian and Turkish translations of al-Ṣūfī, although some astrological works also contain illustrations of the constellations that have recently attracted the attention of historians of Islamic art.

SPHERICAL ASTRONOMY AND SPHERICAL TRIGONOMETRY   Most *zījes* contain in their introductory text the solutions of the standard problems of spherical astronomy, such as, to give only one example, the determination of time from solar and stellar altitude. Rarely is any explanation given of how the formulae outlined in words in the text were derived. There were two main traditions. In the first, the problems relating to the celestial sphere are reduced to geometric or trigonometric problems on a plane. The construction known as the analemma was a singularly powerful tool for solutions of this kind. In the second, the problems are solved by applications of rules of spherical trigonometry. Both techniques are ultimately of Greek origin, and Muslim scholars made substantial contributions to each.

There is some confusion about these contributions in the modern literature. It has been assumed by modern writers that when a medieval writer used a medieval formula that is mathematically equivalent to the modern formula derived by a specific rule of spherical trigonometry, the medieval scholar must have known the equivalent of the modern rule of spherical trigonometry. In fact, however, the medieval formula may have been derived without using spherical trigonometry at all. The first known Islamic treatise dealing with spherical trigonometry independently from astronomy is by the eleventh-century Andalusian Ibn Muʿādh. The contributions to spherical astronomy by such scholars as Thābit ibn Qurra, al-Nayrīzī, Abu l-Wafāʾ al-Būzajānī, al-Khujandī, Kūshyār ibn Labbān, al-Sijzī, Abū Naṣr, are outlined in the recently rediscovered *Keys to Astronomy (Maqālīd)* of al-Bīrūnī, also from the eleventh century.

Already in the work of Ḥabash in the mid ninth century we find a Muslim astronomer at ease with both spherical trigonometric methods and analemma constructions for solving problems of spherical astronomy. In the *zījes* of scholars of the calibre of Ibn Yūnus and al-Bīrūnī we find various methods for solving all of the standard problems of medieval spherical astronomy. The auxiliary trigonometric tables compiled by such scholars as Ḥabash, Abū Naṣr (*c.* 1000) and al-Khalīlī (*c.* 1360) for solving all of the problems of spherical astronomy for any latitude are a remarkable testimony to their mastery of the subject.

Applications of astronomy to aspects of religious practice
The lunar calendar
The Muslim calendar is lunar and the civil months begin with the first sighting of the lunar crescent. The precise determination of the beginnings and ends of the months is particularly important for Ramaḍān, the sacred month of fasting, and various other religious festivals. The legal scholars were content to rely on direct sighting of the

crescent or on alternating 29 and 30-day months to regulate the calendar. But the subject of lunar crescent visibility was generally treated in *zījes* (Fig. 51), and a wide variety of methods and tables were devised to facilitate the solution of this problem. Al-Khwārizmī, for example, compiled a table of the minimum ecliptic elongation of the sun and moon for each zodiacal sign, computed for the latitude of Baghdad, and another early Muslim astronomer, perhaps from Andalusia, compiled a similar table for each of the climates, no doubt inspired by Ptolemy's planetary visibility tables. A few early Islamic astronomers, notably Thābit ibn Qurra and Ibn Yūnus, postulated conditions that appear to be considerably more sophisticated, although they are not yet fully understood. Others like al-Battānī and Ibn al-Zarqāllu, merely played around with numbers, deriving complicated procedures from a basic notion that visibility occurs when the ecliptic elongation is 12° or 1 day's relative motion of the sun and moon after conjunction. Unfortunately no Muslim astronomer has left us any observational data on crescent visibility, and with a few possible exceptions yet to be identified none of the various sets of conditions proposed by the Muslim astronomers appears to be based on such data.

Nowadays there is often confusion about the beginning of Ramaḍān, such as could never have occurred in medieval times. This confusion results from the fact that the crescent may be seen in some locations and not in others, and no less from the reluctance of the religious scholars, who have the final say in announcing the new month, to listen to the astronomers.

51 Extracts from an Egyptian set of tables displaying calculations for the visibility of the lunar crescent. These display for each month of the lunar years 1125 and 1126 Hijra (= AD 1713–14), for sunset on the first day of each month in the civil calendar, the lunar longitude and latitude, angular separation of the sun and moon, and difference in their setting times, followed by a prediction: '[the crescent] will be seen clearly', 'probably' (literally, mostly), 'with difficulty', 'not at all'. In the last case, the month would officially begin the next evening. (Egyptian National Library, Cairo, MS DŞ 155)

The times of prayer

In Islam the times of prayer are astronomically determined. The standard definitions of the five prayer-times, still in use today, are briefly as follows: The Muslim day begins at sunset, and the interval during which the first prayer (*maghrib*) is to be performed lasts from sunset to nightfall. The interval for the second prayer (*'ishā'*) begins at nightfall and lasts until daybreak. The third prayer (*fajr*) is performed in the interval between daybreak and sunrise. The permitted time for the fourth prayer (*zuhr*) begins when the sun has crossed the meridian and ends when the interval for the fifth prayer (*'asr*) begins, namely, when the shadow of an object equals its midday shadow increased by the length of the object. The interval for the fifth prayer may last until the shadow increases again by the length of the object or until sunset. In medieval Andalusian practice the fourth prayer begins when the shadow has increased beyond its midday minimum by one quarter of the length of the object. Isolated instances of the performance of another prayer (*duhā*) in the morning, at a time after sunrise equal to the time remaining till sunset after the *'asr*, are attested in medieval sources. The definitions of the daytime prayers in terms of shadow increases relate the times to the seasonal hours (one-twelfth divisions of the length of daylight), the connection being provided by an approximate Indian formula for time-keeping known to the Muslims in the eighth century. This formula associates the end of the ninth seasonal hour (mid-afternoon) with a shadow increase of 1 gnomon-length, and the end of the tenth hour with a shadow increase of 2 gnomon-lengths.

Clearly the times of the prayers as defined above can be regulated without difficulty by observation, assuming a clear sky. A genre of literature dealing with time-keeping from a non-mathematical point of view was compiled by specialists in folk astronomy, some of whom were also noted legal scholars. These works consist mainly of discussions of Quranic verses and Prophetic statements on these two subjects, embellished with descriptions of simple procedures for time-keeping by shadows and by the lunar mansions. Most surviving examples are of thirteenth-century or later Yemeni and Hejazi provenance, but the trend was set already in the ninth century. Since they involved the religious traditions, they formed part of the corpus of Islamic legal literature, and they were subject to different interpretations by different legal schools. In the first few centuries of Islam, as far as we can tell from the scant evidence available, the times of prayer were regulated by the muezzins themselves, and these were selected as much because of their fine voices as for their ability in folk astronomy.

This notwithstanding, it was found convenient to have tables at hand displaying the length of the prayer times for each day of the year or each degree of solar longitude. The earliest such tables we have are a set for Baghdad displaying the shadow at the *zuhr* and the *'asr* for each 6° of solar longitude, and another, found only in a thirteenth-century Iraqi *zīj*, but probably also dating from the ninth century, displaying the time from sunset to daybreak (based on an approximate Indian formula for time-keeping) and the midday shadow, and the solar altitudes at midday and the beginning and end of the ῾aṣr for each day of the Syrian calendar. About the middle of the tenth century ῾Alī ibn Amājūr compiled two tables displaying the time of day as a function of solar meridian altitude and instantaneous altitude. The first, based on an accurate formula, was computed specifically for Baghdad, and the second, based on an approximate Indian formula, is universal and works quite well for all latitudes. In both cases values are given in seasonal day-hours and minutes for each degree of both arguments.

These early Islamic tables for time-keeping, of which so few examples survive, began a tradition that reached its zenith in thirteenth-century Cairo and fourteenth-century Damascus (Fig. 52). Most of the corpora of tables for time-keeping compiled for these two centres and others such as Jerusalem (Fig. 53), Alexandria, Maragha, Tunis and Taiz, belong to the later period of Islamic astronomy. Some remarkable tables were produced: one, compiled in Egypt in the late thirteenth century, has three arguments (solar or stellar altitude and meridian altitude and half arc of visibility) and displays the time of day or night for any terrestrial latitude; it has over 400,000 entries. It has only recently become apparent that European astronomers in later centuries compiled tables similar in their conception; few of these have been studied yet. Indeed the history of such tables extends from the ninth to the nineteenth century in the Islamic world and from the fourteenth century onwards in Europe and eventually also North America.

It was apparently in Egypt in the thirteenth century that the office of the *muwaqqit* or mosque astronomer responsible mainly for the times of prayer was developed. Most

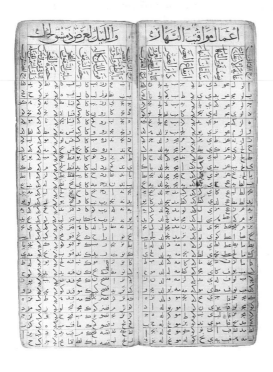

52 An extract from the prayer-tables for the latitude of Damascus, computed in the mid fourteenth century by the *muwaqqit* al-Khalīlī. Twelve functions relating to time-keeping are tabulated across the page for each degree of solar longitude (corresponding roughly to each day of the year); this double page serves the sign of Aquarius. Similar tables have been found for localities between Morocco and Central Asia, Crete and the Yemen. Such tables enabled the *muwaqqit*s to inform the muezzins when the time for each of the five prayers had arrived so that they could summon the faithful to prayer. (Bibliothèque Nationale, Paris, MS ar. 2558)

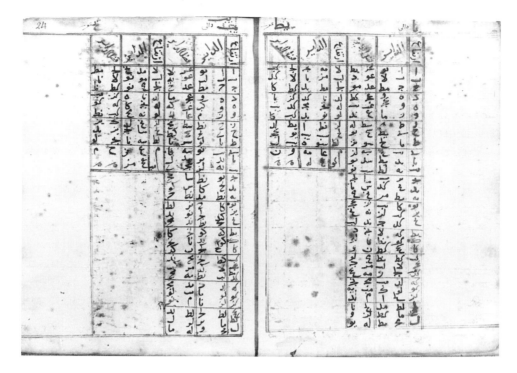

53 An extract from the tables for time-keeping for the sun, compiled for the latitude of Jerusalem by the fourteenth-century *muwaqqit* al-Karakī. For each degree of solar longitude (here Aquarius 11° and 12°, also serving Scorpio 19° and 18°) and for each degree of solar altitude up to the maximum, the time since sunrise and hour-angle are tabulated in equatorial degrees and minutes. The first tables of this kind were prepared in Baghdad in the ninth century and numerous examples for other locations were compiled during the next millennium. (Leipzig University Library, MS 808)

of the Egyptian and Syrian astronomers of consequence from the fourteenth and fifteenth centuries were *muwaqqit*s. The office was particularly important in the Ottoman Empire and in many mosques from Anatolia to the Balkans one can still see the buildings attached to mosques where the *muwaqqit*s kept their books and instruments. There does not appear to have been a similar office in the Muslim East, but we do have evidence that there was a team of *muwaqqit*s in Granada around 1300.

The tables that modern Muslims use to regulate their prayers, published in newspapers, pocket diaries and wall calendars, have a history of over a millennium, only recently documented. The muezzin's call to prayer at five specific times is one of the most distinctive features of modern Islamic life.

The sacred direction
The Quranic injunction that prayer and other ritual acts should be performed facing the Kaaba in Mecca caused no problems for the first generations of Muslims who found

54 An illustration of the Kaaba, with each of the corners and four walls associated with regions of the Islamic commonwealth, and the *qibla* in each region defined in terms of an astronomical horizon phenomenon. From a copy of the cosmography of the fifteenth-century Syrian Ibn-al-Wardī. (Bibliothèque Nationale, Paris, MS ar. 2186)

themselves in localities as far apart as Andalusia and Sind. They had not the means to determine the *qibla*, or direction of Mecca, in the localities they settled, so they simply used the direction of the pilgrim road to Mecca or the cardinal directions; as a result, the earliest mosques and the *miḥrābs* in their *qibla* walls face Mecca only roughly. But the Muslims knew that the Kaaba was aligned in certain astronomical directions (major axis towards the rising of Canopus, minor axis solstitially aligned), and for them the Kaaba was at the centre of their world (Fig. 54). They devised a simple expedient to face the segment of the perimeter corresponding to their location: one should stand in the same direction as one would be standing in if one were actually in front of that segment of the perimeter of the Kaaba. Different schemes of sacred geography were developed by the specialists in folk astronomy and the legal scholars, in which the world was divided into sectors about the Kaaba and the *qibla* in each sector was defined in terms of the rising or setting of the sun or a certain fixed star. This kind of information about the *qibla* was proposed by the legal scholars, and is reflected in the orientations of medieval mosques, only a minority of which are aligned in directions that could have been proposed by the astronomers.

By the early ninth century Ptolemy's list of geographical co-ordinates was available to Muslim scholars and the co-ordinates of Mecca and Baghdad had been investigated by a team commissioned by the Caliph al-Ma'mūn. Further, exact geometrical and trigonometric procedures had been devised for determining the *qibla* from geographical co-ordinates. (The problem of the determination of the *qibla* is easily transformed into a problem of spherical astronomy by considering the zeniths of the localities involved.) al-Bīrūnī's treatise on mathematical geography (*Taḥdīd*), the ultimate goal of which was the determination of the *qibla* at Ghazna, is the most important work of its kind from the medieval period. Since these exact procedures for determining the *qibla* were rather complicated, approximate methods were also derived. Already in the ninth century a table was compiled based on one such method and it displays the direction of Mecca as a function of terrestrial latitude and longitude. Over the centuries several such tables were compiled. We return later to some of the various instruments used for finding the *qibla*.

Only in the eighteenth century did it become possible for the first time to measure longitude differences correctly, and only then could it become obvious that most medieval longitude co-ordinates were incorrect and that even the *qibla*s derived by a correct mathematical procedure but based on these coordinates were also off by a few degrees. Some medieval mosques that are still in use have two *miḥrāb*s indicating the *qibla*, the second being based on modern calculations. Nowadays some Muslims use pocket-compasses provided with lists of *qibla*s for the major cities of the world.

Observation programmes and regional schools of astronomy
Al-Ma'mūn's circle
In the early ninth century the Abbasid Caliph al-Ma'mūn patronised observations first in Baghdad and then in Damascus, gathering the best available astronomers to conduct observations of the sun and moon. Some of the results were incorporated into a *zīj* called *al-Mumtaḥan*, 'tested', although the details of the activities at the two observation posts are somewhat confusing. The *Mumtaḥan Zīj* was apparently compiled in Baghdad by Yaḥyā ibn Abī Manṣūr, but upon his death, according to Ḥabash, the Caliph ordered his colleague Khālid al-Marwarrūdhī to prepare some new instruments and conduct a one-year programme of solar and lunar observations in Damascus in order to compile a new *zīj*. According to Ḥabash this was done, but no such *zīj* is otherwise known to have been prepared before Ḥabash's own *Damascus Zīj*.

These observations, like later ones, were mainly directed towards determining the local latitude and current value of the obliquity, and towards deriving improved parameters for the Ptolemaic planetary models and more accurate star positions. The armillary sphere, the meridian quadrant and the parallactic ruler were known to the Muslims from the *Almagest*, and they added new scales and other modifications, often building larger instruments even when smaller ones would have sufficed. Our knowledge of the instruments used by al-Ma'mūn's astronomers is meagre. An armillary sphere used by Yaḥyā in Baghdad was said to display markings for each 10' of arc, but even contemporary astronomers were not impressed by the precision of the results obtained using it. A mural quadrant made of marble with a radius of about 5 m was used in Damascus, as well as a vertical gnomon made of iron standing about 5 m high. Al-Ma'mūn also patronised measurements of the longitude difference between Baghdad and Mecca (by simultaneous observations of a lunar eclipse) in order to establish the *qibla* at Baghdad properly, as well as measurements of the length of one degree of terrestrial latitude. Most of what is known about the Baghdad and Damascus observatories is provided by Ibn Yūnus and al-Bīrūnī (see p. 163), and the available manuscripts of the *Zījes* of Yaḥyā and Ḥabash await detailed study.

Other observational programmes

Besides the officially-sponsored observations conducted in Baghdad and Damascus in the early ninth century, there are numerous instances of other series of observations conducted in different parts of the Muslim world.

The two brothers called Banū Mūsā made observations in their own house in Baghdad and also in nearby Samarra about 30 years after the *Mumtaḥan* observations. They also arranged for simultaneous observations of a lunar eclipse in Samarra and Nishapur in order to determine the difference in longitude between the two cities. In view of their proficiency in mathematics, it is most unfortunate that neither of the two *zījes* compiled by them has survived.

Al-Battānī carried out observations during the period 887 to 918 in Raqqa in North Syria. He appears to have financed his observational activity himself, and although we have no description of the site where he made his observations, the instruments mentioned in the *zīj* based on his observations include an armillary sphere and mural quadrant, as well as a parallactic ruler, an astrolabe, a gnomon and a horizontal sundial.

The observational activities of the Baghdad family known as the Banū Amājūr were almost contemporary with those of al-Battānī in Raqqa. Father and two sons, and also a freed family slave, all made observations and each compiled a *zīj*, none of which survives. In the accounts of their eclipse observations recorded by Ibn Yūnus it appears that the place where they conducted their observations had some kind of a balcony fitted with slits for observation, but the details are obscure. A particularly interesting account of a solar eclipse in the year 928 that they observed by reflection in water includes a remark that the altitude of the sun was measured on an instrument marked for each third of a degree.

A large mural quadrant was erected at Rayy (near modern Tehran) about the year 950 but we have information only on its use to establish the local latitude and obliquity of the ecliptic. In Shiraz not long thereafter al-Ṣūfī used an armillary sphere with diameter about 5 m to derive the same parameters and to 'observe' equinoxes and solstices. Al-Ṣūfī is best known for his work on the fixed stars, but it seems that this was based more on 'observation' with the naked eye than on 'measurement', looking at the heavens with precision instruments and making estimates of positions. Another contemporary astronomer who conducted observations on which we have no information other than the main parameters of his *zīj* was Ibn al-A'lam. The observations of both al-Ṣūfī and Ibn al'Alam were patronised by the Buwayhid ruler 'Aḍud al-Dawla.

In the late tenth century the distinguished mathematician and astronomer Abu l-Wafā' al-Būzajānī made observations in Baghdad. Most of these appear to have been directed towards the determination of the solar parameters, and the obliquity of the ecliptic and the latitude of Baghdad, although Abu l-Wafā' also collaborated with al-Bīrūnī in Khwarizm (modern Khiva in Uzbekistan) on the simultaneous observation of a lunar eclipse in the year 997. We have no information on the nature of the site where Abu l-Wafā' made his observations, other than its location in a specific quarter of Baghdad.

Contemporaneous with the activity of Abu l-Wafā' was the establishment in 998 of an observatory in the garden of the Baghdad residence of the Buwayhid ruler Sharaf al-Dawla. The organisation of a building and programme of observations was entrusted to Abū Sahl al-Qūhī, a mathematician of considerable standing. We know from contemporary historical records that a special building was erected for the observations, which in turn were witnessed by 'judges, scientists and scholars of note, astronomers, and engineers'. In view of the favourable conditions under which this observatory was established, and the competence of its director, it is somewhat surprising that the two recorded 'observations' that were 'witnessed' by so many dignitaries were the entry of the sun into Cancer and Libra in the year 988. Al-Bīrūnī describes the main instrument which was constructed as a hemisphere of radius 12.5 m on which the solar image was projected through an aperture at the centre of the hemisphere. Activity at the observatory stopped in 989 with the death of Sharaf al-Dawla, so that the institution lasted not much more than a year.

In 994 Abū Maḥmūd al-Khujandī made a measurement of the obliquity using a meridian sextant of about 20 m radius. This instrument was erected in Rayy but al-Khujandī confessed to al-Bīrūnī that it was so large that the centre of the sextant had become displaced from its intended position.

The Egyptian astronomer Ibn Yūnus made a series of observations of eclipses, conjunctions and occultations, as well as equinoctial and solstitial observations. We are extremely fortunate to have not only his reports of these observations but also his citations of earlier observations of the same kind made by individuals such as Ḥabash and the Banū Amājūr. Ibn Yūnus's purpose in making these observations and recording them in the introduction to his *Zīj* is somewhat obscured by the fact that he does not list those observations or present those calculations with which he derived his new solar, lunar and planetary parameters. Neither does he mention any locations for his observations other than his grandfather's house in Fustat and a nearby mosque in al-

Qarāfa. The popular association of Ibn Yūnus with an observatory on the Muqaṭṭam Hills outside Cairo is, as A. Sayılı has shown, a myth. Nevertheless, Ibn Yūnus mentions at least one instrument, probably a meridian ring, that was provided by the Fatimid Caliphs al-ʿAzīz and al-Ḥākim. In a later medieval Egyptian source Ibn Yūnus is reported to have received 100 dinars a day from al-Ḥākim, and it may be that such extremely high payments were made to Ibn Yūnus when he was making satisfactory astrological predictions for the Caliph. Al-Ḥākim made an abortive attempt to found an observatory in Cairo, but this was after the death of Ibn Yūnus in 1009. At some time during his reign there was an armillary sphere in Cairo with nine rings, each large enough that a man could ride through them on horseback.

The observations of al-Bīrūnī were conducted between 990 and c. 1025 in several localities between Khwarizm (modern Khiva) and Kabul. His recorded observations include determinations of equinoxes and solstices, eclipses, and determinations of the obliquity and local latitude.

The corpus of tables known as the *Toledan Tables* was compiled in the eleventh century, based on observations directed by Ṣāʿid al-Andalusī and continued by Ibn al-Zarqāllu. Only the mean motion tables in this corpus of tables are original; most of the remainder were lifted from the *Zījes* of al-Khwārizmī and al-Battānī.

In the thirteenth century there was a substantial observational programme at Maragha. The results are impressive only in so far as theoretical astronomy is concerned (see p. 150). Otherwise the trigonometric and planetary tables in the major production of the Maragha astronomers were modified or lifted *in toto* from earlier sources. This is not a happy outcome for a generously endowed observatory fitted with the latest observational instruments, known to us only from texts. In the early fifteenth century the scene had moved to Samarqand in Central Asia: there a group of astronomers directed by the astronomer-prince Ulugh Beg did impressive work. Only the 40-m meridian sextant survives from the observatory. These men produced a set of tables which, however, it would be foolish to judge before they have been properly studied. The same is true for the short-lived observatory in Istanbul under the direction of Taqi l-Dīn (1577; see frontispiece).

Regional schools of astronomy

After the tenth century there developed regional schools of astronomy in the Islamic world, with different interests and concentrations. They also had different authorities (for example, in the furthest East al-Bīrūnī and al-Ṭūsī, and in Egypt Ibn Yūnus). The main regions were Iraq; Iran and Central Asia; Muslim Spain; Egypt and Syria; the Yemen; the Maghrib; and later also the Ottoman lands. Only recently have the complex tradition of Muslim Spain (tenth–fourteenth centuries), the colourful tradition of Mamluk Egypt and Syria (thirteenth–early sixteenth centuries), the distinctive tradition of Rasulid Yemen (thirteenth–sixteenth centuries), and the staid tradition of the Maghrib (twelfth–nineteenth centuries) been studied. The traditions of Ottoman Turkey and Mogul India are currently being researched.

Astronomical instruments

As noted in the previous section most Islamic observational instruments are lost and known to us only through texts. The state of documentation of the other, smaller Islamic astronomical instruments that do survive leaves much to be desired. Many of the most important instruments are still unpublished, and much that has been written on instruments is on a very amateur level. For these reasons a project is currently underway in Frankfurt to catalogue all Islamic instruments (and European ones) to *c.* 1550 as well as various historically significant later Islamic pieces.

Also the most important writings on instruments have not yet received the attention they deserve. For example, a hemispherical observational instrument for a fixed latitude was devised by the tenth-century astronomer al-Khujandī, the leading instrument-maker of the early period, and this was modified in the twelfth century to serve all latitudes. There are no surviving examples and the available manuscripts have yet to be studied. An important work on instruments was compiled in Cairo *c.* 1280 by Abū 'Alī al-Marrākushī; this has yet to be subjected to a detailed analysis. The author simply collected all of the treatises on instruments known to him and incorporated them into his book. An exciting find of the 1980s was a treatise by the fourteenth-century Aleppo astronomer Ibn al-Sarrāj, the leading instrument-maker of the later Islamic period. In this the author described every kind of instrument known to him as well as those invented by himself. This treatise is currently being studied.

Armillary spheres and globes

In the eighth century al-Fazārī wrote a treatise on the armillary sphere, called in Arabic *dhāt al-ḥalaq*, which means 'the instrument with the rings'. No early Islamic armillary spheres survive, but several other treatises on it were compiled over the centuries. The earliest treatise in Arabic dealing with the celestial globe, called *dhāt al-kursī*, 'the instrument with the stand [literally, the 'throne' on which it sits]' or simply *al-kura*, 'sphere', was written by Qusṭā ibn Lūqā in the ninth century. This treatise by Qusṭā, who was one of the most important translators of Greek works into Arabic, remained popular for a millennium. Of the various surviving celestial globes, which number over 100, none predates the eleventh century. A globe made in Mosul in 1274–5 is preserved in the British Museum (Col. Pl. X).

The spherical astrolabe, unlike the armillary sphere and the celestial globe, appears to be an Islamic development. Various treatises on it were written from the tenth to the sixteenth century, and only one complete instrument, from the fourteenth century, survives. Ḥabash wrote on the spherical astrolabe, the armillary sphere, and the celestial globe, as well as on various kinds of planispheric astrolabes.

Astrolabes

Al-Fazārī also wrote on the use of the astrolabe. The tenth-century bibliographer Ibn al-Nadīm states that al-Fazārī was the first Muslim to make such an instrument; he also informs us that at that time the construction of astrolabes was centred on Harran and spread from there. Several early astronomers, including Ḥabash, al-Khwārizmī and al-Farghānī, wrote on the astrolabe, and introduced features not found on earlier Greek instruments, such as the shadow squares and trigonometric grids on the backs and the azimuth curves on the plates for different latitudes, as well as the universal plate of horizons. Also extensive tables were compiled in the ninth century to facilitate the construction of astrolabes.

Another important development to the astrolabe occurred in Andalusia in the eleventh century, when Ibn al-Zarqāllu devised the single universal plate (ṣafīḥa) called shakkāziyya and the related plate called zarqālliyya with two sets of shakkāziyya markings for both equatorial and ecliptic co-ordinate systems. The latter was fitted with an alidade bearing a movable perpendicular straight-edge (transversal). Several treatises on these two instruments exist in both Western and Eastern traditions of later Islamic astronomy; the Europeans knew of them as the saphea. Ibn al-Zarqāllu's contemporary, ʿAlī ibn Khalaf, wrote a treatise on a universal astrolabe that did not need plates for different latitudes. This treatise exists only in Old Spanish in the Libros del Saber, and was apparently

55 The front of the universal astrolabe of Ibn al-Sarrāj, dated AD 1329. This remarkable instrument not only represents the culmination of Islamic astrolabe-making, but has no equal in sophistication amongst instruments from the European Renaissance. Whereas the standard astrolabe requires a different plate for each latitude, that of Ibn al-Sarrāj has plates that serve all latitudes; indeed, the various components can be used in five different ways to solve all the problems of spherical astronomy for any latitude. IC no. 140 (Benaki Museum, Athens)

not known in the Islamic world outside Andalusia. The instrument was further developed in Syria in the early fourteenth century: Ibn al-Sarrāj devised in Aleppo a remarkable astrolabe that can be used universally in five different ways (Fig. 55).

The astrolabes made by Muslim craftsmen show a remarkable variety within each of several clearly defined regional schools. We may mention the simple, functional astrolabes of the early Baghdad school; the splendid astrolabe of al-Khujandī of the late tenth century, which started a tradition of zoomorphic ornamentation that continued in the Islamic East and in Europe for several centuries; the very different astrolabes of the Andalusian school in the eleventh century and the progressive schools of Iran in the thirteenth and fourteenth centuries; and the remarkable instruments from Mamluk (thirteenth and fourteenth-century) Egypt and Syria. The British Museum possesses a spectacular Mamluk astrolabe by 'Abd al-Karīm al-Miṣrī. After about 1500 the construction of astrolabes continued in the Maghrib, in Iran and in India until the end of the nineteenth century. Many of these, especially those from the Islamic East, were objects of the finest workmanship. Such is the splendid astrolabe made in Isfahan in 1712 for Shāh Ḥusayn, also now in the British Museum (Col. Pl. XI).

Quadrants

Another category of observational and computational devices to which Muslim astronomers made notable contributions was the quadrant, of which we can distinguish three main varieties. First, the sine quadrant with an orthogonal grid. This instrument, in a simpler form, was described already by al-Khwārizmī and was widely used throughout the Islamic period. Some Islamic astrolabes display such a trigonometric grid on the back. The grid can be used together with a thread and movable marker (or the alidade of an astrolabe) to solve all of the standard problems of spherical astronomy for any latitude. Second, the horary quadrant with fixed or movable cursor. This instrument is described already in an anonymous ninth-century Iraqi source and was likewise commonly used for centuries (albeit usually without the cursor, which is not essential to the function of the device). A set of arcs of circles inscribed on the quadrant display graphically the solar altitude at the seasonal hours (approximately, according to an Indian formula). Other Islamic quadrants from the ninth century onwards had markings for the equinoctial hours. The instrument can be aligned towards the sun so that the time can be determined from the observed altitude using the grid. Again, this kind of markings was often marked on the back of astrolabes. Third, the astrolabic quadrant displaying one half of the altitude and azimuth circles on an astrolabe plate for a fixed latitude, and a fixed ecliptic. The effect of the daily rotation is achieved by a thread and bead attached at the centre of the instrument rather than by the movable astrolabe rete. The quadrant with astrolabic markings on one side and a trigonometric grid (Fig. 56) on the other generally replaced the astrolabe all over the Islamic world (with the notable exceptions of Iran, India and the Yemen) in the later period of Islamic astronomy.

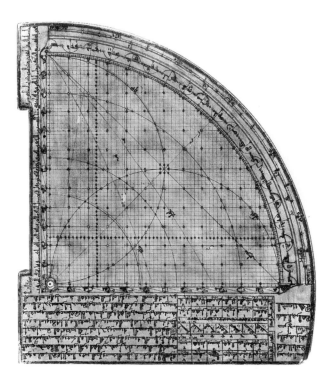

56 A trigonometric grid for solving all the various problems of spherical astronomy without calculation. The grid, a quarter-circle of sexagesimal orthogonal markings like modern graph-paper and with radius 60 units, was developed from simpler varieties first used in Baghdad in the early ninth century. One medieval European quadrant with similar markings has been preserved. This grid, found on an astrolabic quadrant made in Damascus *c.* AD 1800, is remarkable for all the additional markings, lines, arcs of circles and curves, which facilitate the solution of specific problems relating to the astronomically defined times of prayer. (Private collection)

Sundials

We learn from Islamic tradition that the pious Umayyad Caliph ʿUmar ibn ʿAbd al-ʿAzīz (Damascus, *c.* 718) used a sundial, probably a Graeco-Roman one, to regulate the times of the daytime prayers in terms of the seasonal hours. The earliest sundials described in the Arabic astronomical sources are planar, usually horizontal, but also vertical and polar. The mathematical theory for computing the shadow for the seasonal hours at different times of the year and the corresponding azimuths was available from Indian sources, which seem to have inspired the Islamic tradition more than any of the available Greek works. Already the treatise on sundial construction by al-Khwārizmī contains extensive tables displaying the polar co-ordinates of the intersections of the hour lines with the solstitial shadow traces on horizontal sundials for twelve different latitudes. The treatise on sundial theory by Thābit ibn Qurra contains all the necessary mathematical theory for constructing sundials in any plane; likewise impressive from a theoretical point of view is the treatise on gnomonics by his grandson Ibrāhīm.

The earliest surviving Islamic sundial, apparently made in Córdoba about the year 1000 by the Andalusian astronomer Ibn al-Ṣaffār, displays the shadow traces of the equinoxes and solstices, and the lines for the seasonal hours as well as for the times of the two daytime prayers. There is a world of difference between this simple, carelessly constructed piece and the magnificent sundial made in the late fourteenth century by Ibn al-Shāṭir, so devised that it can be used to measure time with respect to any of the

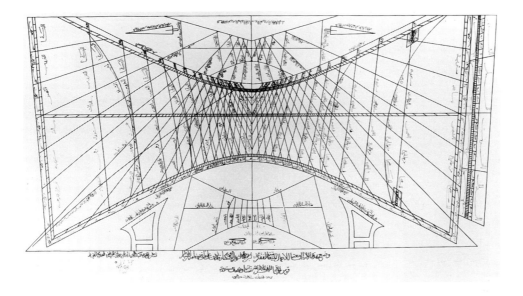

57 The sundial made by Ibn al-Shāṭir for the main minaret of the Umayyad Mosque in Damascus in the year AD 1371–2 was broken by accident in the nineteenth century. The *muwaqqit* who broke it whilst trying to realign it was sufficiently well versed in gnomonics to make this copy, which is still *in situ* on the minaret. Several pieces of the original were discovered in excavations of the drainage system of the mosque in 1958 and are now displayed in the garden of the Archaeological Museum, Damascus. (Alain Brieux, Paris)

five daily prayers (Fig. 57). In the late period of Islamic astronomy a sundial was to be found in most of the major mosques.

Miscellaneous

Several multipurpose instruments were devised by Muslim astronomers. Notable examples are the rule (*mīzān*) of al-Fazārī, fitted with a variety of non-uniform scales for various astronomical functions, and the compendium of Ibn al-Shāṭir, comprising a magnetic compass and *qibla*-indicator, a universal polar sundial, and an equatorial sundial. Of particular interest is a circular brass instrument made in Isfahan *c.* 1700 which is engraved with a world map with Mecca at the centre, and a cartographical grid so devised that the *qibla* for some 150 cities between Spain and China can be read off the outer scale and the distance from Mecca can be read off the non-uniform scale on the diametrical rule (Fig. 58). Muslim interest in projections preserving direction and distance can be traced back several centuries to al-Bīrūnī and Habash.

There are several Islamic treatises on eclipse computers and planetary equatoria (see p. 184) for determining the positions of the planets for a given date. With these the standard problems of planetary astronomy dealt with in *zījes* are resolved mechanically, without calculation. Treatises on eclipse computers are known from the early tenth

58 A cartographical grid with Mecca at the centre, so devised that one can read the direction and distance to Mecca for any locality in the Islamic commonwealth. The instrument, made in Isfahan *c.* AD 1700 but inspired by an earlier tradition, was originally fitted with a magnetic compass and an hour dial adjustable for any latitude for finding the time of day, all now missing. (Private collection)

century, and al-Bīrūnī in the early eleventh describes such an instrument in detail. A newly discovered manuscript (not yet available for research) contains a treatise by the tenth-century Iranian astronomer Abū Jaʿfar al-Khāzin called *Zīj al-Ṣafāʾiḥ*, the *Zīj of Plates*, describing an equatorium. The sole known example of this instrument, made in the twelfth century, is, alas, incomplete: it is in the form of an astrolabe with tables engraved on the mater and additional markings for the foundation of an equatorium. Otherwise the only known early Islamic treatises on planetary equatoria are from eleventh-century Andalusia. The most interesting aspect of the equatorium described by Ibn al-Zarqāllu is the ellipse drawn on the plate for the centre of the deferent of Mercury; it seems that he was the first to notice this characteristic of Mercury's deferent. Al-Kāshī, the leading astronomer of early fifteenth-century Samarqand, has left us a description of a planetary equatorium with which not only ecliptic longitudes but also latitudes could be determined and eclipses calculated.

Transmission to Europe

The Europeans learned of Islamic astronomy through Spain, a region where, because of political problems and the difficulty of communications, the most up-to-date writings were not always available. This explains, for example, how it came to pass that the Europeans came across two major works of Muslim astronomers from the East, al-Khwārizmī and al-Battānī, at a time when these works were no longer widely used in the Islamic East. It also explains why so few Eastern Islamic works became known in Europe. None of the Eastern Islamic developments to Ptolemy's planetary theory were known in Andalusia or in medieval Europe. Al-Biṭrūjī's unhappy attempt to develop planetary models confused Europeans for several centuries; he must be worth reading, they naïvely thought, because he was trying to reconcile Ptolemy with Aristotle. As far as astronomical time-keeping was concerned, this does not seem to have been of much concern to the Muslims in Spain; hence nothing of consequence was transmitted.

On the other hand, some early Eastern Islamic contributions, later forgotten in the Islamic East, were transmitted to Spain and thence to Europe; they have been considered European developments because evidence to the contrary has seemed to be lacking. A good example is the horary quadrant with movable cursor (the so-called *quadrans vetus*), which was invented in Baghdad in the ninth century and (at least in the version with the cursor) virtually forgotten in the Islamic East thereafter; it came to be the favourite quadrant in medieval Europe. What, if any, astronomical knowledge was transmitted through Islamic Sicily remains a mystery, and nothing of consequence is known to have been learned about the subject by the Crusaders.

In the European Renaissance there was no access to the latest Islamic works. So the Europeans contented themselves with new editions of the ancient Greek works, with occasional, almost nostalgic, references to Albategnius (al-Battānī), Azarquiel (Ibn al-Zarqāllu), Alpetragius (al-Biṭrūjī) and the like. A few technical terms derived from the Arabic, such as alidade, azimuth, almucantar, nadir, saphea, and zenith, and a few star-names such as Aldebaran, Algol, Altair and Vega, survived. When the Europeans did come to learn of some of the major Islamic works and to try to come to terms with them it was as orientalists and historians of astronomy, for by this time the Islamic materials other than observation accounts were of historical rather than scientific interest. Thanks to orientalists such as the Sédillots in Paris, works that had been completely unknown to Europeans and mainly forgotten by Muslims were published, translated and analysed. Islamic astronomy was highly respected by such scholars and others, like the historian of astronomy J.-B. Delambre, who, innocent of Arabic, took the trouble to read what his colleagues had written about the subject. But Islamic astronomy, indeed Islamic science in general, received a blow beneath the belt from P. Duhem, a physicist and philosopher ignorant of Arabic, who simply ignored what scholars such as the Sédillots had written. His thesis, that the Arabs were incapable of scientific thought and that whatever merits their science may have had were due to the intellectually superior Greeks, still has many followers, but only amongst those totally ignorant of the research of the past 150 years.

In the period after *c.* 1500 Islamic astronomy declined. All of the problems had been solved, some many times over. Much of the innovative activity had led into a cul-de-sac, from which it would not emerge until modern times, thanks to investigations of manuscripts and instruments. Not that interest in astronomy died down. From Morocco to India the same old texts were copied and studied, recopied and restudied, usually different texts in each of the main regions. But there was no new input of any consequence. Astronomy continued to be used as the handmaiden of astrology, and for the regulation of the calendar and the prayer-times. Where there appears to have been some innovation – such as, for example, in the remarkable device made in Isfahan *c.* 1700 that correctly displays the direction and distance of Mecca for any locality – one can be confident of the existence of an earlier tradition. True some European 'zījes', notably those of Cassini and Lalande, were translated into Turkish and their tables adapted for the longitudes of Istanbul and later Damascus. But the old traditions died hard, and Muslim astronomers for several centuries spent more time copying old treatises and tables than compiling new ones.

Conclusion

The reader will have seen that during the millennium beginning *c.* 750 and especially in the period up to *c.* 1050, although also in the period up to *c.* 1500, Muslim astronomers did first-rate work, most of which was not known in medieval Europe at all. Those few Islamic works from the early period that were transmitted, notably the *Zījes* of al-Khwārizmī and al-Battānī (especially through the *Toledan Tables*) and the banal summary of the *Almagest* by al-Farghānī, convey only an impression of classical astronomy in Arabic garb. But they were in no way representative of contemporary Islamic astronomy in the East, and whilst the Europeans laboured for centuries to come to terms with them, Muslim astronomers were making substantial contributions to their subject that have only been revealed by modern scholarship.

Anyone who leafs through the pages of the two volumes of Fuat Sezgin's *Geschichte des arabischen Schrifttums* listing the manuscript sources for Islamic astronomy and astrology up to *c.* AD 1050, or my survey of the 2,500 scientific manuscripts in the Egyptian National Library, which serves mainly the period after that, will observe the wealth of material relating to this subject that remains untouched by modern scholarship. Very few Islamic astronomical works have been published or have received the attention that they merit. Three out of close to 200 Islamic *zījes* have been published in the optimum way (text, translation and commentary). We have no published edition of the Arabic versions of the *Almagest* (except for the star catalogue), or of any Arabic recensions or commentaries. Many of the published Arabic scientific texts were printed in Hyderabad, most with no critical apparatus. In view of the lack of a published corpus of texts and the improbability that such a corpus will materialise, there is an obvious need for reproduction of manuscripts of particular importance. Otherwise the historian of Islamic astronomy will be forced to continue to rely mainly on microfilms of manuscripts, which some libraries are unable or unwilling to supply. Likewise most of the historically important Islamic astronomical instruments are still unpublished, although the catalogue currently in preparation in Frankfurt promises to make them better known.

In 1845 L. A. Sédillot, whose privilege it was to have access to the rich collection of Arabic and Persian scientific manuscripts in the Bibliothèque Nationale in Paris, wrote: 'Each day brings some new discovery and illustrates the extreme importance of a thorough study of the manuscripts of the East.' Sédillot, like his father before him but like few historians of Islamic science since, also realised the importance of Islamic astronomical instruments. Given the vast amount of manuscripts and instruments now available in libraries and museums elsewhere in Europe, the United States and the Near East, and the rather small number of people currently working in this field, Sédillot's statement is no less true now than it was a century and a half ago.

Bibliography

Dictionary of Scientific Biography, 16 vols. New York: Charles Scribner's Sons, 1970–80. (See especially the articles (in chronological order) al-**F**azārī, **Y**a'qūb ibn **T**āriq, **M**āshā'allāh, **Y**aḥyā ibn Abī Manṣūr, **A**bū Ma'shar, al-**K**hwārizmī, **Ḥ**abash, Ibn **Q**utayba, Banū **M**ūsā, **Th**ābit ibn Qurra, al-**B**attānī, **A**bu l-Wafā' al-Būzajānī, Ibn **Y**ūnus, **I**bn al-Haytham, al-**B**īrūnī, al-**Z**arqālī, al-**B**iṭrūjī (supplement), al-**T**ūsī, **I**bn al-Shāṭir, al-**K**halīlī (supplement), al-**K**āshī, and **U**lugh Beg.)

Encyclopaedia of Islam, new edn, 8 vols to date, Leiden: E. J. Brill, 1960 to present. (See especially the articles on individual astronomers: al-**F**azārī, **Ḥ**abash, al-**B**attānī, **I**bn Yūnus, **I**bn al-Haytham, al-**B**īrūnī, al-**B**iṭrūdjī, al-**M**arrākushī, and on specific topics: '**I**lm al-hay'a [=astronomy]', '**N**udjūm, aḥkām al- [=astrology]', '**M**ınṭaḳa [=ecliptic]' and '**M**ayl [=declination]', '**A**nwā' [=calendrical divisions]', '**K**uṭb [=celestial pole]', '**L**ayl wa-nahār [=night and day]', '**M**adjarra [=Milky Way]', '**M**anāzil [=lunar mansions]', '**N**udjūm [=stars]', '**H**ilāl [=lunar crescent]' and '**R**u'yat al-hilāl [=lunar crescent visibility]', '**M**īḳāt [=astronomical time-keeping]', '**Ḳ**ibla [=sacred direction]', '**M**aṭāli' [= astronomical risings]', and '**M**akka IV: As centre of the world [=sacred geography]'; '**A**ṣṭurlāb [–astrolabe]', '**R**ub' [=quadrant]', '**S**amt [=direction and Mecca-centred world-maps]', '**S**hakkāziyya [= universal projection]' and '**M**izwala [=sundial]'.)

Berggren, J.L. 1986. *Episodes in the Mathematics of Medieval Islam*. Berlin, Heidelberg and New York: Springer-Verlag.

Dalen, B. van 1993. *Ancient and Mediaeval Astronomical Tables – Mathematical Structure and Parameter Values*. Utrecht University.

Goldstein, B.R. 1985. *Theory and Observation in Ancient and Medieval Astronomy*. London: Variorum.

Gunther, R.T. 1932. *The Astrolabes of the World*, 2 vols. Oxford University Press. Reprinted 1976 in one vol. London: The Holland Press (long outdated and full of errors).

Heinen, A. 1982. *Islamic Cosmology* . . . Beirut and Wiesbaden: Franz Steiner.

Kennedy, E.S. 1956. A survey of Islamic astronomical tables. *Transactions of the American Philosophical Society*, new ser., 46. Reprinted, n.d. (c. 1990).

Kennedy, E.S. *et al*, 1983. *Studies in the Islamic Exact Sciences*. Beirut: American University of Beirut Press.

Kennedy, E.S. and Ghanem, I. 1976. *The Life and Work of Ibn al-Shāṭir: An Arab Astronomer of the Fourteenth Century*. Aleppo: Institute for the History of Arabic Science.

Kennedy, E.S. and Kennedy, M.H. 1987. *Geographical Coordinates of Localities from Islamic Sources*. Frankfurt am Main: Institut für Geschichte der Arabisch-Islamischen Wissenschaften.

King, D.A. 1983. *Mathematical Astronomy in Medieval Yemen – A Bio-Bibliographical Survey*. Publications of the American Research Center in Egypt. Malibu, Ca.: Undena Publications.

King, D.A. 1986a. *A Survey of the Scientific Manuscripts in the Egyptian National Library*. Publications of the American Research Center in Egypt. Winona Lake, Ind.: Eisenbrauns.

King, D.A. 1986b. *Islamic Mathematical Astronomy*.

London: Variorum. 2nd rev. edn 1993. Aldershot: Variorum.

King, D.A. 1987. *Islamic Astronomical Instruments.* London: Variorum. Repr. 1995. Aldershot: Variorum.

King, D.A. 1991. Medieval astronomical instruments – a catalogue in preparation. *Bulletin of the Scientific Instrument Society* 31, 3–7.

King, D.A. 1992a. Some remarks on Islamic astronomical instruments. *Scientiarum Historia* 18, 5–23.

King, D.A. 1992b. Some remarks on Islamic scientific manuscripts and instruments and past, present and future research. In *The Significance of Islamic Manuscripts*, ed. John Cooper, pp. 115–44. London: Al-Furqan Islamic Heritage Foundation.

King, D.A. 1993. *Astronomy in the Service of Islam.* Aldershot: Variorum.

King, D.A. and Saliba, G. (eds) 1986. *From Deferent to Equant: Studies in the History of Science in the Ancient and Medieval Near East in Honor of E. S. Kennedy,* Annals of the New York Academy of Sciences, 500.

Kunitzsch, P. 1989. *The Arabs and the Stars.* Northampton: Variorum.

Kunitzsch P. and Smart, T. 1986. *Short Guide to Modern Star Names and Their Derivations.* Wiesbaden: Otto Harrassowitz.

Lorch, R.P. 1995. *Arabic Mathematical Sciences – Instruments, Texts, Transmission.* Aldershot: Variorum.

Mayer, L.A. 1956. *Islamic Astrolabists and their Works.* Geneva: A. Kundig.

Pingree, D. 1973. The Greek influences on Early Islamic mathematical astronomy. *Journal of the American Oriental Society* 93, 32–43.

Ragep, F. Jamil 1993. *Naṣīr al-Dīn al-Ṭūsī's Memoir on Astronomy (al-Tadhkira fī 'ilm al-hay'a),* 2 vols. Berlin, Heidelberg and New York, Springer-Verlag.

Sabra, A.I. 1987. The appropriation and subsequent naturalisation of Greek science in medieval Islam. *History of Science* 25, 223–43.

Saliba, G. 1991. The astronomical tradition of Maragha: a historical survey and prospects for future research. *Arabic Science and Philosophy* 1, 67–99.

Saliba, G. 1994. *A History of Arabic Astronomy: Planetary Theories during the Golden Age of Islam.* New York University Press.

Samsó, J. 1992. *Las ciencias de los antiguos en al-Andalus.* Madrid: Mapfre.

Samsó, J. 1994. *Islamic Astronomy and Medieval Spain.* Aldershot: Variorum.

Savage-Smith, E. 1985. *Islamicate Celestial Globes – Their History, Construction, and Use.* Washington, D.C.: Smithsonian Institution Press.

Sayılı, A. 1960. *The Observatory in Islam.* Publications of the Turkish Historical Society, ser. VII, no. 38.

Sezgin, F. 1974, 1978, 1979. *Geschichte des arabischen Schrifttums,* V: Mathematik, VI: Astronomie, VII: Astrologie, Meteorologie und Verwandtes. Leiden: E. J. Brill.

Storey, C.A. 1958. *Persian Literature: A Bio-Bibliographical Survey,* vol. II, pt 1. London: Luzac and Co.

Suter, H. 1900. Die Mathematiker und Astronomen der Araber und ihre Werke. *Abhandlungen zur Geschichte der mathematischen Wissenschaften* 10 (1900), and 14 (1902), 157–85. Both reprinted 1982. Amsterdam: The Oriental Press.

Toomer, G.J. 1968. A Survey of the Toledan Tables. *Osiris* 15, 5–174.

Van Brummelen, G. 1991. The numerical structure of al-Khalīlī's auxiliary tables. *Physis,* new ser., 28, 667–97

Varisco, D.M. 1993. *Medieval Agriculture and Islamic Science – The Almanac of a Yemeni Sultan.* Seattle, Wa.: University of Washington Press.

Vernet, J. 1978. *La cultura hispanoárabe en Oriente y Occidente.* Barcelona: Ariel. German language edn 1984: *Die spanisch-arabische Kultur in Orient und Okzident.* Zurich and Munich. French language edn 1985: *Ce que la culture doit aux Arabes d'Espagne.* Paris.

Wright, R.R. 1934. *The Book of Instruction in the Elements of Astrology by . . . al-Bīrūnī.* London: Luzac & Co.

OLAF PEDERSEN

European Astronomy in the Middle Ages

Introduction

The 1,000 years of the Middle Ages were not a static period in the intellectual and scientific development of Europe. Not least the history of cosmology and astronomy shows how the initial cultural setback was overcome so that radical new ideas on the structure of the universe could emerge at the same time as the more technical description of the behaviour of the celestial bodies reached a degree of accuracy and sophistication equalling that of the Greek founders of these disciplines.

The initial conditions were poor. The collapse of the Roman empire extinguished the essentially bilingual civilisation in the West where the Roman schools in the former provinces disappeared and the Greek language passed into oblivion. The *Almagest* became a closed book. Most other fundamental works of Greek science met with the same fate. Only a small number of Latin encyclopaedias, by authors such as Pliny, Macrobius and Martianus Capella, were available. They summarised more or less faithfully the results of ancient science but were silent about the methods by which such results had been achieved. In consequence, they were useless as guides to further research.

Monastic learning

As the only centres of learning in the early period, the new monastic schools had to satisfy themselves with such limited and second-hand sources. Moreover, there was no strong stimulus to do research or make regular observations at these institutions where scientific knowledge was mainly regarded as a helpful tool for the proper understanding of the Bible. Nevertheless, there are at least two reasons why these 'dark' centuries deserve a place in any account of the history of science in general and astronomy in particular.

First, the propagation of the Christian faith with its belief in one God as the creator of heaven and earth gradually changed the common attitude to nature, paving the way for a scientific approach to its phenomena by rejecting the multitude of gods and spirits previously thought to be responsible for them. Already at the beginning of the seventh

59 Medieval and Renaissance Europe.

century this movement bore fruit in the monumental *Twenty Books of Etymologies*, in which Bishop Isidore of Seville demonstrated how all natural phenomena and all human affairs could be described in a language that was completely devoid of mythological features. Among other things this led to an important distinction between astronomy and astrology. The former was concerned with phenomena caused by the daily rotation of the heavens and the course of the sun, moon and planets. The latter tried to predict human events by nativities presupposing a correlation between the twelve signs of the zodiac and the organs of the body. It was branded as mere superstition with the result that the ancient belief in astrology was now on the decline.

Second, the Christian liturgy made new demands on time-reckoning. The Julian calendar maintained its position for civil purposes. It was also used for defining 'fixed' ecclesiastical events such as Christmas, Epiphany and the Annunciation. On the other hand, the 'movable' feasts of Easter and the derived dates of Pentecost and the beginning of Lent were determined by rules going back to the lunar calendar of ancient Israel. Thus Easter Sunday had to be the first Sunday after the first full moon after the vernal equinox. The determination of this date in the civil calendar was a rather difficult problem which found no general solution until the beginning of the sixth century. All these questions were exhaustively dealt with in a great work, *On the Theory of Time-reckoning*, written in 725 by the Benedictine Bede of Yarrow. Unsurpassed in clarity and comprehensiveness it remained for centuries the most impressive monument of monastic science. It was also Bede who first made consistent use of the Christian era in his *History of the English Church and People*, in which the years are numbered from the birth of Christ.

Less attention was paid to the problem of determining the time of the day. In monastic communities the seven 'canonical' hours of prayer were determined by simple sundials which seem to have been the only astronomical instruments of this period. In some places water-clocks were used to find the hour at night as in the ancient world. Also the ancient practice of defining one hour as the twelfth part of the day or the night survived with the result that the hours of the day were longer than the hours of the night in summer, and the hours of night were longer during winter. The custom of using such seasonal hours prevailed until the fourteenth century when the newly invented mechanical clock with its uniform motion gradually made people in general familiar with the equal or equinoctial hours which had been used by astronomers since antiquity.

The astronomy of the schools

From Carolingian times onwards (*c.* AD 800) the monastic schools, usually situated in the countryside, were supplemented by a growing number of cathedral schools situated in the cities and open to both lay and clerical pupils. Here the secular part of the curriculum was taught within the traditional framework of the seven 'liberal arts', one of which was astronomy. This was still a predominantly literary activity based on the meagre Latin sources; but the introduction of new instruments in the form of simple sighting tubes testifies to an increased interest in observing the details of the starry heavens more closely, and records of spectacular phenomena such as comets and eclipses were frequently entered into calendars and historical annals.

Another outcome of the teaching of the schools was a number of specialised manuals and textbooks in which the traditional material was presented in a more unified and systematic way. The greatest attention was still paid to the elementary phenomena of spherical astronomy whereas planetary phenomena were more summarily described. However, Hipparchus's theory of the motion of the sun was usually presented in a qualitative way. It was often built into a more comprehensive geo-heliocentric system derived from Martianus Capella and propagated by John Scot Eriugena in the ninth century (Col. Pl. XII). Here Venus and Mercury are supposed to be moving in circles around the sun which has an eccentric orbit around the earth, while the moon and the superior planets (Mars, Jupiter and Saturn) move in concentric orbits. Outside the planetary region is the firmament, a spherical shell or 'sphere' containing the fixed stars and carrying all other celestial bodies around in its daily revolution. Because of the biblical reference to the 'waters above the heavens' the firmament was sometimes supposed to be surrounded by a 'crystalline' sphere of water frozen to ice. Further away was an 'empyrean' sphere (mentioned already by Martianus Capella). Being invisible these two spheres were of no interest to astronomy as such (as later underlined by Nicole Oresme) although they remained an integral part of the popular picture of the world throughout the Middle Ages.

At the same time as the potential of the Latin sources was being exhausted in the West, the Muslim world had recovered most of the heritage of Greek science, making its principal works available in Arabic translations. Already before the turn of the millennium Latin scholars began to be attracted by the rumours of the scientific treasures possessed by the Moors in Spain. Over the next three centuries an increasing number of students visited this country. Quite a few settled there to collaborate with Muslim or Jewish scholars. Others returned to their schools provided with new books and instruments and familiar with the Arabic language. Before the year 1200 this movement had made most of the extant Greek works in any field of learning available in Latin, together with a number of treatises of more recent date originating within the pale of Islam itself. The general effects of this cultural interchange were both numerous and profound, and in fact decisive for the future course of European science as a whole.

The impact of this movement on astronomy was felt early in the eleventh century when the introduction of the plane astrolabe (described on pp. 166–7 and 233–4) initiated a new development. About 1050 its construction and use were described by the Benedictine Herman the Lame of Reichenau in two small writings which are the first treatises on scientific instruments to emerge on Latin soil. The astrolabe was a highly versatile instrument by means of which altitudes could be measured, the time of the observation being determined both by day and by night, and a number of problems in spherical astronomy solved by mechanical manipulation of its parts. In consequence, spherical astronomy was transformed from a qualitative discourse on diurnal phenomena into a precise geometrical science. Moreover, the new possibility of producing numerical data from observations called for a better knowledge of the motions of the sun, moon and planets, a demand which was met in 1126 when Adelard of Bath translated into Latin al-Khwārizmī's great collection of astronomical tables with their reminiscences of Indian procedures of calculation (see pp. 137 and 151). At the same time appeared a Latin version of the Ptolemaic *Toledan Tables* (see p. 164) with their *Canones*. They were quickly adapted to many other localities such as Pisa, Marseilles, Toulouse and London, as a testimony to the enthusiasm with which this new mathematical astronomy was welcomed. When finally Ptolemy's *Almagest* appeared in Latin the underlying theory behind the tables was revealed, and medieval astronomy was in principle able to continue where ancient astronomy had left off.

University astronomy

The general explosion of knowledge during the twelfth century was more than the individual schools and single masters could cope with. In some cases this led to the emergence of specialised schools with several masters involved in the same general subject. Thus Bologna became famous for law, Montpellier for medicine, Paris for theology, and Chartres for natural philosophy. Out of this situation grew the medieval *studium generale*, or university, in which several specialised schools were united in a common structure under various forms of corporate government. Everywhere the access to the 'higher' faculties of law, medicine or theology presupposed a degree from the introductory faculty of arts which everywhere provided the basic education. To this faculty belonged the teaching of astronomy as an integral part of that natural philosophy which formed, together with moral philosophy and metaphysics, the standard components of the curriculum. An interesting consequence of this system was the fact that no medieval student could pass his master's degree without being acquainted with at least the rudiments of astronomy and cosmology.

On the other hand, the medieval university never gave a prominent position to astronomy. There was no higher faculty of science, and the subject was placed among the introductory arts taught to young boys who had to study several other disciplines at the same time. In consequence the teaching had to be of an elementary character. The *Almagest* was available, but much too difficult and had to yield to briefer and more easily digestible expositions. In particular, an *Astronomical Compilation* by the ninth-century Baghdad scholar al-Farghānī (Alfraganus of the Latins) was widely used in a translation by Gerard of Cremona (d.1187), who had also produced the standard version of the *Almagest*.

Even more popular was a series of textbooks produced by the university teachers themselves, especially the works of John of Sacrobosco, who went from England to Paris where he taught until his death about 1236. He was one of the small number of professors who devoted a lifetime to science without coveting a more lucrative job in one of the higher faculties. His books comprised: *Tractatus de Sphera* containing the elements of spherical astronomy and a few paragraphs on eclipses and the motions of

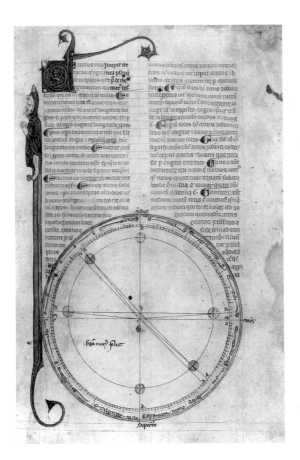

60 The first page of the standard textbook of planetary theory of the late Middle Ages, the anonymous *Theorica planetarum*. The figure shows the eccentric circular orbit of the sun around the earth according to Ptolemy and Hipparchus. (British Library, MS Harl. 13, f. 3r)

the sun and moon (Col. Pl. XIII); *Compotus* with a concise exposition of the principles of time-reckoning, including a perspicacious remark on the shortcomings of the calendar and expressing the need for a reform; and finally *Algorismus* explaining how to perform calculations with the new Arabic numerals. The exposition is always clear and to the point, containing nothing which a student had better forget at a more advanced stage of his education.

Sacrobosco's laconic remarks on planetary theory were supplemented by a manual, *Theorica planetarum* (Fig. 60). Many works with this title appeared from different hands, the most widely used being the work of an unknown author of the first half of the thirteenth century. It described the various kinematic models of Ptolemaic astronomy, with clear definitions of their parameters and variables, but omitting much numerical data which could be found in astronomical tables and their *canones*. In numerous extant codices we find Sacrobosco's three books bound together with *Theorica planetarum*, a calendar and a set of tables, as a *Corpus astronomicum*, which provided the basic equipment for teachers of astronomy until the sixteenth century, when several such volumes enjoyed a prolonged life in printed editions.

University cosmology

Thus astronomy entered into university teaching as an essentially mathematical discipline, in agreement with the Archimedean tradition from antiquity, according to which it is the purpose of science to discover mathematical relationships between the phenomena of nature, which are considered as 'given' facts derived from experience and observation. But as a result of the twelfth century translations of all the works of Aristotle this concept was immediately confronted with the idea of science as a quest for the causes of the phenomena. This metaphysical philosophy of science was behind Aristotle's book *On the Heavens*, which entered the university curriculum as the standard manual on cosmology, providing answers to questions about both the physical and the metaphysical structure of the universe which were conspicuously absent from Sacrobosco's textbooks and similar works.

On the Heavens aimed at describing the universe in terms of 'physics', defined as the science of motion and rest. In the middle of everything is the spherical and immovable earth in the neighbourhood of which 'light' and 'heavy' bodies composed of the four elements can perform 'natural' motions up and down in order to come to rest at their 'natural places'. This 'elementary' part of the world fills a 'sub-lunary' sphere outside which we find the moon and the other heavenly bodies. They are never at rest, their natural motions being circular and eternal in sharp contrast to the bodies in the lower world. In consequence they must consist of a special 'fifth element' or 'ether' that is unknown here below. This region is subdivided into seven spherical shells or 'spheres' each of which carries a planet around from west to east with a definite period. Further out is the eighth sphere or the 'firmament' which contains the fixed stars. Each day it performs a uniform rotation from east to west in which also the interior spheres of the planets take part. Since infinite velocities are unthinkable, the firmament must have a finite size. Its outer surface marks the boundary of the universe. Outside there is nothing, not even an empty space.

The existence of two alternative descriptions of the universe presented a dilemma that gave rise to several schools of thought. At one end of the spectrum were the strict Aristotelians who followed the twelfth-century 'Commentator' Ibn Rushd of Córdoba (Averroës of the Latins) in a very literal interpretation of the works of 'The Philosopher' (i.e. Aristotle). They found that the many eccentric circles and epicycles of Ptolemy were incompatible with the homocentric principles of their master, just as they were attracted by the apparent simplicity of a system that could be understood without hard mathematical brainwork. Others followed Ibn Rushd's contemporary and fellow citizen al-Biṭrūjī (see p. 150), whose book on *The Motions of the Heavens* (translated by Michael Scot in 1217) tried to develop a mathematical theory without eccentric circles; this proved to be a failure.

At the other extreme were the mathematical astronomers who were less interested in the structure of the universe than in being able to calculate planetary positions. They made short shrift of any theory that was unable to do so by means of tables for practical use, even if it were pronouncedly 'physical'. A widely accepted kind of mean position was indicated by Campanus of Novara in his great *Theorica planetarum*, datable to about 1258. It showed how each one of the Ptolemaic models could be imbedded in a 'physical' mechanism comprising several ethereal bodies. Thus the epicycle became a (true) sphere rotating inside a spherical cavity in a spherical shell. All this was placed within a larger shell or 'sphere' with concentric surfaces. This was a medieval version of the mechanical universe which Ptolemy had described in his *Planetary Hypotheses*, but carefully excluded from the *Almagest*. Assuming that the various spheres were just large enough to accommodate the respective kinematic models, and that there were no empty spaces between them, Campanus was able to repeat Ptolemy's calculation of the size of the universe, or at least the radius of the inner surface of the firmament, which he found to be a little more than 70 million 'miles', equivalent to about 120 million kilometres.

The astrological connection

The thirteenth century did not only revive Aristotelian cosmology and mathematical astronomy. It also witnessed the return of astrological beliefs which had been kept at bay since the time of Isidore and Bede. There were several reasons for this change. Already twelfth-century writers such as Hildegard of Bingen and Bernard Silvestre had resurrected the Platonic view of the human being as a microcosm whose bodily organs were correlated to the macrocosmic signs of the zodiac (Col. Pl. XIV). Another stimulus came from the wave of translations. It is significant that Ptolemy's astrological manual, *Tetrabiblos*, was translated into Latin several decades before the *Almagest*. It was followed by a great number of works of Arabic origin, among which *Introduction to Astronomy* (i.e. astrology) by Abū Maʿshar, or Albumasar, became immensely popular.

Learned churchmen such as John of Salisbury made strong objections to this movement, which was, nevertheless, strengthened in the thirteenth century by Aristotelian cosmology. Aristotle had placed the 'first mover' in the eighth sphere from which causal 'influences' flowed through the planetary spheres to the elementary world, where they were ultimately responsible for all changes as instruments of the irresistible necessity which permeated the cosmos as a whole. This had immediate consequences for the practice of medicine, since, for example, phlebotomy and other cures must be performed at times when the stars were in favourable positions; and manuals on astrological medicine became increasingly common. The ultimate result of all such ideas was the belief that the fate of individual men and women was totally determined by the configuration of the planets at the time of birth as recorded in personal 'nativities' (or horoscopes).

Such extreme views were incompatible with Christian theology and provoked ecclesiastical intervention. Best known is the promulgation in 1277 of no less than 219 condemnatory propositions by the Bishop of Paris. Here several Aristotelian doctrines, such as the eternity of the world, were branded as heretical, and at the same time a blow was dealt to astrology: whatever influence the stars may have here below, they do not suspend free will and moral responsibility. However, the debate went on, and astrology went from strength to strength during the fourteenth century when the great calamities of famines, wars and pestilence turned anxious minds to ask what the future had in store, as if the astrologers could provide the answer.

Sometimes, ordinary astronomy was seen as a mere introduction to the 'great science' of astrology which 'links natural philosophy with mathematics', as maintained by the anonymous author of the thirteenth-century *Speculum astronomiae*. Other writers branded astrology as a mere superstition without scientific foundation, and also dangerous to society when princes listened more to their court astrologers than to their ordinary political advisers. But no agreement was obtained and the Middle Ages came to an end with astrology as part and parcel of everyday life at all levels.

Technical developments

Astrology was a mixed affair. On the one hand, it was based on a traditional set of rules for interpreting celestial phenomena in terms of terrestrial events. These rules were arbitrary and without any scientific foundation. On the other hand, the relevant phenomena had to be presented as numerical values of planetary positions at a given time, which only astronomy could provide. In this way the spread of astrological beliefs gave rise to an increasing demand for people with an expert knowledge of technical astronomy. This had consequences in more than one area.

In the social sphere it created a new civil profession. A scholar wishing to dedicate his life to astronomy would find very few opportunities within the universities. Now he might be able to earn his living by practising astrology at the same time as cultivating his scientific interests, either as court astrologer to a prince or as a salaried 'mathematician' of a city; in the latter case he would also act as surveyor of land and controller of weights and measures. Kepler's career shows how this system outlived the Middle Ages.

The technical work of the astrologer comprised several mathematical operations. One of them was to delineate or 'cast' the twelve 'houses' or sectors in which the sky had to be divided at a given time. This could be done in a variety of ways leading to different results. This gave rise to various schools of astrologers according to the preferred method; but essentially the choice of method was arbitrary. The next step was to ascertain the positions of the planets (including the sun and the moon) at a given time – in the past if a birthday nativity was ordered by a client, or in the future if the purpose was the 'election' of a favourable date for a planned or foreseen event. This was done by a set of astronomical tables according to procedures described in the accompanying *canones*. This explains the proliferation of tables in late medieval manuscripts. Most of them belong either to the Toledo family already mentioned, or derive, in the later period, from the so-called *Alfonsine Tables* compiled about 1272 by a team of Spanish astronomers working under the patronage of King Alfonso X of Castile. The latter are marked by a consistent use of the sexagesimal number system. However, the original version is lost, and only a Latin translation from about 1320 has been preserved.

A set of tables was characterised by a number of fundamental parameters. These included for each planet a *radix*, i.e. its position at the epoch on which the tables were based, and a value of the mean motion of the planet per hour, day, month and year which together with the *radix* made it possible to calculate the mean position of the planet at a given time. Finally, a number of auxiliary tables of 'equations' made it possible to calculate the actual deviation of the planet from its mean position so that its true position would appear within one of the twelve 'houses'. This completed the scientific part of the work, the next step being a judgement concerning what influence the location of a given planet in a particular 'house' would have. All this explains why astrology contributed to make the science of tables a predominant part of late medieval astronomy.

The actual determination of the position of a planet was a cumbersome and long-winded affair, since the calculation of the 'equations' involved several multiplications of numbers with four or more sexagesimal digits. Already Muslim astronomers had facilitated the procedure by means of special calculating instruments, which about 1260 turned up in Europe under various names such as *theorica*, *volvella*, or *equatorium*. The first type was described by Campanus of Novara; it was soon followed by other models, some of them of great ingenuity and sophistication. They were all aimed at reducing the number of tables, in particular by doing away with the tables of equations and the corresponding calculations, which were to be replaced by the mechanical manipulation of the graduated scales of an instrument. The details of the equatoria are described on pp. 169–70. Here we shall only notice that the idea of mechanical calculation was first conceived by astrologers who wanted to facilitate and speed-up their business.

Of more fundamental importance than the techniques of manual or mechanical calculation was the problem of the theories and parameters on which the tables were based. Here we meet with another example of cultural interchange with the Muslim world. For whereas the Ptolemaic models were never seriously challenged in the West, the numerical parameters of the *Almagest* were sometimes changed in consequence of new observations made more or less systematically at one of the Muslim centres of learning, where there were real observatories of a type and size unknown in Europe. For example, in the *Almagest* the eccentricity of the orbit of the sun was assumed to be $\frac{1}{24}$, or 2; 30 in sexagesimal notation (see pp. 83–5). Seven hundred years later al-Khwarizmī in Baghdad changed it to 2; 14. The *Toledan Tables* (see p. 164) reduced it to 1; 59; until it was changed back to 2; 30 in the Latin version of the *Alfonsine Tables*. Thus there was no steady improvement towards the modern value of about 2; 00. This was the result of the great veneration with which both Muslim and Christian astronomers preserved old observations. It can be illustrated by the theory of precession which played an overwhelming role in late medieval astronomy.

For a given time the tables provided planetary positions in a co-ordinate system based on the ecliptic with its point of origin (0°) at the vernal equinoctial point where the ecliptic intersects the celestial equator. Of course, it was important to know where the fixed stars are at the given time. Now already Hipparchus (see p. 81) had realised that, relative to the equinoctial point, the stars perform a slow, eastward motion at a rate which Ptolemy (see p. 87) assumed to have a constant value of at least 1° per century. This 'motion of the eighth sphere' was re-examined by al-Battānī (or Albategnius) in the ninth century and found to be about 1° in 66 years (cf. p. 153). However, his successors did not conclude that the ancient value was wrong, but that the rate of precession had changed since Ptolemy's time. So instead of being constant it was a function of time, which had to be determined in a way which accounted for all the observational material. Several such functions were proposed. In the *Alfonsine Tables* the precessional motion of the firmament was resolved into two components, one of them being a uniform rotation with a period of 49,000 years, and the other an oscillatory movement to and from the equinox lasting 7,000 years. A similar and equally intricate theory of the 'trepidation of the equinoxes' (see pp. 101, 149–50 and 153) was even developed by Copernicus, who was as convinced as his medieval predecessors of the validity of the historical observations – a belief that was first abandoned when Tycho Brahe decided to 'restore' astronomy by means of a completely new set of observations.

Cosmological innovations

Along with the development of theoretical astronomy, radically new cosmological ideas emerged within the framework of university teaching. This happened in the wake of the promulgation of 1277 (see p. 182) which branded several Aristotelian doctrines as wrong at the same time as it provoked philosophers to reconsider the notion of necessity. The indubitable necessity in nature was now viewed as 'contingent' upon the will of the Creator who might have made a different world. Together these two ideas led to a critical attitude towards the established cosmology and physics, resulting in a wealth of new and fertile hypotheses.

Thus the Oxford scholar Thomas Bradwardine (d. 1349 as Archbishop of Canterbury) attacked the Aristotelian doctrine that there could be no empty space 'outside the world', arguing that God's infinity implied an infinite universe. The finite firmament still defined the 'place' of the world; but it is surrounded by an infinite, void 'space' of a divine nature. At about the same time, his Paris colleague Jean Buridan used the omnipotence of God to criticise the traditional picture of the world inside the firmament. Aristotle had taught that the planets perform their ordained motions because their respective spheres were ensouled. This rather animistic concept had already been abandoned in the thirteenth century in favour of the assumption of 'separate intelligences' as governing planetary motions. Now Buridan proposed that all the spheres move because of a purely physical 'impetus' which was implanted in them at their creation and has kept them going ever since. That he also explained terrestrial motions in terms of an impetus of varying forms is a testimony that the traditional separation of the sublunary world from the heavens was now beginning to crack.

In Paris Buridan's pupil Nicole Oresme (d.1382 as Bishop of Lisieux) adopted both the impetus theory and the idea of an infinite space. In a major commentary on Aristotle's *On the Heavens*, written in French in 1377, he also used both thought experiments and mathematical arguments to destroy the doctrine of 'natural places'. This did away with the notion of 'levity' as a particular quality on a par with gravity; and gravity was no longer understood as a tendency to move towards the centre of the world, but (as already earlier scholastics had maintained) as a general tendency in all bodies to 'unite with their likes'. Finally he subjected the usual arguments for the immobility of the earth to a painstaking criticism, proving among other things that all celestial phenomena would appear in exactly the same way if the earth were supposed to perform a daily rotation about its axis, although he could not persuade himself that it really does so.

In the fifteenth century, all these new ideas came together in Cardinal Nicholaus of Kues (d.1464), who described an infinite universe containing stars everywhere, with the earth as a 'noble star' among others. Each star rotates about its particular axis with the result that an observer on any of them would see the universe as we do, only with different axes and poles. It is natural to regard this intimation of what has been called the general cosmological principle as the last word of medieval cosmology on the threshold of the Renaissance.

Bibliography

Duhem, P. 1913–59. *Le Système du Monde*, vols I–X. Paris.

Grant, E. (ed.) 1974. *A Source Book in Medieval Science.* Cambridge, Mass.: Harvard University Press. [Contains English translations of many of the texts cited in this chapter.]

North, J. D. 1988. *Chaucer's Universe.* Oxford University Press.

North, J. D. 1994. *The Fontana History of Astronomy and Cosmology.* London: Fontana. Published in the USA as *The Norton History of Astronomy and Cosmology.* New York: W. W. Norton.

Pedersen, O. 1993. *Early Physics and Astronomy: A Historical Introduction.* Cambridge University Press. [Detailed bibliography and a lexicon of ancient and medieval astronomers.]

Current research in medieval astronomy is published in specialised journals, in particular the *Journal for the History of Astronomy* (ed. M. A. Hoskin), Science History Publications, Cambridge.

N. M. SWERDLOW

Astronomy in the Renaissance

The period from the late fifteenth century to the early seventeenth saw a transformation of astronomy more fundamental and more extensive than anything in its history since its development in antiquity from the Babylonians to Ptolemy. While it is not possible to write a continuous history of astronomy in antiquity, owing to the absence of most sources between the Babylonians and Ptolemy, the sources for the Renaissance and early modern period survive in such quantity that each step, however small, is documented, and the problem for the historian is one of selection, understanding, and interpretation of the mass of material. There are, however, clearly three factors that lie at the foundation of this transformation, respectively mathematical, physical, and observational, and four principal figures, Regiomontanus, Copernicus, Tycho Brahe and Kepler. The story begins with the recovery and exposition by Regiomontanus of Ptolemy's mathematical and observational methods, and some of the additions made by Arabic astronomers, still contained within the physics of rotating spheres as the cause of motions in the heavens. With the same physical foundation, Ptolemy's methods were applied faithfully by Copernicus, who also extended their application to the heliocentric theory. A more radical change came with Tycho Brahe, whose improved instrumentation and vast quantity of observations of an unheard of degree of accuracy greatly expanded Ptolemy's procedures, while his evidence against spheres in the heavens removed the physical foundation on which astronomy had rested since antiquity. With the necessity of finding a new physical foundation and with Tycho's observations, both of which required the development of new mathematical techniques, Kepler at last went far beyond Ptolemy's methods, and discovered entirely new principles for the precise description of the motions of bodies in the heavens based upon an entirely new physics. It is the purpose of this chapter to describe this history, not in the detail it deserves, but at least to convey some understanding of the extraordinary ideas and discoveries that brought the science of the heavens from the ancient and medieval to the modern world.

Regiomontanus

> I cannot wonder at the indolence of common astronomers of our age, who, just as
> credulous women, receive as something divine and immutable whatever they come
> upon in books either of tables or their canons, for they believe in writers and make
> no effort to find the truth.

With these polite remarks Johannes Müller of Königsberg (1436–76), known as
Regiomontanus, began a devastating criticism of contemporary astronomy in a letter to
Giovanni Bianchini (*c.* 1400–*c.* 1470), himself one of the most accomplished astronomers
of the age. Bianchini was the chief fiscal officer to the Estensi of Ferrara, and was trained
in the law, but he was passionately devoted to astronomy and astrology, especially to the
works of Ptolemy, and in his free time rested his mind from business by computing
astronomical tables. He had computed tables for spherical astronomy, eclipses and
planetary theory, and had written a large introduction to and commentary on the
Almagest as far as Book VI called *Flores almagesti*. In the summer of 1463 he had sent
Regiomontanus as a test a problem in spherical astronomy. This began a correspondence
extending into the spring of 1464 in which each told of his interests and posed problems
in mathematics and astronomy for the other to solve, some of considerable difficulty.
Bianchini, then in his sixties and perhaps past his prime, could solve few; Regiomontanus,
then all of twenty-seven and the most proficient mathematician in Europe, could solve
all of Bianchini's and all of his own. He came from the small town of Königsberg in
Franconia. Something of a prodigy, in 1447, when only eleven years old, he entered the
University of Leipzig and computed a daily ephemeris for the sun, moon and planets
for the year 1448. He soon exhausted the resources of Leipzig, whatever they may have
been, and in 1450 enrolled at the University of Vienna.

When Regiomontanus arrived, there was one person with a serious interest in
astronomy, Georg Peurbach (1423–61), who entered the university in 1446 but only
began lecturing on astronomy in 1454, after he received his master's degree, when he
presented his *Theoricae novae planetarum* (*New Theories of the Planets*), the earliest copy of
which is probably one made by Regiomontanus. The *Theoricae novae* was intended to
replace the standard school text on planetary theory, the old *Theorica planetarum* attributed
to Gerard of Cremona (see p. 179); and it did, for there were more than fifty printed
editions, some with extensive commentary, through the early seventeenth century, long
after it had outlived its usefulness. What Peurbach did in this work was to describe
Ptolemy's solar, lunar and planetary theory that underlay the *Alfonsine Tables* (see p. 183).
He did this first physically, describing complete, solid spheres, then geometrically as
circles, explaining the technical terms of astronomy and astronomical tables (Figs 61a
and b). Among Peurbach's other astronomical, calendrical and mathematical works the
most important was the *Tabulae eclipsium* (*Tables of Eclipses*), about 100 pages of closely
written columns of numbers for computing eclipses of the sun and moon.

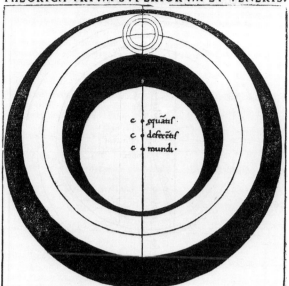

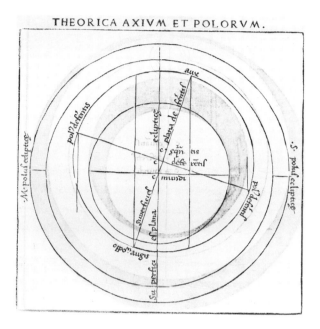

61 Peurbach, *Theoricae novae planetarum* (Nuremberg, *c.* 1473), ff. 6r and 6v. Figure 61a (*left*) shows a cross-section of a spherical planetary model. The centre of the eccentric sphere carrying (*deferentis*) the epicycle is marked c[*entrum*] *deferentis*, and the sphere rotates uniformly about the c[*entrum*] *aequantis*, the centre of the equant circle (not shown). Figure 61b (*right*) shows the axes and poles of the spheres. The axis of the sphere carrying the epicycle, with its end points labelled *polus deferentis*, passes through the c[*entrum*] *deferentis*. (British Library)

The new light of Italian humanism and Greek learning came to Vienna in 1460 when Cardinal Johannes Bessarion (1403–72) arrived as legate of Pope Pius II to enlist the aid of Emperor Friedrich III and the German princes for a crusade to recapture Constantinople from the Turks. Bessarion, originally from Trebizond and a student of the Platonising Gemistus Pletho in Mistra, brought with him to Italy when he came to the Council of Ferrara–Florence (1438–42) a large collection of Greek manuscripts, including the mathematical sciences, to which he continued to add until, in the last years of his life, he presented them to the city of Venice where they formed the foundation of the Biblioteca Marciana. He was a great patron of Greek learning, particularly of translations into Latin, and saw in Peurbach and Regiomontanus talent in the mathematical sciences exceeding any of his circle in Italy. He had begun a translation of the *Almagest* himself, but was too busy to complete it. Now he asked

Peurbach, who knew the book almost by heart, but in Latin only, to prepare an abridgement and accompany him on his return to Italy. Peurbach requested that Regiomontanus be allowed to join them and began working on the abridgement, but fell ill in April of 1461 after completing only the first six books, and shortly before his death asked his younger colleague to carry the work to completion.

In the autumn Regiomontanus journeyed to Rome with Bessarion, and got right to work. It happened that in December there was a lunar eclipse and close approaches of Saturn, Mars and Venus to each other and to stars in Capricorn. Regiomontanus observed these, estimating the separation of the planets by comparison with the separation of the fixed stars given in the Alfonsine star catalogue. He computed the positions of the planets with the *Alfonsine Tables*, finding notable errors, and continued such comparisons sporadically into the following year, the errors forming the basis of the criticism of current practice in astronomy in the letter to Bianchini. But he did a good deal more. He completed Peurbach's abridgement of the *Almagest*, probably revising his teacher's own contribution, and in 1462 or '63 dedicated the *Epitome of the Almagest* to Bessarion. The *Epitome* is far more than an abridgement of the *Almagest*, for it recasts the work in the form of propositions in which geometrical theorems are proved, models, instruments and methods of observation described, and procedures for deriving parameters from observation explained, all with a precision and clarity that for the first time gave Europeans a reliable guide to the full range of Ptolemy's mathematical techniques in spherical astronomy, eclipses and planetary theory. Additional materials were drawn from Arabic works translated into Latin. The *Epitome* is the great advanced treatise on astronomy of the Renaissance of letters, of the recovery of Greek learning, and was studied into the early seventeenth century as a preparation for the *Almagest* and even in place of the *Almagest*. Copernicus used it extensively in his work and, as we shall see, it may have provided a crucial step on the way to his own principal innovation in astronomy.

Thus, when Regiomontanus wrote his criticism of contemporary astronomy to Bianchini in 1464, he was in a better position than anyone to know what was wrong and what, within the limits of Ptolemy's methods, should be done to correct it. He begins his examination with the sphere of the fixed stars concerning which there is disagreement both on the rate of its motion with respect to the equinoxes and on the inclination of the equator to the ecliptic. But worse still, when one considers the hypotheses for the motion of the fixed stars, one attributed (incorrectly as it turns out) to Thābit ibn Qurra, the other implicit in the *Alfonsine Tables* – probably according to Peurbach's description in the *Theoricae novae* – a variation of the obliquity of the ecliptic results that is flatly contradicted by the observations of all ages from Ptolemy, through al-Battānī to Peurbach and Regiomontanus and their contemporaries Paolo Toscanelli and Leon Battista Alberti. Likewise, the *Alfonsine Tables* mislocate the apogee of the sun at the time of Ptolemy, and when the sun is in the beginning of Aries in the fixed ecliptic it will have a declination of 6°. Moving on to the planets, Saturn and Jupiter pose no problems, but Mars is found to have errors of as much as 2°, indicating possible errors in its eccentricity or the radius of its epicycle, and Venus errors of $\frac{3}{4}°$ in longitude and an unspecified amount in latitude. (The errors were found from the observations of December 1461, and both planets can in fact show much larger error.) In addition, the eccentricities and epicyclic radii of these planets indicate a variation in distance that should produce a great variation in apparent size that is not observed. (It took Galileo's observations with the telescope to show that the variations in size do occur, but cannot be detected with the unaided eye.) According to the tables, Mercury should frequently be visible at our latitude although it is seldom or never seen. And in the case of the moon the errors are so large and frequent that 'even ordinary people tear at this divine science of the stars with a sharp tooth'. The end of the eclipse of 17 December 1461 occurred an hour earlier by observation than by computation, other eclipses show great errors in duration and magnitude, and the lunar eccentricity and epicyclic radius imply a variation of apparent size of four to one, which is never observed.

The errors pointed out by Regiomontanus were serious, in fact more serious than he could know, for their correction ended up requiring the observations of Tycho, the theory of Kepler, and the telescope of Galileo, in short, the total reform of astronomy in the late sixteenth and early seventeenth centuries. And Copernicus's heliocentric theory was certainly of help along the way. No-one would have been more surprised than Regiomontanus at the form these corrections took, for he surely imagined that a consistent application of Ptolemy's methods would be adequate to bring the science to perfection. He asked Bianchini to work with him to remove the rust from the heavenly spheres and guide them back to the royal roads, perhaps an allusion to Ptolemy, at one time mistakenly identified as one of the Egyptian kings of that name. But this is the last letter, or the last surviving letter of the correspondence, and as far as is known Bianchini made no further contributions to astronomy. Regiomontanus, however, was only beginning, and was to spend the rest of his life attempting to carry out this great project and more besides. Here we can but briefly mention his contributions, which amount only to the initiation of his plan, but were essential if the science were to advance beyond where Ptolemy and his Arabic successors had left it.

First came trigonometry, the foundation of all computation. After completing the *Epitome*, he undertook, in reverse order as he put it in the dedication to Bessarion, the more basic *De triangulis omnimodis* (*On Triangles of Every Kind*), eventually extending to five books on plane and spherical trigonometry. He also calculated two enormous sine tables to every minute of arc — it appears as though he was a lightening calculator — the first with a unit radius of 6,000,000, the second with a radius of 10,000,000, the first modern seven-place sine table containing only 13 errors greater than ±1 in its 5,400 entries. The importance of such a table at the dawn of modern science should not be underestimated. High precision in calculation is not possible without precise trigonometric tables, and this table is the first to attain a degree of precision exceeding any application at its time. It formed the basis of trigonometric tables for the next century or so, or even to the present day, for what is a modern seven-place sine table if not Regiomontanus's table with every error corrected?

Next was spherical astronomy, the geometry and trigonometry of the circles of the celestial sphere, essential for the determination of time, of risings, culminations and settings, of transformation of co-ordinates, resolution of components of parallax, reduction of observations, as well as the essential astrological applications of computing limits of houses, directions of significators, and projections of rays. For all these Regiomontanus computed his largest set of tables and, along with the *Epitome*, his most important work for the next century and a half, the *Tabulae directionum*, completed after he left Italy in 1467 to take up a professorship at the newly founded University of Pressburg (Bratislava). They were dedicated to his new patron, Johann Vitez, Archbishop of Gran (Esztergom) and Chancellor of Hungary, to whom Peurbach had earlier dedicated his *Tabulae eclipsium*. Also completed in Hungary, and dedicated to King Matthias Corvinus, were the *Tabulae primi mobilis*, about which he had earlier written to Bianchini. These tables of ninety pages give direct solutions to right spherical triangles at 1° intervals, for which he sets out no less than sixty-three problems showing their application.

But he did not stay long in Hungary. In 1471, perhaps on rumours of a plot by Vitez to remove Matthias as king, the faculty of the university dispersed as rapidly as it had earlier assembled, and Regiomontanus, with his collection of manuscripts, made his way to Nuremberg, the great commercial centre of Southern Germany and of trade between Northern Europe and Italy. Shortly after his arrival, he wrote another of his long letters filled with mathematical problems to one Master Christian, 'the most outstanding of mathematicians', who has been identified with Christian Roder, the Rector of the University of Erfurt. He told of being with the king and nobles of Hungary when he was shown contradictory and mistaken astrological predictions from Italy for the previous year. The incompetence of these excited such wonder that he explained that the errors came, not from the art of astrology, but from relying upon poorly translated or corrupt texts and inaccurate calculation of the motions of the sun, moon and planets. What is necessary, he explained, is a restoration of astronomy by co-ordinating the work of several observers making frequent observations, and he asks Christian to take part in this warfare carried out, not on horseback with spears, but with books and new, large astronomical instruments that he is constructing. He asks him to send any investigations or observations he may have, and he in return will send his own. And when this co-operative venture has advanced sufficiently, he will correct the tables and compute ephemerides for thirty and more years that he intends to *print*.

The improvement of astronomy was only one part of Regiomontanus's plan; the other, about which he also informed Christian, was for the general improvement of mathematics by setting up the first printing firm devoted to scientific publishing, a

192

62 Regiomontanus, *Kalendarium* (Nuremberg, *c.* 1474), f. 15v, giving dates, times, magnitudes, durations and illustrations of solar and lunar eclipses, 1479–82. (British Library)

remarkably far-sighted idea scarcely 20 years after the invention of typography. Regiomontanus appears to have been the first person to grasp the importance of printing for the improvement of the sciences – for as yet not one serious work in the mathematical sciences had been printed – and was certainly the first to act on it. He also grasped how essential it was that printed texts and translations be accurate since a bad printing of hundreds of copies could do much more damage than a single bad manuscript. His first completed publications were the *Theoricae novae planetarum* of Peurbach, notable for its many hand-coloured woodcuts (Figs 61a and b), and the first edition of Manilius's *Astronomica*, an astrological poem of the early first century. He next printed his own *Kalendarium*, in both Latin and German versions with different type fonts for each, containing the ecclesiastical calendar along with an analysis of its errors for 1475 to 1531, tables and instruments for finding the true positions of the sun and moon, and illustrations of solar and lunar eclipses for 1475–1530 (Fig. 62). This was followed by his most ambitious undertaking, *Ephemerides* for 32 years (1475–1506), giving daily positions of the sun, moon and planets along with their aspects to the moon and to each other, a book of nearly 900 pages (Fig. 63) which set the pattern for ephemerides continued by generations of astronomers well into the seventeenth century and, with increasing precision but no astrological aspects, to our own day.

At about the same time he printed a prospectus of his intended publications, namely, most of the mathematical works of antiquity in new or improved translations, a few from the Middle Ages, and many of his own compositions, nearly fifty titles in all. Among the ancient works were Euclid's *Elements*, the surviving works of Archimedes, the *Conics* of Apollonius, Ptolemy's *Geography*, *Almagest*, *Tetrabiblos* and *Harmonics*, and Theon's commentary on the *Almagest*, and among his own the *Epitome of the Almagest*, *Problems Pertaining to the Almagest* (lost), *Tabulae directionum*, *Tabulae primi mobilis*, *De triangulis omnimodis*, and a large commentary on Ptolemy's *Geography*. He also intended to print a map of the inhabited world and individual maps of Germany, Italy, Spain, France and Greece, all before any maps had yet been printed.

Unfortunately, little of this plan was carried out. His last printing was his dialogue on planetary theory, *Disputationes contra Cremonensis deliramenta* (*Disputations against the Cremonese Nonsense*), a dialogue on the faults of the *Theorica planetarum* attributed to Gerard of Cremona with an amusing preface in which he defends his intention to print new and improved translations and accurate texts against critics who had charged him with being an arrogant young German too far from the culture of Italy to do anything of value. In the summer of 1475 he journeyed to Rome, summoned by Sixtus IV to assist in a reform of the calendar. There he died at the age of forty in July of 1476, perhaps of plague which was epidemic in Rome that year. His death was a great loss to astronomy and the mathematical sciences. Considering his extraordinary capacity for work and the level of productivity reached by printers in the 1470s and '80s of two to three thousand page settings per year, there is reason to believe that he could have published a substantial number of the works listed in his prospectus in the next twenty or so years. Still, his *Kalendarium* and *Ephemerides* were frequently reprinted, the *Tabulae directionum* appeared in 1490 and the *Epitome* in 1496. In the first half of the sixteenth century, texts of Archimedes, Ptolemy, Theon and al-Battānī were printed from his manuscripts, as were a number of his own works, particularly under the supervision of Nuremberg's distinguished astronomer, astrologer and geographer Johann Schoener (1477–1547), who revered Regiomontanus's memory as an ornament of the mathematical sciences and of his adopted city. Regiomontanus's works were essential to the development of astronomy through the sixteenth century, and their use and authority was greater a century after his death than at any time during his life.

The planned reform of astronomy never got off the ground, but the observations had a different fate. Regiomontanus himself barely made a start, but after his departure from Nuremberg his associate Bernhard Walther (1430–1504) undertook the first systematic observations in Europe. Using a modification of Ptolemy's parallactic rulers by which the chord of the zenith distance is measured, he observed hundreds of solar zenith distances in the meridian from 1475 to 1504. Likewise, using a *rectangulum instrumentum* or *radius astronomicus*, an instrument like a cross-staff for finding the chord of arcs of up to about 30°, he measured a few hundred angular distances between planets and fixed stars until 1488. In that year he acquired an armillary with which he made

63 Regiomontanus, *Ephemerides, 1475–1506* (Nuremberg, *c.* 1474). Opening for November 1475, showing on the left-hand page the longitudes of the sun, moon and planets for each day, and on the right-hand page the aspects of the moon to the sun and planets, and of the sun and planets to each other. These aspects were used for weather prognostication, determining favourable times for blood-letting and taking medication, and other activities of everyday life. (British Library)

another few hundred direct measurements of longitude and latitude of planets until 1504, which seems to have been the busiest of all his years of observation (he died on 19 June). Three of his observations, of Mercury, were used by Copernicus, and following the publication of the entire corpus along with the observations of Peurbach and Regiomontanus by Schoener in 1544, Walther's observations were used by Tycho and Kepler, who wrote but did not publish an examination of many of them. The observations were republished by Willibrord Snell in 1618, and in that form have remained the earliest European observations of use to modern astronomy.

Copernicus

After the brilliance and energy of Regiomontanus, Nicholas Copernicus (1473–1543) may seem rather colourless. Neither in mastery nor extension of Ptolemy's methods did he equal his predecessor, but he was industrious and honest in applying what he learned from Ptolemy and Regiomontanus, and brought forth, not only a rival to the *Almagest*, but the most famous single discovery, or theory, in the entire history of astronomy. He was born to a prosperous merchant family in Torun in Royal Prussia (now part of Poland), and following the early death of his parents, was raised in the household of his maternal uncle Lucas Watzenrode, a sour and gloomy man who became Bishop of Warmia (Ermland) in 1489 and intended a similar ecclesiastical career for his nephew. He attended the University of Cracow from 1491 to 1495, where courses were offered in astronomy, astrology and mathematics, but left without taking a degree. At this time his uncle acquired for him a canonry of the Cathedral Chapter of Warmia and later other benefices which assured his living. From 1496 to 1501 he studied canon and civil law at the University of Bologna, then from 1501 to 1503 medicine at the University of Padua, and finally took a degree in canon law from the University of Ferrara in 1503. While at Bologna he worked with the professor of astronomy Dominico Maria di Novara (1454–1504) – his earliest observations date from this period – and in 1500 he gave a lecture on mathematics in Rome. When he returned to Warmia, 'this remote corner of the earth', as he later called it, he lived with his uncle in Heilsberg, serving as his personal physician, but in 1510 moved to Frauenberg, the headquarters of the Chapter, where he spent most of the rest of his life carrying out administrative duties.

It appears as though his new theory came to him about the time he left his uncle's service, if indeed he did not leave and give up further ecclesiastical advancement in order to devote more attention to astronomy. After working out the basic principles of his theory, he wrote a description that circulated in manuscript before 1514 and in 1575 came into the hands of Tycho Brahe, who had copies made with the title *Memoir (Commentariolus) by Nicholas Copernicus on the Hypotheses of the Motions of the Heavens that He Invented*. The theory of the *Commentariolus* is similar to that of *De revolutionibus*, but has deficiencies that Copernicus recognised at the time and errors that he came to recognise later. Among the former were that little evidence is given for the heliocentric theory – it is introduced in a series of postulates – and among the latter that the parameters of the hypotheses or models for the motions of the sun, moon and planets were extracted from the *Alfonsine Tables*, of doubtful reliability. Copernicus must soon have decided to write a far more comprehensive work proving, or at least giving convincing evidence for, the heliocentric theory, and deriving improved parameters from observation according to Ptolemy's methods as presented in the *Almagest* and Regiomontanus's *Epitome*. But this was an ambitious undertaking, for from about 1512 to 1529, the dates of the earliest and latest observations of the planets in *De revolutionibus*, he must have devoted his effort principally to making observations, after which there still remained the great labour of carrying out the derivations and computing tables. He had accomplished something of this sort by 1535 when he prepared an ephemeris based on his new theory and parameters which was sent off for publication. Unfortunately, the plans for publication fell through, and the ephemeris has disappeared.

By this time he was also writing *De revolutionibus orbium coelestium* (*On the Revolutions of the Heavenly Spheres*), but became reluctant to complete or publish it. Part of his hesitancy came from the realisation that he was unable to prove the heliocentric theory, but only present arguments for it based upon the internal consistency of his hypotheses, which he feared, rightly as it turned out, few would find convincing. And even after more than 20 years, the derivation of new parameters remained uncertain and he now had doubts about the reliability of the ancient observations he was of necessity using. He considered the expedient of publishing only tables, like the *Alfonsine*, in which his theory would be neither stated nor justified, but his closest friend Tiedemann Giese (1480–1550) argued that he should rather follow Ptolemy in setting out the reasons for his theory and the demonstrations of its parameters.

So matters stood until 1539 when Georg Joachim Rheticus (1514–74), a professor of mathematics at Wittenberg, came to learn of Copernicus's new theory, the general outlines of which were already known to some astronomers. Rheticus must immediately have won Copernicus's confidence, for he gave him the manuscript of *De revolutionibus* to read and gave him permission to write an account to stimulate interest in the book that he was now determined to complete. This Rheticus did splendidly in the *Narratio prima* (Danzig, 1540), the 'first account' of Copernicus's book in the form of a letter to Johann Schoener whom Rheticus had earlier visited in Nuremberg. Copernicus continued to revise his book, perhaps with assistance from Rheticus, and by May of 1542 printing in Nuremberg by Johann Petreius began under Rheticus's supervision. However, in October Rheticus left to take a new appointment at the University of Leipzig, and supervision was turned over to Andreas Osiander (1498–1552), who added an unsigned preface stating that Copernicus's new theories, like all astronomical hypotheses, were intended only for computation and were not to be taken as true or even probable. This was not Copernicus's opinion, and neither was it that of Rheticus or Giese, who, after the book appeared, petitioned the town council of Nuremberg to compel Petreius to issue a corrected edition with new front matter, which the council rejected. Copernicus's own reaction is not known, if indeed he had any, for he had suffered a stroke in December of 1542 and, after remaining comatose for several months, only received a copy of the book on the day of his death, 24 May 1543.

There are two records of Copernicus's early planetary theory. One is the *Commentariolus* and the other an earlier page of notes containing the parameters of the *Commentariolus* with preliminary steps showing how they were derived from the *Alfonsine Tables* and providing evidence for the analysis by which Copernicus arrived at the heliocentric theory. The planets display two inequalities in their motion, the first or zodiacal inequality, the planet's own non-uniform motion about the sun, represented in Ptolemy's theory by the motion of the centre of the epicycle, and the second or solar inequality, the displacement in the apparent position of the planet due to the motion of the earth, represented geocentrically by the motion of the planet on the epicycle. The preceding description applies to the superior planets; for the inferior the function of the inequalities is reversed, the motion of the centre of the epicycle representing the motion of the earth and the motion of the planet on the epicycle the heliocentric motion of the planet. Copernicus's original concern in planetary theory was the violation of uniform circular motion in the hypothesis for the first inequality, but we shall reverse the order of his work by first considering the analysis through which he derived the heliocentric theory.

The evidence comes from the page of notes containing the parameters of the *Commentariolus* which shows that he was using an alternative model for the second inequality of the planets mentioned briefly by Ptolemy at the beginning of the treatment of retrogradations in *Almagest*, 12.1 and more fully explained by Regiomontanus in

Epitome, 12.1–2. In this alternative model, the epicycle is replaced by an eccentricity that rotates such that it always lies in the direction of the mean sun. The epicyclic and eccentric models for the superior planets, Mars, Jupiter and Saturn, are superimposed in Fig. 64a in which the earth is *O*, the centre of the epicycle *C*, and the planet *P*. Complete parallelogram *OCPN* and *N* will be the centre of an eccentric of eccentricity *e* equal to the radius of the epicycle *r*. And since in the epicyclic model *r* is always parallel to the direction from the earth to the mean sun, *e* will always lie in the direction *OS* to the mean sun *S*, and *P* will move on the eccentric circle about *N*. Copernicus must have been considering this model, for in his notes the number corresponding to the radius of the epicycle is called an 'eccentricity'. What if it is assumed that the mean sun lies at the centre of the eccentric, if *S* coincides with *N* as in Fig. 64b? It can be seen that the planet now moves in a circle about the mean sun, which in turn moves about the earth, and the radius *R* of each planet's orbit can be measured in units of the fixed distance *e* from the earth to the mean sun. This too is shown in Copernicus's notes by a list of parameters under the heading 'Proportion of the celestial spheres to an eccentricity of 25 parts', the same units used to measure the radii of the spheres in the *Commentariolus*.

What we have reached so far is not the Copernican but the Tychonic theory, the planets move about the mean sun which in turn moves about a fixed earth. But Regiomontanus also describes an eccentric model for the inferior planets, Mercury and Venus, that leads instead to the Copernican theory with the earth moving about a fixed mean sun. The superimposed epicyclic and eccentric models are shown in Fig. 65a, identical to the models for the superior planet except that the centre of the epicycle *C* always lies in the direction from the earth *O* to the mean sun *S*. Again completing parallelogram *OCPN*, *N* is the centre of an eccentric of eccentricity *e* equal to the radius of the epicycle *r*, and now the radius of the eccentric *NP* will always be parallel to the direction *OS*. And indeed in Copernicus's notes the number for the radius of Mercury's epicycle is called an 'eccentricity'. One could, of course, transform the epicyclic model to a Tychonic form by assuming that *S* coincides with *C*, thereby fixing the distance from the earth to the mean sun for both inferior planets. But this is not the transformation of the eccentric model, and is not what Copernicus did. Rather, as shown in Fig. 65b, the sun moves to the fixed centre *S*, and the planet and earth move parallel through the distance *R*, and again the radius of the planet's sphere can be measured in units of the fixed distance from the earth to the mean sun.

The eccentric models for the superior and inferior planets have led respectively to the Tychonic and Copernican theories, and there is no evidence in Copernicus's notes for how he decided between these alternatives. There is, however, a compelling reason for his decision remarked on by Kepler in a draft for his defence of Tycho's originality against a charge by N.R. Ursus that Tycho had been anticipated by, among others, Copernicus. According to Kepler, 'Copernicus could not see [the Tychonic theory]

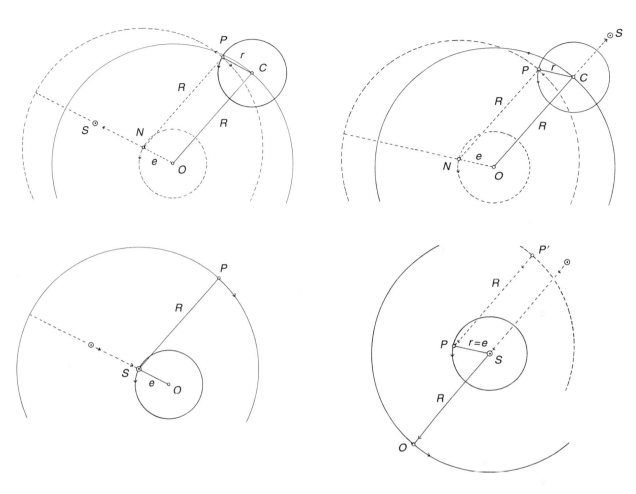

64 (*above*) Copernicus: a) (*top*) the equivalent
epicyclic and eccentric models for the second
inequality of the superior planets, and b) the
conversion of the eccentric model to the Tychonic
theory by moving the sun to the centre of the
eccentric.

65 (*above*) Copernicus: a) (*top*) the equivalent
epicyclic and eccentric models for the second
inequality of the inferior planets, and b) the
conversion of the eccentric model to the
Copernican theory by moving the sun, planet and
earth in parallel.

because he believed in the reality of the spheres'. What Kepler was referring to is a
peculiarity of the Tychonic theory that affects only Mars and was also a problem for
Tycho for some time, as we shall see. The Tychonic system is shown in Tycho's own
illustration in Fig. 71. Note in particular the circular paths of Mars about the mean sun
and of the mean sun about the earth. Since the radius of Mars's sphere is about $\frac{3}{2}$ the
radius of the sphere of the mean sun, the paths must cross, meaning physically that the
two spheres must intersect. This would not be a problem if there were no solid spheres,
as Tycho later decided, but since, as both Tycho and Kepler remark, Copernicus still

adhered to the solidity of the spheres, in no way could he admit such an intersection and so was forced to conclude that only the motion of the earth around the sun was physically possible.

The preceding analysis shows *how* Copernicus may have been led to the heliocentric theory, but not *why* he believed it to be true, the reasons for which he gives in the sections of the *Commentariolus* on the second inequality and in *De revolutionibus*, 1.10. Above all is the creation of a unified planetary system in which each body has its place and nothing can be changed. The heliocentric theory uniquely determines the order and relative distances of the planets, which are arbitrary or require additional assumptions in geocentric theory, in which each planet has its own eccentric and epicycle and there is no common measure of the whole system as the sphere of the earth provides in the heliocentric. The apparent retrograde motion of the planets is explained adequately in geocentric theory, in which in the lower part of the epicycle, near opposition for a superior planet and inferior conjunction for an inferior planet, the backward motion of the planet exceeds the forward motion of the centre of the epicycle. Copernicus, however, believed the heliocentric explanation superior, namely, that a superior planet appears to move retrograde when the earth passes it and an inferior planet when it passes the earth, for this also shows just why the retrogradation of the superior planet occurs only near opposition and of an inferior planet only near inferior conjunction. In addition, other unexplained properties of geocentric theory were seen as obvious consequences of the conversion from the heliocentric. It was evident why Mars and Venus had larger epicycles, longer retrograde arcs, and longer periods of invisibility than Jupiter and Saturn, simply because the speeds and distances of Mars and Venus were much closer to those of the earth than were those of the more remote planets. Further, the most obvious distinction of superior and inferior planets in the geocentric theory, that the radii of the epicycles of the superior planets are always parallel to the direction from the earth to the mean sun while the centres of the epicycles of the inferior planets lie in the direction of the mean sun, was seen as a direct consequence of the conversion from the heliocentric to the geocentric theory.

These are all good arguments for the heliocentric theory – and they were Copernicus's own reasons for adopting it – but on the other side were two serious problems, namely, the effect of the motion of the earth on the most distant part of the universe, the fixed stars, and on the closest, the earth itself. If the sphere of the fixed stars were at a distance of about 20,000 terrestrial radii, the value from Ptolemy's *Planetary Hypotheses* that had become canonical, then the earth moving in a circle about the sun at a distance of some 1,200 terrestrial radii should produce unequal divisions of the celestial sphere by the horizon and obvious annual displacements in the positions of stars of nearly 7°, but nothing of the kind was observed. To address this problem, Copernicus supposed that the sphere of the fixed stars was at a distance so great that the annual motion of the earth would produce no observable effects. But this in turn raised further problems remarked on by Tycho in his *Astronomical Letters*. First, if the stars were so distant that the annual displacement were 1', there would be a vast space of nearly eight million terrestrial radii between the outermost planet and the stars, filled with nothing and serving no purpose, which Tycho called 'absurd'. Second, if, as was believed, the apparent diameters of bright stars were 1' or 2', as great or greater than their annual displacements, then the stars must be as large as or larger than the earth's orbit around the sun, which also seemed impossible. This objection was only answered when Galileo found with the telescope that the apparent diameters of stars were much smaller than previously imagined, so they could be removed to great distances without being any larger than the sun.

The second problem concerned the earth. How could the earth, the centre toward which heavy things move downward in straight lines, either be carried in an annual motion about the sun or, more remarkably still, possess a rapid daily rotation about its own centre, a necessary consequence of its motion about the sun? For the physical problem of the rotation of the earth, which should produce notable effects on the motion of projectiles and falling bodies, Copernicus proposed in *De revolutionibus* what he considered to be minimal alterations of Aristotelian theory of natural motion compatible with the motion of the earth. The natural motion of a sphere, he assumed, is to rotate uniformly in place by virtue of its form, that is, the spherical form is the cause of its own rotation, no force is required. This natural motion holds both for the celestial spheres that carry the earth and planets about the sun, thereby accounting for their motions and fixed orbits, and also for the spherical body of the earth, comprising earth, water and air, for which a daily rotation, just as being at rest, will produce no detectable effects. *Gravitas* (heaviness), the descent of heavy bodies toward the centre of the moving earth, is explained as a 'natural striving placed in the parts' to come together to form a globe, as is also shown by the spherical form of the sun and moon.

It must be emphasised that Copernicus proposed what he believed to be small modifications of Aristotelian physics, and we must be careful not to attribute to him its later repudiation, based in part on the motion of the earth, by Galileo and Descartes in the following century. Contrary to what is usually said, he himself did not in the slightest eliminate the distinction between the celestial region of unchanging uniform circular motion and the terrestrial region of rectilinear motion and change. The heavenly bodies, including the earth, are still carried in circles by uniformly rotating spheres, the terrestrial elements still have their natural rectilinear motions, which are simply added to the natural rotation of the earth, and the principal modification is granting to any natural spherical body, including the earth, a natural rotation.

The heliocentric theory is Copernicus's most important, and only lasting, contribution to astronomy, but it was not his motivation for taking up planetary theory in the first place, which was really more traditional. He explains in the *Commentariolus* that his original concern was the violation of uniform circular motion in Ptolemy's hypothesis for the first inequality of the planets, which is physically or mechanically impossible if the motions of the heavens are brought about by the rotation of spheres. Thus, in Fig. 61a from Peurbach's *Theoricae novae* showing a cross section of a spherical *theorica* or hypothesis, the sphere carrying the epicycle, which is eccentric to the *c[entrum] mundi*, the centre of the world, has its centre at the point marked *c[entrum] deferentis*, the centre of the sphere 'carrying' the epicycle, but rotates with uniform angular motion about the *c[entrum] aequantis*, the centre of the 'equant' circle, the circle on which the uniform motion takes place. The problem becomes clearer, or worse, in Fig. 61b, showing the axes and poles of the spheres. The axis of the sphere carrying the epicycle is the line passing through the *centrum deferentis* with its end points marked *polus deferentis*. How can the sphere rotate on this axis passing through its centre and yet have a uniform angular motion around the off-centre *centrum aequantis*?

This is the problem with which Copernicus begins the *Commentariolus*, and he was hardly the first to be concerned with it for he was anticipated by a number of Arab astronomers. In the eleventh century Ibn al-Haytham (see p. 148) raised it as one of many physical problems in Ptolemy's astronomy in his *Doubts concerning Ptolemy*, and it was later taken up by astronomers associated with the observatory of Maragha founded in 1259 in North-western Persia (see pp. 150 and 164). The first generation of the 'Maragha school', as it has come to be called, included the philosopher and mathematician Naṣīr al-Dīn al-Ṭūsī (1201–74) and Mū'ayyad al-Dīn al-'Urḍī (d. 1266), who supervised the construction of the observatory. They were followed by al-Ṭūsī's student Quṭb al-Dīn al-Shīrāzī (1236–1311; Fig. 49), and in the next century in Damascus by Ibn al-Shāṭir (1304–75). Their method was to decompose Ptolemy's single motion with respect to the centres of the eccentric and equant into uniform circular motions produced by spheres rotating uniformly about axes passing through their centres. The transmission of their inventions from Arabic in the East to Latin in the West is obscure. Yet Copernicus's lunar and planetary theory in longitude in the *Commentariolus*, right down to the additional complications for Mercury, is that of Ibn al-Shāṭir in nearly every detail, except for the heliocentric arrangement and the extraction of parameters from the *Alfonsine Tables*, and it is hard to believe in light of so many and such complex identities that Copernicus was entirely without knowledge of his predecessors (see p. 150).

The principles underlying Copernicus's modification of Ptolemy's hypothesis for the first inequality of the superior planets are shown in a heliocentric form in Fig. 66. The centre of the earth's sphere, the mean sun, is S, the centre of the eccentric M, the centre of the equant E, and the two eccentricities e are equal. According to Ptolemy, the planet P moves on a circle of radius R about M such that its angular motion k is uniform about E. Copernicus lets the centre C_1 of a large epicycle of radius r_1 move in a circle of radius R about S through k while the centre C_2 of a smaller epicycle of radius r_2 moves in the opposite direction through k so that r_1 is always parallel to the apsidal line SA. The planet P' then moves in the same direction as C_1 through $2k$, where r_1

66 Copernicus, *Commentariolus* (*c.* 1510–12). Ptolemy's and Copernicus's hypotheses for the first inequality of the superior planets superimposed. The planet is *P* in Ptolemy's hypothesis and *P'* in Copernicus's hypothesis in the form using two epicycles; both *P* and *P'* move uniformly about the equant point *E* and very nearly coincide.

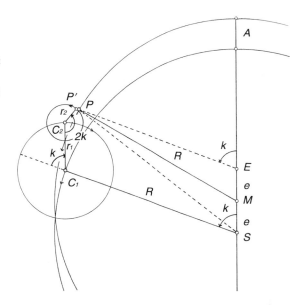

coincides with r_2 when C_1 is in the apsidal line. Provided that $r_1 = \frac{3}{2}e$ and $r_2 = \frac{1}{2}e$, *P'* will always lie on the line *EP*, thus moving uniformly about *E*, and the apparent directions *SP* and *SP'* will nearly coincide, the greatest difference, for Mars, amounting to about 3'. The path of *P'* has, in Kepler's words, an 'exorbitation' from the circle, that is, it lies just outside of and nearly coincides with the circle described by *P*.

It cannot be too strongly emphasised that Copernicus's hypothesis 1) preserves uniform motion about the equant, but as a result of uniform circular motions that can be produced by rotating spheres, and 2) is observationally indistinguishable from Ptolemy's hypothesis. The common belief that Copernicus did away with the equant, or wished to do away with the equant, is entirely false, for it was as fundamental to his planetary theory as to Ptolemy's. And there was, even at the time of Tycho and Kepler, no observational distinction between the hypotheses of Ptolemy and Copernicus; the motivation for Copernicus's hypothesis was purely physical. Kepler remarks in the *Astronomia nova* that Copernicus's hypothesis is physically justified only if solid spheres are granted; once Tycho did away with solid spheres the hypothesis lost its purpose, and Kepler began his own investigations with Ptolemy's equant motion pure and simple.

In *De revolutionibus* the hypothesis is changed to the form in Fig. 67, which shows the complete hypothesis for both inequalities. The larger epicycle is replaced by an equal eccentricity $e_1 = r_1 = \frac{3}{2}e$, which has the same effect since r_1 is always parallel to e_1. The second inequality is accounted for by the motion of the earth *O* in a circle of radius *s* about the mean sun through the mean anomaly *a*, uniformly with respect to a line *FG* which is drawn parallel to *EP*. The true position of the planet as seen from the earth is found through two corrections, c_1, the equation of centre, the difference in direction

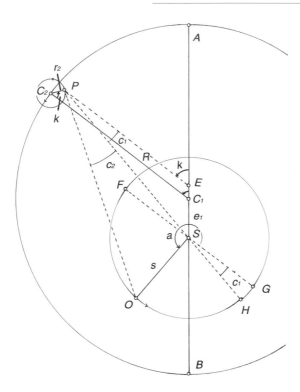

67 Copernicus, *De revolutionibus* (1543). The complete hypothesis for both inequalities of a superior planet in the form using a fixed eccentricity and a single epicycle. The planet *P* moves uniformly about the equant point *E* and non-uniformly about the mean sun *S*, and the earth *O* moves uniformly in a circle about *S*, which is the centre of its sphere.

between the uniform motion of the planet about the equant – again, note that the equant is essential – and the non-uniform motion about the mean sun, and c_2, the equation of the anomaly, or as Copernicus calls it, the parallax of the sphere (*parallaxis orbis*), which reduces the heliocentric direction of the planet *SP* to the geocentric direction *OP*. The corrections are found from trigonometric formulae nearly identical to those used by Ptolemy, and Copernicus's tabulations of the corrections are also nearly identical to Ptolemy's.

For practical purposes, Copernicus's planetary theory therefore amounts to a heliocentric transformation of Ptolemy's with uniform motion about the equant preserved. This was immediately recognised by those proficient enough to read Copernicus's book with understanding, and it had the good fortune of making his work useful even if the heliocentric theory was not accepted. Erasmus Reinhold (1511–53), Rheticus's fellow professor of mathematics at Wittenberg, who did not believe the heliocentric theory and may himself have invented something like the Tychonic theory, wrote a large commentary on *De revolutionibus* in which he used both Ptolemy's and Copernicus's hypotheses for the sun, moon, and first inequality of the planets, and derived all their parameters over again from Copernicus's observations with meticulous (and superfluous) precision. These parameters were then the basis of his *Prutenic Tables* (1551), which became the basis for derivative tables and most ephemerides until about the middle of the seventeenth century when Kepler's *Rudolphine Tables* at last came into common use. In this way, Copernicus's work, along with Regiomontanus's tables for spherical astronomy, formed the foundation of practical astronomy and astrology for the better part of a century. It is also worth noting that the Gregorian calendar reform was based upon the mean length of the tropical year in *De revolutionibus* and the *Prutenic Tables*, which is essentially the same as the tropical year in the *Alfonsine Tables*, and on the mean synodic month of the *Prutenic Tables*.

We remarked earlier that Regiomontanus saw the reform of astronomy as primarily a correct application of Ptolemy's methods. Copernicus, even with the heliocentric theory and the attempt to represent planetary motion strictly through the uniform rotation of spheres, was of the same mind. In a sense, it was Copernicus who carried out the plan envisioned by Regiomontanus for the improvement of astronomy through an application of Ptolemy's methods, an application based upon the *Almagest* and the *Epitome*. The most famous evaluation of Copernicus is Kepler's remark: 'Copernicus, ignorant of his own riches, took it upon himself for the most part to represent Ptolemy, not nature, to which he had nevertheless come the closest of all'. This is entirely just. Copernicus built upon the foundation of Ptolemy's descriptions of the apparent motions of the heavens, Ptolemy's observations, and Ptolemy's methods of deriving parameters from observation. Thus, he did, as Kepler said, represent Ptolemy rather than nature. But there was little else he could do, for to represent nature it was first necessary to discover what nature showed, and that required a new approach based upon observations more extensive, diverse and accurate than any yet made, and upon an analysis of planetary motion that abandoned all previous methods to seek, if necessary by trial and error, entirely new ones. Work of this kind did not begin until nearly half a century after Copernicus, and then it took the ingenuity and labour of Tycho, Kepler and nearly half a century to accomplish.

Tycho Brahe

Of all the advantages that may be conferred by fortune, the advantage of birth is not the least. Tygge Brahe (1546–1601), Latinised as Tycho, was born into an ancient family of the Danish nobility whose members regularly served in the Council of the Realm, and unlike other artists and philosophers, he never stood hat in hand before the great. Where others could only beseech, he could demand, which he did, at times imperiously, through his turbulent life. He was raised by his uncle, who intended for him a career in the Danish administration, to which end, beginning in 1559, he attended universities in Copenhagen, Leipzig, Wittenberg and Rostock, studying the liberal arts and law in preparation for a career he never wanted. For from an early age his passion was astronomy, which he pursued in secret, educating himself from books and taking observations with improvised instruments. He found that neither the *Alfonsine* nor the *Prutenic Tables* were sufficiently accurate, something shown in particular by a great conjunction of Jupiter and Saturn in August 1563 in which the former were in error by a full month and the latter by a few days, following which he began to record his observations and seek improved instruments. In 1569 he settled for a year in Augsburg where he made a pair of compasses 1.5 m in radius with a scale for measuring angles up to 30° and a gigantic wooden quadrant 4.5 m in radius which could, with difficulty, be rotated in azimuth for measuring altitudes in any vertical circle. He also read widely in astrology and carried out alchemical experiments, interests he maintained for the rest of his life.

He returned to Denmark in 1570 and not long after came into an inheritance that made him financially independent. The following year he moved to Herrevad Abbey where on 11 November 1572 occurred an event that will be forever celebrated in the history of astronomy. On returning from his alchemical laboratory in the evening, he noticed an unfamiliar star in Cassiopeia, brighter than any he had ever seen, even brighter than Venus. This, as we now know, was one of the few supernovae ever seen in our galaxy, and the first to be studied with care (see pp. 339–41 and Fig. 116). Tycho observed the star from night to night – by May 1573 it was of the second magnitude and remained visible into 1574 – measuring its distance from other stars to determine whether it showed any motion or parallax. He found that it showed neither, and concluded that it was located above the moon, above the planets, in the region of the fixed stars. It was, he later wrote, a special creation of God, a star of enormous size and brightness, preceding the anticipated change of triplicity of the next great conjunction of Jupiter and Saturn by ten years, and thus portending momentous changes of empires, kingdoms and religions, especially in Northern Europe (all of which certainly came to pass in the following century). Kepler later remarked that if nothing else the new star foretold the appearance of a great astronomer. Indeed, it was above all the new star that made Tycho resolve to devote his life to astronomy, and in 1573 he wrote a small book on the star, the first of his innovative publications.

In 1575 Tycho again travelled south as he was considering leaving Denmark to settle in some place more conducive to his astronomical research and closer to those with similar interests. He stopped in Kassel to visit Wilhelm IV (1532–92), Landgrave of Hesse, who for some years had been conducting observations himself with instruments of very high quality, a visit that marked the beginning of a correspondence that Tycho later published in his *Astronomical Letters* (1596). From Kassel he went to Frankfurt, Basel, Venice, and then returned by way of Augsburg, Regensburg, where he attended the coronation of his future patron Rudolph II as King of the Romans (i.e. Germans, he became Emperor in 1576), Nuremberg and Wittenberg. As soon as he reached Denmark, he began to plan his move to Basel, on account of the excellence of its university and learned men, not to mention its healthy climate, when, to make a long story short, in February 1576 King Frederick II offered him the island of Hven, in the Danish Sound (since 1658 part of Sweden), and generous financial support to build facilities for and maintain his research. Tycho hesitated only a few days before accepting the offer, and by spring he was planning an institute for scientific research, some forty years before Bacon's description of Salomon's house in the *New Atlantis*.

Tycho called the building that would house his family, assistants, instruments, laboratory and library Uraniborg, and it was by any standard extraordinary, planned according to Vitruvian and Palladian principles of proportion (*symmetria*). Its construction took 5 years, from 1576 to 1581. Most of what is known of it comes from the description and woodcuts in Tycho's *Astronomiae instauratae mechanica* (1598), as it was torn down for bricks in 1623 and entirely demolished before the middle of the century. In the plan of Fig. 68, *D* is the heated winter dining room, *E, F, G* are bedrooms, *H* the kitchen, *K* (in its centre) a deep well with a pump and pipes to distribute water to different rooms, and *T* the library with a large brass celestial globe at *W*. In the elevation *N, O, R, S* are the observatories containing both mounted and portable instruments, and each panel of the roofs could be removed individually to give access to any part of the sky. Finally, underground at the left side is the alchemical laboratory containing no less than sixteen furnaces, and at the right the well running up to the kitchen. The building was surrounded by a garden and a massive wall with outbuildings for a print shop and servants quarters. Tycho later needed more space for instruments, and built between 1584 and 1587 an observatory outside the walls called Stjernborg, which was erected with the roofs at ground level and the instruments solidly mounted in underground pits. The joys of observing on Hven during the Baltic winters are not to be imagined.

Tycho drove his assistants, perhaps as many as sixty over a period of twenty years, and the unfortunate peasants of Hven hard, but he was also mercilessly critical of his own work, and where precision is the goal, only merciless criticism will do. This can be seen above all in his innovations in instrumentation. None of Tycho's instruments survive, but there are detailed descriptions of their construction and use in the *Mechanica*, in which he does not hesitate to evaluate their problems and inaccuracies. His instruments are distinguished by their large size, precise manufacture, and ingenious use, but as one studies the descriptions, one can only despair at the difficulty of using such contrivances and wonder at the astonishing results, as shown by modern analysis, that Tycho achieved. The instruments fall into three broad classes. First are quadrants for measuring altitude that may either be fixed in the meridian or free to rotate in azimuth. Col. Pl. XV shows the most famous of Tycho's instruments, the great mural quadrant some 2 m in radius, graduated by diagonal transversals to sixths of a minute. The plate also shows other instruments, a clock, assistants at work in Uraniborg, and Tycho with his dog. Meridian altitudes give the declination co-ordinate directly, but a second co-ordinate is harder to find. In principle, right ascension can be found by timing meridian transits, but Tycho never found a reliable clock, for an error of only 1 minute of time produces an error of 15' of right ascension, where 24 hours correspond to 360°.

Hence it was necessary to turn to the second class of instruments, armillaries, graduated rings for the direct measurement of either ecliptic or equatorial co-ordinates. Tycho first made a zodiacal armillary for measuring longitude and latitude with a

ORTHOGRAPHIA
PRÆCIPVÆ DOMVS ARCIS VRANIBVRGI
IN INSVLA PORTHMI DANICI VENVSIA *vulgo* HVENNA, ASTRONOMIÆ INSTAV-
RANDÆ GRATIA CIRCA ANNVM 1580 à TYCHONE BRAHE
EXÆDIFICATÆ.

ICHNOGRAPHIA ET EIVS EXPLICATIO

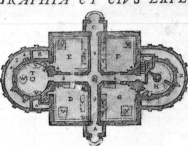

A Ianua Orientalis. C. Oc-
gulos rectos concurrentes, qui
Cænaculum hybernum sive hypocau-
angulo post fornacem, parvum quod-
gyricum esset, in quo tamen quinq.
ptiis ad manus isthic operi Pyrono-
jus illud descendendum foret. B
qui aquas hinc inde cùm lubuit, in
culum illud hybernum. E. F. G.
pro ascensu in superiorem contigna-
cementitius 40. ulnas profundus,
quas per siphones hinc inde occultè
Cameras tam superiores quam infe-
descensu in Laboratorium Chymi-
bus magnus Orichalcicus num.   exhibitus. V. Quatuor Mensæ pro Studiosis, 4.
in quatuor angulis conclavium. Y. Lecti in iisdem conclavibus, hinc inde dispositi.
discernet. Intelligenda autem sunt hæc omnia in eâ quantitate, veluti fundamento majoru domus supra depictæ quadrare poterunt: Li-
cet hic coarctationis loci gratiâ in duplo quasi minori formâ exhibeantur.

cidentalis. Q. Transitus 4. ad an-
tamen posteà in tres redacti sunt, ut
flum D. ampliaretur, atq, in ejus
dam & secretum Laboratorium spa-
distinctim erant furni, qui prom-
mico inserviebant, nè semper in ma.
Fons aquarium volubilem rotans,
sublime eiaculabatur. D. Cæna-
Cameræ pro hospitibus. L. Gradus
tionem. H. Coquina. K. Puteus
artificio hydraulico serviens & a.
per murum transeuntes in singulas
riores distribuens. P. Gradus pro
cum. T. Bibliotheca. VV. Glo-
Camini tàm è laboratorio inferiori ascendentes, quàm
Cætera acutus inspector propriâ intentione facilè

diameter of 1.2 m, but the weight of its rings caused them to move out of their correct plane. He greatly preferred equatorial armillaries of his own design for measuring right ascension and declination, which could be larger since they had fewer rings, and he made several including a gigantic equatorially mounted declination ring, shown in Fig. 69, fully 2.7 m in diameter and graduated to fourths of a minute. The third class is sextants and similar devices for measuring angular separations as large as 60° in any direction, as between two stars. Tycho had used such instruments since he was in Augsburg, and he had several made including the ingenious bipartite arc shown in Fig. 70, which reached a precision of better than 1'. Precision came not only from size, but from refinements, as the transversal scale, splitting minutes into smaller units, and an ingenious sight with adjustable parallel slits, used in pairs to determine when an object is exactly centred, both of which are shown in Col. Pl. XV. The increase in precision Tycho achieved was extraordinary. Various statistical evaluations of his observations have been made with different results, but it is safe to say that fundamental stars and solar

69 (*left*) Tycho Brahe, *Astronomiae instauratae mechanica* (1598), f.C3v. Great equatorial armillary, 2.7 m in diameter. The complete circular ring is for measuring declination, and the half-circular ring below it is for measuring right ascension. (British Library)

70 (*right*) Tycho Brahe, *Astronomiae instauratae mechanica* (1598), f.C4v. Bipartite arc 1.5 m in length, for measuring separations between stars of up to 30° in any direction. Each arc, *FC* and *FD*, is 15°. Tycho believed it accurate to less than 1'. (British Library)

altitudes were measured to considerably less than 1' and Kepler considered reduced observations of planets to be safely within 2'.

Just as important as the accuracy of Tycho's observations is their quantity, four large volumes in J.L.E. Dreyer's edition of Tycho's works, many thousands of observations of stars, sun, moon, and planets, and Tycho not only made the observations, but used them. He established the positions of fundamental zodiacal stars by repeatedly measuring their distances from α Arietis, whose position had previously been established with great care, and then computing their differences of right ascension, repeating the procedure until the differences around the zodiac added up to 360°. Measured distances of other stars from the fundamental stars were then used to build up a catalogue, first of 777 stars published in the *Progymnasmata*, and later of 1,000, distributed in manuscript in 1598 and finally published by Kepler in the *Rudolphine Tables*. He also found that the common motion of the stars, or the precession of the equinoxes, was 51" per year and uniform. From meridian altitudes of the sun, and many other observations, he found an improved value of the obliquity of the ecliptic, a much improved solar theory, and made a table of atmospheric refraction of the sun. Unfortunately there is a consistent problem in all of these because he believed the sun had a horizontal parallax of 3', the canonical value since Ptolemy, and his corrections for this false parallax introduced systematic errors. Likewise, he made countless observations of the moon – altitudes, declinations and right ascensions, distances from stars and, of course, eclipses – which led to his improved lunar theory and his discovery of two new inequalities each in longitude and latitude. The planets were observed extensively in all parts of their orbits and every configuration with the sun, not merely the oppositions used by Ptolemy and Copernicus for deriving parameters. In the account of his work in the *Mechanica*, Tycho claims to have derived new elements of the motions of the planets in both longitude and latitude from 25 years of observations, so that all that remained was to compute tables which could then be the basis of new ephemerides. Here he was surely too optimistic about his own progress, although these observations later formed the foundation of Kepler's investigation of Mars in the *Astronomia nova* and derivation of the elements of all the planets for the *Rudolphine Tables*.

Tycho's most famous discovery, or invention, is connected with a great comet that he first saw on 13 November 1577 while he was catching fish for dinner in one of his ponds. It was a spectacular sight, as brilliant as Venus, with a blue–white head 7' or 8' in diameter and a reddish tail 22° in length. Tycho's immediate object was to determine its motion and distance, by no means an easy task since, as we know, its apparent motion is really a result of the motion of the earth and the comet. He observed the comet until January and found no perceptible parallax, indicating that it was well above the moon, and concluded that it was moving *non-uniformly and retrograde about the sun* at an inclination of 29; 15° to the ecliptic just outside the path, but within the sphere, of Venus, which he already considered to move about the sun along with Mercury. Since his analysis provided the best evidence yet found that comets were in the heavens rather than phenomena of the upper air, the common opinion following Aristotle, he decided to publish a book on the subject, but it took more than 10 years until *De mundi aetherei recentioribus phaenomenis* (*On More Recent Phenomena of the Celestial World*) appeared in 1588. During that long period the consequences drawn from the comet became even more remarkable, although the course of events is not entirely clear.

In the summer of 1580 a skilled mathematician named Paul Wittich (*c.* 1545–87), whom Tycho may earlier have met at Wittenberg, came to Hven to work as an assistant. He remained only four months, and left during October when Tycho required his help in observing another comet that had just appeared. Later Tycho came to suspect him of revealing to the Landgrave as his own inventions technical details of instruments, including transversals and parallel sights, that Tycho would have preferred to keep secret or at least receive credit for inventing. But initially Wittich brought information of his own that was of great interest to Tycho. The first was what are called the prosthapharetic formulae for reducing multiplication of trigonometric functions to addition and subtraction, which save vast amounts of time and reduce errors in calculation. The second, although less certain, may have been several variants of Copernicus's planetary models, most in a geocentric form although not the Tychonic theory, that have recently been found in a copy of *De revolutionibus* annotated by Wittich. How much Tycho learned of these at the time is not clear, and at least for the moment he did nothing with them.

Then, in 1582 Tycho had a clever idea for testing the geocentric and heliocentric hypotheses. In Ptolemy's theory, Mars is always located beyond the sun, and since the sun has a parallax of about 3', the parallax of Mars must always be less. On the other hand, in the Copernican theory, Mars at opposition is closer than the sun, although owing to its large eccentricity there is a considerable variation in its distance. At the end of 1582 Mars was due to reach opposition and, if closer than the sun, would show a parallax of about $4\frac{1}{2}'$, which Tycho attempted to find by taking measurements of its distance from fixed stars. Initially he believed he found no parallax at all, but he later reconsidered and concluded that Mars did display a parallax showing that it was closer than the sun. For some time Tycho may almost have been an unwilling Copernican, for he could not believe the motion of the earth physically possible (and he also took seriously scriptural objections to a moving earth). By sometime in 1583 or '84, he saw his way to the theory that bears his name – even if he was not the first to work it out – but at first it appeared to have a problem that made it just as impossible as the movement of the earth.

The Tychonic System is shown in Fig. 71 in its most famous illustration from *De mundi aetherei recentioribus phaenomenis*: the five planets encircle the sun while the sun and moon move about the earth, which Tycho considered to be absolutely at rest, so each day the heavens, including the fixed stars just beyond Saturn at a distance of about 14,000 terrestrial radii, rotate about the central earth (see also Col Pl. XVI). In determining the

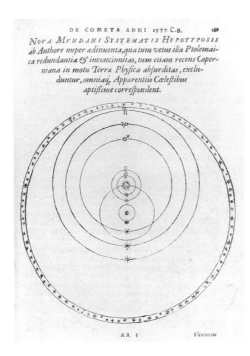

DE COMETA ANNI 1577 C.8.

Nova Mvndani Systematis Hypotyposis ab Authore nuper adinuenta, qua tum vetus illa Ptolemaica redundantia & inconcinnitas, tum etiam recens Coperniana in motu Terra Physica abfurditas, excluduntur, omniaq, Apparentiis Cœleftibus aptiffime correfpondent.

AA 3   *Flexiosen*

71 Tycho Brahe, *De mundi aetherei recentioribus phaenomenis* (1588), p. 189. The Tychonic System, in which the moon and sun move about a fixed earth, which does not rotate, and the five planets move about the sun. The fixed stars are just beyond the circle of Saturn (♄). Note the intersection of the circles of Mars (♂) and the sun. Tycho decided such an intersection was possible after he determined that there were no solid spheres. (British Library)

order and distances of the planets and accounting for all the unexplained features of geocentric models, it is fully the equal of the Copernican theory, from which it is observationally indistinguishable. And of course it entirely eliminates the physical problems of the earth's motion. But it does have its peculiarities. One, which may appear subtle but is worth remarking on, is that measured from the direction from the earth to the sun, the heliocentric motions of inferior and superior planets take place in opposite directions, Venus and Mercury in the positive sense, Saturn, Jupiter and Mars in the negative. The other, which is more striking and far better known, is the intersection of the circles of Mars and the sun. We had noted earlier that this intersection could well have motivated Copernicus's rejection of the Tychonic theory, and it was a problem for Tycho too, for as long as he believed in the solidity of the celestial spheres he could not bring himself to believe his theory physically possible owing to such an interpenetration of the spheres of Mars and the sun.

The solution came through a further analysis of the distance and motion of the comet of 1577. Already in 1579 Tycho was aware of Michael Maestlin's belief that the comet had passed the earth at a distance of 155 terrestrial radii in the direction of the sphere of Venus and continued moving beyond the sun to a distance of 1,495 terrestrial radii. By the beginning of 1587, perhaps aided by a communication from Christopher Rothmann pointing out that the motion of comets is evidence against the solidity of the planetary spheres, Tycho again evaluated his observations and decided that the comet had moved from 173 to 1,733 terrestrial radii, evidently passing directly through the 'spheres' of Venus and the sun. The conclusion was inescapable: there are no 'solid' celestial spheres and the 'ridiculous' penetration of the spheres of Mars and the sun is really not a problem after all. This in itself was one of Tycho's most important discoveries, for the elimination of solid spheres totally transformed the physical basis of astronomy, and in so doing both raised problems and created new possibilities essential to Kepler's work, for something now had to take their place to account for the motion of heavenly bodies. Tycho published his theory of the planetary system in the book on the comet. He was very proud of it, and spent a good part of the rest of his life defending the originality of his discovery.

The Tychonic theory was, like the Copernican, a consistent explanation of the second inequality of the planets, but with regard to the first inequality Tycho followed Copernicus in preserving Ptolemy's equant motion by reducing it to uniform circular motions, although in the double-epicycle form since the motion of the centre of the system, the mean sun, about the earth made equivalent eccentricities holding fixed directions from the movable centre seem less plausible (although it really makes no difference). And we have noted that Tycho's claim in the *Mechanica* to have derived new elements for the planets seems premature. With the far more complex theory of the moon, however, he had great success, improving lunar theory in every way and discovering two new inequalities in both longitude and latitude. The lunar theory was Tycho's latest contribution to astronomy; most of the work was done between 1594 and 1600. He was aided by his most loyal and proficient assistant Christian Sørensen Longomontanus (1562–1647), who was at Hven from 1589 to 1597 and rejoined Tycho in Prague for seven months in 1600, when he worked out the final lunar theory, which in justice should be credited to both of them.

The lunar theory was published posthumously in the *Astronomiae instauratae progymnasmata* (*Introductory Studies of the Restored Astronomy*) of 1602, the composition and printing of which had been in progress for years. In addition to lunar theory, it contains accounts of solar theory, the fixed stars and the new star of 1572, and was intended as one of a series of volumes, including the book on the comet, that would treat definitively all aspects of astronomy. But it was not to be. Owing to various demoralising disputes concerning Hven and other property, some of his own making, and the apparent loss of favour with King Frederick's successor, Christian IV, Tycho left Hven and Denmark forever in 1597. After two years in Germany, in 1599 he entered the service of Emperor Rudolph II, who granted him an estate in Benatky, a few hours from Prague, that had the possibility of becoming a new Uraniborg, along with a generous salary and a promise of continued support for his work, although these were not exactly dependable. His instruments, at least the portable ones, were being brought to his new residence. Longomontanus returned, and soon a promising new assistant from Graz, Johannes Kepler, arrived to take over work on Mars so that Longomontanus could devote himself to the completion of the lunar theory. Tycho was not old, and was driving everyone as hard as ever, but the days of designing new instruments and making observations were over, and what remained was the reduction of the observations and their application to the improvement of the theory of the planets. When Tycho died unexpectedly after a brief illness on 24 October 1601, this task had barely begun. But by good fortune it was in the hands of the one person whose tenacity and originality exceed Tycho's own. For it was Kepler, basing his work upon the foundation of Tycho's observations, who transformed astronomy more in a decade than in the fifteen centuries since Ptolemy.

Kepler

Johannes Kepler (1571–1630) was the first astronomer to come of age a Copernican. He had the good fortune to attend the University of Tübingen where from Michael Maestlin (1550–1631), himself a Copernican, although not in his publications, he learned of the heliocentric theory. Kepler never doubted it for a moment, and in student disputations regularly defended the motion of the earth. However, he had no intention of pursuing astronomy, for he was studying for the ministry, devoting the greater part of his attention to theology, an interest that never left him and on which his views were always idiosyncratic. But before he completed his theological studies, in early 1594 he was offered the position of mathematics teacher at the Protestant Stiftsschule in Graz, and was convinced by his professors to take it. In addition to teaching the mathematical sciences and various humanistic subjects, he was also the district mathematician with the responsibility of preparing an annual almanac predicting the weather, harvest and notable events of the year. In his first year, 1595, he was quite successful, predicting an unusually cold winter and an invasion of the Turks into Austria, both of which came to pass.

At about this time he began to ponder the question that was to occupy him for the rest of his life, namely, what plan did God have in mind when he created the universe? He asked this question in a very particular form. The Copernican theory gives the order, distances, and periods of the planets with certainty. Why did God set these out exactly as they are? He tried a variety of relations between period or speed and distance, looking for some kind of pattern, but to no avail. Then on 19 July 1595, a day he was never to forgot, he was explaining to his class the great conjunctions of Jupiter and Saturn, which occur just over 240° apart so that when they are marked on a circle and connected in order by straight lines a series of nearly equilateral inscribed triangles is formed, their sides enclosing an area in which another circle may be inscribed. And the radii of the outer and inner circles are in the ratio of $2:1$, quite close to the ratio of the distances of Saturn and Jupiter of about $9\frac{1}{2}:5$. But even this promising beginning failed when he found that other inscribed polygons had no relation to the distances of other planets, and the number of polygons was without limit while there were only six planets. Then it struck him that there were five and only five regular polyhedra, and that the inscription of these in a series of spheres gave something very close to the distances of the planets from the sun and also showed why there were six planets, neither more nor less. Here indeed was God's plan.

Kepler wrote in great excitement about his discovery to Maestlin, asking for his advice and for more accurate distances than those he had extracted from Copernicus. Maestlin complied, and Kepler began writing the first of his great works, the *Mysterium cosmographicum*, or, to give its complete title, which pretty well sums up its contents, 'Introduction to the Dissertations on the Universe, containing the Mystery of the Universe concerning the wonderful proportion of the celestial spheres and the true and appropriate reasons for the number, magnitude, and periodic motions of the heavens demonstrated from the five regular geometrical bodies', which appeared in Tübingen in 1596. There are principles set out in the *Mysterium* that were to form the foundation of much of Kepler's later work, in light of which he issued a new edition in 1621 with supplementary notes. The first principle was that the plan of the universe was based upon geometry, in this case the regular polyhedra that determined the distances of the planets from the sun, as in the famous picture in Fig. 72, although the agreement with the Copernican distances found by observation was only approximate. But the discrepancies were also of interest to Kepler as they meant, not that the polyhedral theory was wrong, but that there was yet another factor to be taken into account that could be of interest, and he set about looking for it.

It was here that he reached his second important principle. The distances from Copernicus had been measured from the centre of the earth's sphere, the mean sun, except for earth which were taken from the true sun in order to give the earth a greatest and least distance. But is it possible that all the distances should be taken from the true sun? With help from Maestlin, who calculated the distances, eccentricities and directions of the apsidal lines from the *Prutenic Tables*, Kepler compared the greatest and least distances from the mean sun and true sun respectively with those from the polyhedra. The results were not conclusive. Nevertheless, he was by now certain that distances and eccentricities should be measured from the true sun, one reason being that while the eccentricities of Mars and Venus from the mean sun had changed between Ptolemy and Copernicus, those from the true sun had remained constant. He therefore set out to find a relation between distances from the true sun and periods, reasoning that, in the absence of solid spheres, some force from the sun must move the planets and, as the speeds of the outer planets are slower, this force must weaken with distance. Since the length of the orbit increases and the force decreases with distance, the distance has a double, additive effect on the period, that is, the increment in the period is twice the increment in the distance. Using this relation, Kepler again calculated the distances, this time mean distances from the true sun. The results were not better than from the polyhedra – they were, he admitted, a bit worse – but he was sure that some day the two methods would be reconciled (and in a note in the 1621 edition he points out that Book V of the *Harmonice mundi* does just that).

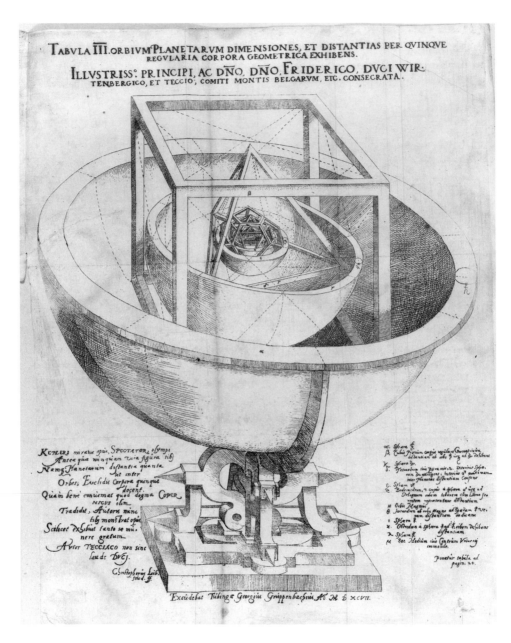

72 Kepler, *Mysterium cosmographicum* (1596), tab. 3. Illustration of a wooden model in the form of a large bowl, apparently never built, of the placement of the five regular polyhedra between the spheres of the planets. The polyhedra determine the distances of the planets from the sun and show why there are exactly six planets. (British Library)

Kepler's speculation about the double effect of distance on speed gave him another important insight, namely, the physical principle underlying Ptolemy's equant. Either through knowledge of Tycho's work or on his own, Kepler had also done away with solid spheres. Consequently, there was no longer any objection to returning to Ptolemy's hypothesis for the first inequality with its bisected eccentricity and violation of uniform circular motion, which Kepler now considered physically essential. In this hypothesis, the planet moves in its eccentric in equal times through a smaller arc near aphelion and a larger arc near perihelion. But the effect of the eccentricity is double, for 1) the distant arc appears smaller from the sun than the closer arc and 2) the force from the sun moving the planet weakens with distance such that the planet really moves slower near aphelion than near perihelion. Combining the two effects, it is as though the planet moved uniformly through equal arcs on a circle about an equant point at twice the eccentricity from the sun as the centre of the eccentric. The demonstration is hardly rigorous, but Kepler says he knows the reader will rejoice at learning the cause of the equant. Nevertheless, problems still remain for according to Copernicus the speed of the inferior planets vary, not with their own distances from the sun, but with the earth's, and the earth itself, which moves uniformly in a circle about the mean sun, has no equant. This last question was to become of great importance in Kepler's later work.

Kepler sent copies of the *Mysterium* to various mathematicians for their opinions of his great discoveries. Galileo, who received two copies that Kepler simply had delivered to astronomers in Italy – neither had heard of the other – responded within hours of receiving the book, after reading the preface. He told Kepler that he had come to Copernicus's opinion many years ago, and from it could explain the causes of many natural effects inexplicable by the common hypothesis, although he does not dare to publish for fear of ridicule. Kepler guessed that Galileo was referring to the tides, and he was probably right. Tycho had already seen the book, and despite some misgivings at such *a priori* theorising, was also encouraging. He pointed out that the eccentricities he had derived would provide a more accurate test of the distances, and invited Kepler to join him as he had just left Denmark and was without adequate assistance. Kepler also believed that Tycho's observations could help him, and after any number of complications he did join Tycho in Prague in October 1600.

He soon began the work that was to occupy the better part of his time for the next six years, his 'War on Mars', as he called it in the account of his research in the *Astronomia nova* (1609). Already when he arrived he found that Longomontanus had worked out a double-epicycle model for Mars, geocentric of course, equivalent to a heliocentric equant model with the eccentricities, from the mean sun to the centre of the eccentric and from the centre to the equant, in the ratio of 5:3. It could nicely reproduce ten oppositions from 1580 to 1600 to an accuracy of 2' of arc, although it failed badly for longitude outside of opposition and for latitude everywhere. Guided by his speculations in the *Mysterium*, Kepler suspected that the problem was in taking the apsidal line and the oppositions with respect to the mean rather than the true sun. His first task was therefore to find the time and longitude of oppositions to the true sun in Mars's inclined circle properly reduced to the ecliptic. This required an investigation of the inclination and nodal line of Mars's orbit, and the results were themselves of the highest interest. For it turned out that the nodal line passed through the true rather than the mean sun and, when accurate observations of Mars's latitudes were used rather than the faulty extremal values relied upon by Ptolemy and, following him, Copernicus, the plane of the orbit no longer oscillated, but was fixed. At once Kepler had corrected the two most serious problems in latitude theory.

With the oppositions to the true sun correctly reduced to ecliptic longitude, Kepler took up finding the eccentricities in an equant model like Ptolemy's, but without assuming a bisected eccentricity. The problem was of great difficulty, requiring four oppositions, and as there was no direct geometric solution, he had to use iterations based upon trial and error. After some 70(!) trials, Kepler found that where the radius of the eccentric is 100,000, the eccentricity from the sun to the centre of the eccentric is 11,332, from the centre to the equant 7,232, and the total 18,564. Note that the ratio of the total to the eccentricity from the sun to the centre is about 8 : 4.88 while Longomontanus found 8 : 5, and Ptolemy 2 : 1. In recomputing twelve oppositions, eight agreed to less than 1' and the worst error was just over 2', certainly a great success. However, when tested by the calculation of latitudes at opposition and longitudes outside of opposition, there were large errors. In fact, both of these indicated that the eccentricity should be bisected as Ptolemy had found. But when oppositions were computed with a bisected eccentricity of 9,282, the errors rose to about 8' (near octants, 45° and 135°). Kepler remarks that this error would not have been significant to Ptolemy, who considered his observations accurate only to 10', but with Tycho's observations such an error is significant, and indeed the entire reform of astronomy will rest upon those 8'. What Kepler now realised is that his model gave correct *directions* of the planet, which are all that matter at opposition when one is observing on a straight line from the sun to the planet, but incorrect *distances*, which affect longitudes observed outside of opposition, when the earth is not on the line from the sun to the planet, and for the same reason all latitudes. The model arrived at thus far, which Kepler called the 'vicarious hypothesis', while flawed in distance, was nevertheless useful for finding the theoretically correct direction of Mars, and Kepler so used it as his standard for evaluating other models.

Having initially failed to conquer Mars directly, Kepler turned to the motion of the earth, specifically to the second inequality of Mars produced by the motion of the earth. If the eccentricity of Mars should be bisected, what about the earth, the only planet with a simple eccentric according to Ptolemy, Copernicus and Tycho? Already in the *Mysterium*, he had suspected that the earth may also have an equant, since only in that way could its true speed vary with distance from the sun in accordance with his physical theory. He therefore set out to investigate the hypothesis of the earth, not, as had always been done, by observing the sun, which depends only upon direction and cannot distinguish a bisected from a simple eccentricity as small as the earth's, but by examining the effect of the earth's location on the apparent position of Mars, which can isolate the contribution of the distance between the sun and earth. Since the heliocentric directions of Mars were known correctly from the vicarious hypothesis, its apparent directions could then be used to find the eccentricity of the earth. Here it was fortunate that Tycho had observed Mars for many years throughout its orbit, for Kepler now required observations with Mars and the earth in a variety of configurations. From several different examinations, Kepler found that where the radius of the earth's eccentric is 100,000, the eccentricity is about 1,800, half of Tycho's eccentricity of 3,584. Immediately he applied this to show that in computing the second inequality of a planet, here of course Mars, the motion of the earth, or the equivalent motion of the sun in the Tychonic theory or of the planet on the epicycle in the Ptolemaic, must be taken as having a bisected eccentricity and equant. This by itself corrected the largest remaining theoretical error in planetary theory, reaching about 45' for Mars, as Tycho had already discovered, far larger than the errors of a few minutes to which Kepler was to devote the rest of his investigations.

Now, Kepler knew that an equant, however accurate its effect, was only a geometrical consequence of a more basic physical cause of planetary motion. In the *Mysterium* he identified this as a relation of motion to distance from the sun, which he now wishes to quantify and explain. He defines a quantity he calls 'delay' (*mora*), which is the inverse of velocity, that is, delay is time per unit distance where velocity is distance per unit time. The distances here are always circular arcs, so delay = time/arc, and the arcs are usually small. He then shows, purely from the geometry of Ptolemy's equant model, that the time required by a planet to traverse equal small arcs at aphelion and perihelion is nearly proportional to the distance from the sun, a relation he believes to hold everywhere, that is, the delay in each arc is directly, or the velocity inversely, proportional to the distance from the sun. (The relation is in fact exact at the apsides, and for small eccentricities may be extended with little harm. Far too much has been made of this error, which Kepler later corrected anyway.)

The underlying physical cause of the motions of the planets, and of this relation of velocity or delay to distance, Kepler believed to be a 'motive virtue' (*virtus motrix*) that extends outward from the sun as an immaterial image (*species*) of the body of the sun, just as light is an image of the fire of the sun. Like light, at any distance from the sun the virtue, propagated instantaneously, forms a spherical surface, and as such surfaces are larger at greater distances, in fact as the square of the distance, so both light and the virtue attenuate with distance. Light illuminates only the surface of bodies, but the virtue, which comes from the entire body of the sun, penetrates the entire body of a planet and moves the planet by its own motion of rotation in a circle produced by the rotation of the solar body itself. Kepler here inferred by reason of his physical theory that the sun must rotate several years before it was discovered with the telescope from the movement of sunspots, and he guessed that its period is about three days, $\frac{1}{30}$ the period of Mercury, just as the earth rotates in about $\frac{1}{30}$ the period of the moon. Both the earth and the sun, in order to maintain their rotation, have a 'motive soul' (*anima motrix*). Like light, the virtue spreads out from the sun in straight lines in all directions, but since it is the *circular* rotation of the virtue, the image of the rotation of the sun, that moves the planets, Kepler holds that its action to move the planets spreads out *circularly* (but *not* in a single plane) and decreases linearly with distance. This is an obscure argument – and here it has been much condensed – but it is what Kepler requires to account for the linear proportionality of delay to distance.

The virtue rotates with the same angular speed at all distances, but the planets are not carried around so fast, both because the virtue weakens with distance and because the planets are material and are thus inclined to rest. Here Kepler holds the Aristotelian principle of 'inertia' (a term he used later in the *Epitome of Copernican Astronomy*), or resistance to motion, that the natural state of a body is rest and a force is required to keep it in motion, which in turn is proportional to the force. All of Kepler's physics depends upon this principle – every motion requires a force proportional to the motion – and thus it differs from Newtonian physics, in which inertia is a resistance to change of state, of rest or of motion in a straight line, and a force produces an acceleration, a change of motion. He also compares the power of the sun to move the planets to a magnet, not here as an attracting force, which he takes up after he determines that the path of the planet is not a circle, but as a directing force, for which he cites William Gilbert's demonstration of terrestrial magnetism. It is through this directing force that the rotation of the solar body and the virtue cause a planet to move in a circle, just as also the rotation of the earth causes the moon to move.

With his physical theory in hand, Kepler next sets out to compute its effect on the motion of a planet, here the earth, in an eccentric circle. He has established that the delay of a planet in each small arc is proportional to its distance from the sun, and thus the delay in a larger arc, equal to the sum of the delays in its constituent smaller arcs, is proportional to the sum of the distances to the smaller arcs. He first attempted to compute the delays by summing the distances from the sun to the planet at the beginning of each 1° arc and dividing them by the sum of all the distances in the circle, proportional to the periodic time taken as 360°, to give the mean anomaly proportional to time in degrees. Thus, in Fig. 73, the sum of the distances from the sun S, as $SP_a + SP_1$, is to the sum of all distances as the mean anomaly in arcs $P_aP_1 + P_1P_2$ is to 360°. But this method was extremely tedious as finding the mean anomaly, the sum of the delays, for any arc necessitated computing and summing all the distances in that arc, which were strictly innumerable.

So Kepler hit on an expedient. Since the distances SP contained in the plane of the eccentric are infinite in number, and since the areas of triangles with equal bases are proportional to their altitudes, he substituted for the distances and sums of distances the areas and sums of areas of small sectors with their vertices at S, as $P_aSP_1 + P_1SP_2$, and equal bases P_aP_1, P_1P_2. (One can see, by the way, that the use of areas follows more intuitively from considering delays in equal arcs than velocities in equal times.) *If the areas of the sectors are proportional to the distances from the sun, to which the delays are proportional, the delays will in turn be proportional to the areas.* This, quite simply, is the origin of the area law. Note that Kepler did *not* consider an area to be a sum of distances, but just that a larger area contains more distances. As he put it, rather cryptically, 'in this area is a measure of the capacity (*facultatis*) by which the distances added together are effective for accumulating the delays'. However, he was also aware that *SP* is generally *not* the altitude of the sectors – since it is only perpendicular to the eccentric in the apsidal line – and is in fact longer than the altitude. Imagine that P_2 is itself the base of a tiny sector P_2SP_2; its altitude, perpendicular to the circle, must pass through the centre M and is thus $P_2Q_2 < P_2S$; and the same holds for altitude $P_4Q_4 < P_4S$. Consequently, the areas of the sectors are not proportional to *SP*, but to *PQ < SP*, and the sums of the sectors fall short of the sums of distances *SP* although they are proportional to the sums of distances *PQ*. Kepler later calls *PQ* the 'diametral distance', and it will prove very important.

Even though the areas are (as yet) imperfectly equivalent to distances for measuring delays, Kepler shows how to apply them. The area of any sector *ASP* is the mean anomaly, proportional to the sum of the delays in arc *AP* and thus to time. It may be divided into sector *AMP* from the centre, called the eccentric anomaly *E*, to which angle *E* at centre *M* is proportional, and triangle *SMP*, called the physical equation. Angle *SPM* is called the optical equation, and angle *ASP* the true or equated anomaly. Now consider sector ASP_m, where AMP_m is a right angle, which we divide into sector AMP_m and right triangle SMP_m. The area of sector AMP_m is E_m, and the area of triangle SMP_m is $\frac{1}{2}eR$ in radians, or in degrees $e° = (180/\pi)eR$, the maximum physical equation. Thus, the mean anomaly is $M = E_m + e°$. The altitude of any other triangle, as SMP_2, is $P_2N = R \sin E$, and since triangles on the same base, here *e*, are proportional to their altitudes, the area of SMP_2 is $e° \sin E$. Hence, the area of any sector, as $ASP_2 = AMP_2 + SMP_2$, or the mean anomaly, is $M = E + e° \sin E$. For later reference, the distance $PQ = R + e \cos E$.

With the second inequality due to the motion of the earth more or less under control – the solar equations were good to about 30″ – Kepler resumed his attack on the first inequality of Mars. If now the true method of computing equations and true elements have been found, they must produce true equations, meaning equations in agreement with the vicarious hypothesis. But the results were otherwise, for computing the equations by areas produced errors of about ±8′ near octants such that the planet was made too fast near the apsides and too slow near mean distances. The equant with bisected eccentricity also produced errors of ±8' near octants, but in the opposite directions, so again an error of 8' was significant. Kepler first suspected that the substitution of areas for distances in the circle might be at fault, but a careful analysis showed that the effect of the substitution was too small and in fact made the planet a bit faster near mean distances while the distances would make it slower still and increase the error. He analysed the problem further by comparing three distances used in an earlier, faulty derivation of elements, that had suggested that the planet was not necessarily located on a circle, with computed distances to the eccentric, and found that the original distances fell significantly short of the distances to the eccentric. His conclusion was now certain: the path of Mars is not a circle, but falls within the circle, less near the apsidal line and more near the mean distances, which was later confirmed by many more observations. This would also explain the errors in the equations, for if the areas of circular sectors made the delays too long, that is, the planet too slow, near mean distances, the areas of sectors of a curve within the circle would make the delays shorter and thus the planet faster, as it should be.

The discovery that the planet's path was not a circle has been greatly celebrated – mostly under the mistaken belief that it freed astronomy from burdensome restrictions – but it was not celebrated by Kepler, who arrived at it only after much difficulty and many failures and knew that his troubles were only beginning. For a circle is located uniquely by three points, but it was not obvious what sort of curve was contained within the circle nor how to go about finding it. So Kepler returned to his physics. He had earlier inquired into the physical cause of motion in an eccentric, for which he employed an epicyclic model, not because the epicycle was real, but in order to analyse the physics geometrically. He now resumes the model and suggests that the centre of the epicycle moves according to the area law through the eccentric anomaly while the planet moves on the epicycle in the opposite direction uniformly through the mean anomaly. The result is that the planet approaches toward and withdraws from the sun, describing some kind of curve inside the eccentric circle which he calls 'oval', meaning egg-shaped.

Now began Kepler's greatest headache. What was the shape of this curve? How does one measure arcs, distances, areas and equations in it? Ten chapters, fifty long pages are spent investigating this oval figure and methods of computing equations with it, all of which failed in one way or another. In one, to measure the area of the oval he substituted a nearly coincident ellipse, and found that, in Fig. 73, when AMP_m is a right angle, when $E = 90°$, the distance $P_mP'_m$ is 858, the maximum width of the 'lunule' between the oval and the eccentric circle. In computing equations from this ellipse by areas, he found errors of about ±6' near octants, but now such that the planet was made too slow near the apsides and too fast near mean distances, opposite to applying areas to the circle. After much frustration, a comparison of many distances derived from observation showed that the oval fell too far within the circle so that, as Kepler says, his physical causes went up in smoke. Note that this would account for the errors in the

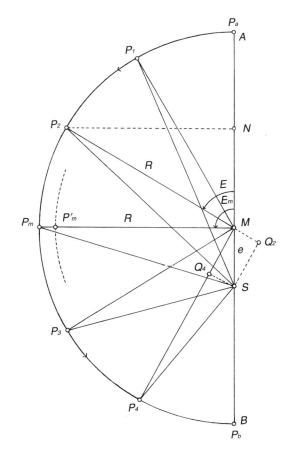

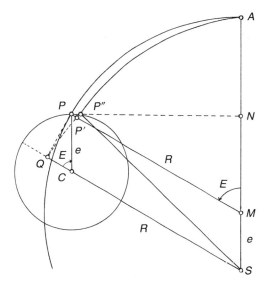

73 (*left*) Kepler, *Astronomia nova* (1609). Discovery of the area law. Kepler first tried to compute delays in small arcs, as $P_aP_1 + P_1P_2$, by summing distances, as $SP_a + SP_1$. But this was very laborious, so he substituted the areas of sectors, as area $ASP_2 =$ sector P_aMP_2 + triangle SP_2M.

74 (*above*) Kepler, *Astronomia nova* (1609). Discovery of the elliptical path. The distance of the planet from the sun is $SQ = R + e \cos E$, and its direction SP'' where P'' lies on an ellipse. Kepler found that also $SP'' = R + e \cos E$, showing that the planet must in fact be at P'' and thus move on the ellipse.

equations, since the areas of sectors near mean distance will now be too small, making the delays too short and the planet too fast.

Kepler grasped that the areas and distances in the circle were too large and in the oval too small, that the distances given by observation fell about half way between, and thus that the maximum width of the lunule $P_mP'_m$ should be about half of 858 or 429. But just what this meant was not obvious. Then, when worrying over the cause of the reduction of the lunule, he happened to notice that at $E = 90°$ the secant of the optical equation, 5; 18°, is 100,429, that is, in Fig. 73 the secant of angle SP_mM, or distance SP_m, is 100,429. It was, he wrote, as though he were awakened from sleep and saw a new light. He reasoned that the correct distance of the planet was not 100,429, but $R = 100,000$, less by the correct width of the lunule. This would result if, in Fig. 74 showing an epicyclic model used to produce an eccentric, the distance of the planet, although not its direction, is given, not by SP to the epicycle, but by SQ to the foot Q of a perpendicular from P, which 'librates' on the diameter of the epicycle through $CQ = e \cos E$. The distance from the sun, the 'diametral distance', as Kepler now calls it, is thus $SQ = R + e \cos E$, which the attentive reader will recall is the distance PQ in Fig. 73 that was correctly proportional to the area of each sector of the eccentric from S and thus to the delay in each arc. A comparison with the distances from observation used earlier to check the oval showed reasonably good agreement, so Kepler concluded that at last he had a model that correctly represented distances.

He devotes a long chapter to the physical cause of the libration, which we shall consider later in its final form in the *Epitome of Copernican Astronomy*, and then, on the verge of success, runs into yet another problem. 'Galatea, wanton girl, pelts me with an apple, and, hoping to be seen first, flees to the willows' (Virgil, *Ecl.* 3.64). In Fig. 74, the eccentric anomaly E is again AMP and the distance correctly given by $SQ = R + e \cos E$. With S as centre draw an arc of radius SQ. Where is the planet? Kepler's first guess, reasonably enough, was at P' on line MP showing the eccentric anomaly. But this introduced errors of $5\frac{1}{2}'$ and $4'$ at octants and produced a figure, broader near aphelion, that he called 'cheeky' (*via buccosa*). So he put aside the diametral distances, and reasoned as follows: Computing equations by areas of the circle produced errors in excess while the ellipse, used in place of the oval, with a lunule 858 in width produced (nearly) equal errors in defect. The correct curve must lie midway between, and since the only (symmetrical) figure between an ellipse and a circle is another ellipse, the correct curve is an ellipse cutting off a lunule with a width of 429. This is a rather mysterious step, made even more mysterious when Kepler says he was so sure it was right that he did not even have to compute equations from it.

What did concern him, however, until he almost went mad, was why, when the distances produced by the librations were so obviously true, as shown by the observations, the planet should follow an ellipse, as shown by the equations (which he had not computed). Put another way, the ellipse gave the right *directions*, or so it appeared, and the librations the right *distances*. He was nearly back to where he started with the vicarious hypothesis and the bisected eccentricity! The answer came when he realised that the figure produced by the librations was none other than the very ellipse he wanted. 'O foolish me, as though the libration on the diameter could not be the way to the ellipse.' It is easy enough to locate the planet on the ellipse. In Fig. 74 draw the ordinate PN from the circle to the apsidal line. An ellipse cuts all ordinates in the ratio of the minor to major axes, and this ratio is given by the eccentricity, so a point P'' of the ellipse can be constructed. What Kepler then shows is that $SP'' = R + e \cos E$, that is, P'' is also constructed by the libration. Earlier it was shown that areas of sectors of the circle from S proportional to the delays in the circle are also proportional to the distances $R + e \cos E$. And he now shows that the same holds for the ellipse, that is, the delays in the circle (*not in the ellipse*) are likewise proportional to the areas of sectors and to distances in the ellipse. The qualification 'not in the ellipse' is important, for arcs in the ellipse are in general not proportional to arcs in the circle, and in any case Kepler cannot measure arcs or delays in the ellipse. Later, in the *Epitome of Copernican Astronomy*, he shows that delays in small arcs perpendicular to the direction from the sun to the planet, as arc P_1P_1' in Fig. 75, the instantaneous circular motion of the planet due to the circumsolar virtue alone without the libration, are proportional to distances and areas of sectors in the ellipse. This is true because the distance from the sun to the planet, as SP_1, is the altitude of a small sector P_1SP_1' formed by an arc perpendicular to that distance, to which the area of the sector is proportional.

We have deferred consideration of the physical explanation of the libration that accounts for the change of distance between the planet and the sun such that the planet describes an ellipse. Kepler worked this out after a fashion in the *Astronomia nova*, but refined it in the *Epitome of Copernican Astronomy*, and it is the refined theory we examine here. Now, the planets are carried about the sun by the rotation of the circumsolar 'virtue', which has a 'directing force' like a magnet, and this by itself would produce a circular path. However, just as the poles of two magnets also attract and repel each other, an analogous property in the sun and planets produces a variation of distance. The body of the sun is like a giant magnet, one pole of which is its centre and the other its entire surface; it is the rotation of this giant magnet that causes the virtue to rotate. The body of the planet is also a magnet, containing within it immaterial straight lines or 'fibres' that tend to maintain a fixed direction while the planet moves about the sun and rotates on its axis just as, in Kepler's analogy, the needle of a compass continues to point north when you walk around a castle with it. One end of the fibres, related to the centre of

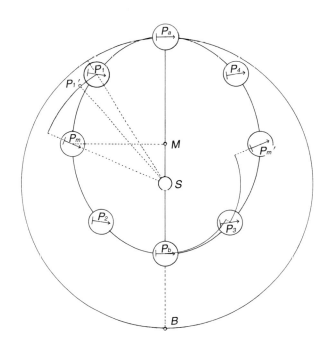

75 Kepler, *Epitome astronomiae copernicanae* (part 2, 1620). Generation of the elliptical orbit from librations. Magnetic fibres in the planet are attracted and repelled by the sun, the point of the arrow attracted in semicircle P_aP_b and the back repelled in semicircle P_bP_a. Kepler shows that if the fibres incline such that they are directed at the sun at ±90° of eccentric anomaly, at P_m and P_m', the planet will move in an ellipse.

the sun, is 'solipetal' (*solipeta*, sun-seeking) or 'friendly' to the sun, and the other, related to the surface, is 'solifugal' (*solifuga*, sun-fleeing) or 'unfriendly' (cf. Newton's *vis centripeta* and *vis centrifuga*, surely a coincidence of nomenclature). According to Kepler's principle of inertia, every motion requires a force. Thus, as the circumsolar virtue causes the planet to move about the sun, the action of the sun on the fibres attracts or repels the planet with a force, like the virtue, inversely proportional to the distance, and causes the planet to approach toward or withdraw from the sun at a rate that depends primarily on the alignment of the fibres with respect to the direction from the sun to the planet. The attraction and repulsion also cause a slight deflection of the fibres by which Kepler quantifies the libration.

The motions are shown qualitatively in Fig. 75. When the planet is at aphelion P_a and perihelion P_b, the fibres, shown by the arrow, are perpendicular to the direction to the sun, so the planet is neither attracted nor repelled. As the planet is carried around the sun, from P_a to P_b the solipetal end of the fibres, the point of the arrow, now turned toward the sun, is deflected toward the sun and the planet is attracted inward. Then from P_b to P_a the solifugal end of the fibres, the back of the arrow, now facing the sun, is deflected away from the sun and the planet is repelled outward. A slight delay in restoring the perpendicular alignment of the fibres at P_a and P_b will cause the apsidal line to advance, which does take place very slowly. In quantifying the libration, Kepler shows that if the deflection of the fibres is always equal to the optical equation, so that at $\pm 90°$ of eccentric anomaly, P_m and $P_m{'}$, the fibres are directed exactly at the sun, the libration is equal to $e\,(1 - \cos E)$. Since the distance at aphelion $SP_a = R + e$, the distance at any point $SP = R + e - e\,(1 - \cos E) = R + e \cos E$, exactly the distance of the planet in the ellipse.

All that now remains is to show how to compute equations. In Fig. 76, the ellipse is APB, its major auxiliary circle $AP{'}B$ with centre M, the sun is at S, and the eccentricity $SM = e$. The mean anomaly M in the circle, area $ASP{'}$ = sector $AMP{'}$ + triangle $SMP{'}$, or $M = E + e \sin E$, locates point $P{'}$ at an eccentric anomaly $AMP{'} = E$. Draw the ordinate $P{'}N$ to the apsidal line, which meets the ellipse at P, the location of the planet.

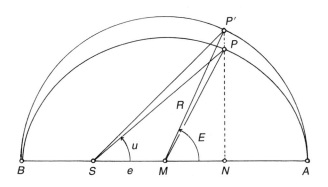

76 Kepler, *Astronomia nova* (1609). Determination of the position of the planet P in an ellipse with centre M and the sun at S. From the eccentric anomaly $E = AMP{'}$, one may easily find the mean anomaly M = area $ASP{'}$ and the true anomaly $u = ASP$. But to find the eccentric anomaly E from the mean anomaly M geometrically is not possible, although it may be approximated by iteration or interpolation.

Because of the constant proportionality of $P'N$ to PN, area ASP has the same ratio to the area of the ellipse as area ASP' has to the area of the circle, and thus area ASP' in the circle may be used to measure area ASP in the ellipse. The true or equated anomaly, ASP, is u, and the distance $SP = R + e\cos E$. Given the eccentric anomaly E, it is straightforward to find the mean anomaly M and the equated anomaly u. And given u, E may be found with somewhat more trouble, and then M is straightforward.

But to find E from M, the eccentric anomaly from the mean anomaly, which of course is what always must be done, is quite another matter. For the relation $M = E + e\sin E$, known as 'Kepler's equation', equivalent to area ASP' = sector AMP' + triangle SMP', has no geometrical solution for E. If we are given arc $AP' = E$ and point S on diameter AB, we may construct triangle $SMP' = e\sin E$, and the sum $E + e\sin E = M$. But if we are given an area M of part of a circle and point S on diameter AB, we must construct a line from S to a point P' on the circle such that SA and SP' contain the area M. For then we could easily divide sector ASP' into E and $e\sin E$ by drawing MP' from the centre. Kepler does not know how to construct SP'. Therefore, he calls upon geometers to solve the problem: 'Given the area of part of a semicircle and a point on the diameter, to find the arc, and the angle at that point, by the sides of which angle and by which arc the given area is contained. Or, to divide the area of a semicircle in a given ratio from any point on its diameter.' Whoever shows me how to do this, he says, will be my great Apollonius – after Virgil, *Ecl.* 3.104, *eris mihi magnus Apollo* – a pun on *Apollonius Gallus* (1602), in which François Viète wrote disparagingly of the mathematical abilities of astronomers in general and Copernicus in particular, as Kepler had earlier noted, making the same pun, after describing his seventy iterations for the vicarious hypothesis. The problem, as Newton among others was later to show, is not solvable, but there are various ways of finding E from M by iteration or interpolation, as Kepler himself was the first to do.

The *Astronomia nova*, while probably the most important of Kepler's works, marks, not the end of his astronomical researches, but, appropriate to its title, a new beginning. Everything now remained to be done, and done differently than anything that had been done before, specifically, the new methods had to be applied to all the other planets to derive the elements of their orbits and compose tables for calculation. In addition, the theory of the moon, although greatly improved by Tycho, had yet to be brought within the new methods and accounted for by correct physical principles. The work took many years and had many interruptions, including Kepler's move from Prague to Linz in 1616, and finally appeared in two large publication. The first was the *Epitome astronomiae copernicanae* (*Epitome of Copernican Astronomy*, 3 pts 1618–20–21), a comprehensive treatise on the new astronomy written, so Kepler intended, for the general reader, although it is not easy going by any standard. The second was the *Rudolphine Tables* (1627), a total reform of all earlier tables and the culmination of more than fifty years of work by Tycho, his assistants, and Kepler. But already in 1617 Kepler began publishing annual ephemerides based upon the methods and parameters of the *Rudolphine Tables*, so by the time they appeared their application was established, although few could appreciate their superiority to all other tables, and for some years because of their strangeness they were neglected in favour of more traditional tables that are now known to be far less accurate.

We cannot leave Kepler without giving some attention to one of his most remarkable works, which he himself may have considered his finest, the *Harmonice mundi* (*Harmony of the World*) of 1619. The question he had asked in 1595 that began all of his research was, what plan did God have in mind when he created the universe? Part of the answer was the five regular polyhedra, which showed why there were just six planets and determined their distances. But the distances from the polyhedra agreed with the observed distances only roughly, and the polyhedra by themselves did not explain why the orbits should be eccentric to the sun and with a variety of eccentricities. Clearly God could have made the distances agree perfectly with the polyhedra and the orbits concentric circles had he wished, but that would have been less interesting, less diverse than the plan he chose. What was that plan? In searching for an answer, Kepler turned to the ratios of musical intervals. These were, he believed, close to fundamental relations since he also believed that the foundation of harmonic ratios, as for the octave, fifth, fourth, etc., was the divisions of the circle formed by the sides of the constructible regular polygons. Hence, like the distances from the polyhedra, Kepler's harmonics is founded in geometry.

In searching for harmonic ratios among the planets, he eliminated distances, periods, delays, and path lengths, and decided upon apparent angular velocities as seen from the sun, which for each planet are nearly in an inverse proportion to the square of its distances. These will be perceived directly, not by ourselves on the moving earth, but by beings on the sun, who are of a more spiritual nature, but are also simpler, and see only the moving directions of the heavenly bodies and have no way of learning their distances. He finds a number of harmonic ratios between the extremal velocities, aphelial and perihelial, of each planet and of any two adjacent planets, and he develops these into a yet greater number of harmonic relations throughout the planetary system. Then, from the extremal velocities of each planet, he determines its greatest and least relative distances from the sun, and from these its eccentricity. The eccentricities found in this way agree quite well with those derived from observation, and confirm that God gave the planets their eccentricities just so that they will display the harmonic ratios in their extremal motions.

When he had nearly finished writing all of this, he discovered a new relation between the distances and periods of the planets that is known as Kepler's third or harmonic law, that the squares of the periods of the planets are proportional to the cubes of their (mean) distances from the sun, that is, $P^2 \propto r^3$ or $P \propto r^{3/2}$. Newton showed that this relation is a direct consequence of the inverse-square law of gravity, indeed, is the most important evidence for the inverse square. For Kepler it was a way of extending his harmonic speculations beyond the eccentricity of each planet to the distances of all the planets, for with this relation he could find the distances of the planets from their periods or mean motions, which compared very favourably with the observed distances. This at last solved the problem of the relation of period and distance that had occupied him since he wrote the *Mysterium*. Then, combining the distances found in this way with the eccentricities derived from the harmonic ratios, he found the greatest and least distances, and compared these with distances derived from observation, and again the agreement was very good. Finally, he compared these distances with the distances from the regular polyhedra, finding this time a reasonably good agreement. His conclusion was that the polyhedra served for the rough form of the universe while the harmonies provided its perfection. He had at last understood God's plan. At the end he says a prayer, thanking God for all the delight he takes in His work, and asking that if he has made any mistakes in showing the glory of His works to men, God will inspire him to correct them.

Bibliography

General

Dreyer, J.L.E. 1906. *A History of Planetary Systems from Thales to Kepler*. Cambridge University Press. Reprinted 1953 as *A History of Astronomy from Thales to Kepler*. New York: Dover Publications.

Koyré, A. 1973. *The Astronomical Revolution*. Trans. R.E.W. Madison. Ithaca: Cornell University Press.

North, J.D. 1994. *The Fontana History of Astronomy and Cosmology*. London: Fontana. Published in the USA as *The Norton History of Astronomy and Cosmology*. New York: W.W. Norton.

Pannekoek, A. 1961. *A History of Astronomy*. New York: Interscience. Reprinted 1989. New York: Dover Publications.

Taton, R. and Wilson, C. 1989. *Planetary Astronomy from the Renaissance to the Rise of Astrophysics, Part A: Tycho Brahe to Newton. The General History of Astronomy*, vol. 2. Cambridge University Press.

Van Helden, A. 1985. *Measuring the Universe. Cosmic Dimensions from Aristarchus to Halley*. University of Chicago Press.

Regiomontanus

Gerl, A. 1989. *Trigonometrisch-astronomisches Rechnen kurz vor Copernicus: Der Briefwechsel Regiomontanus–Bianchini*. Boethius 21. Stuttgart.

Hamann, G. 1980. *Regiomontanus-Studien*. Sitzungsberichte der Österreichischen Akademie der Wissenschaft. Phil.-hist. Kl. 364. Vienna.

Kremer, R.L. 1980. Bernard Walther's astronomical observations. *Journal for the History of Astronomy* 11, 174–91.

Swerdlow, N.M. 1990. Regiomontanus on the critical problems of astronomy. In *Nature, Experiment, and the Sciences*, ed. T.H. Levere and W.R. Shea, pp. 165–95. Dordrecht: Kluwer.

Zinner, E. 1968. *Leben und Wirken des Joh. Müller von Königsberg, gennant Regiomontanus*. 2nd edn. Osnabrück: O. Zeller. English trans. E. Brown 1990: *Regiomontanus, His Life and Work*. Amsterdam: North Holland.

Copernicus

Armitage, A. 1957. *Copernicus, the Founder of Modern Astronomy*. New York: Thomas Yoseloff.

Copernicus, N. *Complete Works II. On The Revolutions, Complete Works III. Minor Works*. Trans E. Rosen. 1978 and 1985. Warsaw: Polish Scientific Publishers.

Copernicus, N. *On the Revolution of the Heavenly Spheres*. Trans. A.M. Duncan. 1976. Newton Abbot: David & Charles; New York: Barnes & Noble.

Neugebauer, O. 1968. On the planetary theory of Copernicus. *Vistas in Astronomy* 10, 89–103.

Rheticus, G. J. *Narratio Prima*. Trans. and ed. H. Hugonnard-Roche and J.-P. Verdet with M.-P. Lerner and A. Segonds. 1982. Studia Copernicana, 20. Wroclaw: Polish Academy of Sciences.

Rosen, E. 1971. *Three Copernican Treatises*. 3rd edn. New York: Octagon.

Swerdlow, N. M. 1973. The derivation and first draft of Copernicus's Planetary Theory. A Translation of the Commentariolus with Commentary. *Proceedings of the American Philosophical Society* 117, 423–512.

Swerdlow, N. M. and Neugebauer, O. 1984. *Mathematical Astronomy in Copernicus's De Revolutionibus*. New York: Springer-Verlag.

Zinner, E. 1988. *Enstehung und Ausbreitung der copernicanischen Lehre*. Ed. H.M. Nobis and F. Schmeidler. 2nd edn. Munich: C.H. Beck.

Tycho

Brahe, T. *Description of His Instruments and Scientific Work as given in Astronomiae Instauratae Mechanica*. Trans. and ed. H. Raeder, E. Strömgren and B. Strömgren. 1946. Copenhagen: Munksgaard.

Dreyer, J.L.E. 1890. *Tycho Brahe. A Picture of Scientific Life and Work in the Sixteenth Century*. Edinburgh. Reprinted 1963. New York: Dover Publications.

Gingerich, O. and Westman, R.S. 1988. The Wittich Connection: conflict and priority in late sixteenth-century cosmology. *Transactions of the American Philosophical Society* 78 (7).

Hellman, C.D. 1971. *The Comet of 1577. Its Place in the History of Astronomy*. 2nd edn. New York: Columbia University Press.

Schofield, C. 1981. *Tychonic and Semi-Tychonic World Systems*. New York.

Thoren, V. 1990. *The Lord of Uraniborg*. Cambridge University Press.

Westman, R.S. 1975. Three responses to the Copernican theory: Johannes Praetorius, Tycho Brahe, and Michael Maestlin. In *The Copernican Achievement*, ed. R.S. Westman, pp. 285–345. Berkeley: University of California Press.

Kepler

Casper, M. 1959. *Kepler*. Trans. C.D. Hellman, London: Abelard-Schuman. Reprinted 1993. New York: Dover Publications.

Field, J.V. 1988. *Kepler's Geometrical Cosmology*. University of Chicago Press.

Grafton, A. 1973. Michael Maestlin's account of Copernican planetary theory. *Proceedings of the American Philosophical Society* 117, 523–50.

Kepler, J. *Mysterium Cosmographicum. The Secret of the Universe*. Trans. A.M. Duncan, with introduction and commentary by E. J. Aiton. 1981. New York: Abaris Books.

Kepler, J. *The New Astronomy*. Trans. W. H. Donahue. 1992. Cambridge University Press.

Small, R. 1804. *An Account of the Astronomical Discoveries of Kepler*. London. Reprinted 1963. Madison: University of Wisconsin Press.

Stephenson, B. 1987. *Kepler's Physical Astronomy*. New York: Springer-Verlag. Reprinted 1994. Princeton University Press.

Stephenson, B. 1994. *The Music of the Heavens: Kepler's Harmonic Astronomy*. Princeton University Press.

G.L'E. TURNER

Later Medieval and Renaissance Instruments

Astronomy is the systematic study of the heavens, and is the foundation of all the sciences. It was necessary for time-telling, for keeping a record of the seasons, and for guidance in travel by sea and land. In view of the universally visible marvels of the heavenly bodies in the night sky, it is not surprising that all early human societies believed that the stars influenced people's fate. Therefore, alongside what we today would see as the scientific study of the heavens, and closely related to it, there existed prognostication, or astrology, the study of the same phenomena for the purpose of casting horoscopes and predicting future events. The first practitioners of astronomy and astrology used simple instruments and methods of computation. The earliest observatories were monumental in scale, using natural features allied with stone obelisks as foresights and backsights. As the requirement for greater precision grew, instruments of a more sophisticated kind gradually came to be designed and used.

The true watershed for serious astronomical observation came, however, at the end of the fifteenth century when Christopher Columbus set sail from Spain to the West in 1492. The discovery of the Americas created the economic requirement for accurate astronomical observation, particularly related to some means of finding the longitude at sea when out of sight of land. For over two centuries this constituted an intractable problem, the final solution of which had to wait till the design of the marine chronometer by John Harrison (1693–1776) in the mid eighteenth century. But from the end of the fifteenth century, it became of great practical importance to make accurate astronomical observations, and for this purpose instruments were devised for use in the observatory, by navigators and by land surveyors.

Though scientific instruments of engraved brass were made from the thirteenth century in Europe, it was not until the second half of the fifteenth century that the trade began to develop into a serious and lucrative one. A major event that contributed to the change in the approach to astronomical observation was the invention of the printing process in the 1450s. The skills in calligraphy and the engraving that this revolutionary process generated, associated with the economic pressures that resulted from the age of discovery, formed a base for the precision instrument-making trade in Europe. The pioneers were Gemma Frisius (1508–55) and Gerard Mercator (1512–94), of Louvain in the Low Countries, whose influence in globe, map and instrument making spread rapidly across the continent of Europe and to England.

To summarise, there occurred, in the period under discussion, a transformation in world view that is embodied in the whole meaning of the word 'Renaissance'. This profoundly affected the whole concept of cosmology, and so the way in which the heavens were observed and studied. From the point of view of instrumentation, this change produced an overriding requirement for greater accuracy, so that more, and above all, more precise instruments were needed.

Astronomical instruments

The armillary sphere and celestial globe

The armillary sphere was the earliest of demonstration and teaching instruments associated with astronomy. The concept of a spherical earth and sky is by no means self-evident, but it was generally accepted in the Greek world by the time of Aristotle (382–322 BC). The Ptolemaic world system placed the earth in the centre of a series of nested spheres representing the planets and the starry sky. It was this system that the armillary sphere reproduced. Its basic form was a series of metal or wooden rings or circles representing the equator, ecliptic, tropics, polar circles, solsticial colure, and other meridian circles, generally with the polar axis extended to form a handle. Its purpose was to represent the position and motion of those circles, and to explain how they were related to one another. In its later, more elaborate form, the armillary was supported in a stand carrying a horizon ring, into which the meridian ring slotted, and against which it was adjustable. Early versions, showing the handle, are illustrated in thirteenth and fourteenth-century manuscripts, while elaborate and elegant examples on stands, from the sixteenth century, survive. It is interesting to note that the armillary was used as a device or badge by Manoel I, who came to the throne of Portugal in 1495. It was during his reign that astronomical navigation was first seriously promoted in a country noted for its seafaring, with the result that the sea route to India was discovered, and the Portuguese empire founded. The instrument was both a potent symbol and a practical teaching device, by means of which students could be shown the workings of the heavenly bodies.

The armillary is, of course, a close relative of the celestial globe, of which the design and mounting were first described by Ptolemy in the *Almagest*. But though numerous celestial globes of Islamic origin are known, it was not until the end of the fifteenth century that globes were produced as part of an established craft in Europe. The celestial globe made by Hans Dorn in 1480 marks the start of this development. It coincided with the great voyages of exploration, and the increased concern both with deep-sea navigation and with map-making, and its most famous exponent was the great

77 European astrolabe, *c.* 1200. Brass, diam. 11.3 cm. This is one of the earliest known European astrolabes, with a rete inspired by Hispano-Moorish designs. There are 13 star pointers, none of them named. On the back of the instrument are scales for 360°, the zodiac and the calendar. Crudely drawn is a shadow square and an unequal hour diagram. The three plates are marked out for latitudes from 23° to 48°. IC no. 161. (British Museum, MLA 1961, 12–1, 1)

map-maker, Gerard Mercator. His celestial globe of 1551, following the production of a terrestrial globe ten years earlier, was the largest printed globe made until that time, and embodied the latest astronomical knowledge.

Astrolabe

In many ways the planispheric astrolabe is the archetypal scientific instrument because of its antiquity and remarkable sophistication. It is a flat, circular brass instrument, which embodies a stereographic projection of the globe of the earth and of the hemisphere of the heavens – a sort of flattened armillary sphere (Figs 77, 78). The origin of the stereographic projection is nearly always the South Pole, and the plane of the projection the equator. Although it can be used for time-telling either by the stars or by the sun, the astrolabe has an additional function as an analogue computer, important for solving problems the mathematics of which would be too daunting to solve in other ways.

The European university arose from the cathedral schools during the twelfth century, and by 1350 their number had reached thirty. Science was taught in the arts faculties as part of the *quadrivium*: arithmetic, geometry, astronomy, music. The astrolabe had, at that time, an obvious role in teaching astronomy, which encompassed astrology, since in the medieval cosmological view, the planets ruled the life of mankind, and so prognostications were vitally important. By 1500 the number of universities had increased to seventy, which explains the fifteenth-century increase in astronomical texts and in the making and use of astrolabes and associated instruments. A number of examples have

78 English astrolabe, *c.* 1295. Brass, diam. 46 cm. The rete has 40 star pointers, of which 33 are named in Arabic and 7 in Latin. The mater has a stereographic projection for 42° (ROMA), while three separate plates are marked out for latitudes 48°30', 51°, 52° (LVNDONIARVM), 53°, 54°, 55°. The back bears comprehensive calendrical information, including a table of epacts in the medieval form beginning with 0. This astrolabe belonged to Sir Hans Sloane and formed part of the foundation collection of the British Museum. IC no. 290. (British Museum, MLA Sloane 53)

survived of fairly simple and small astrolabes that were made at that period in Germany, and it has been postulated that these were made for the use of scholars at the growing number of universities founded at that time in the Germanic states. These were used largely for teaching. During the sixteenth century some very large astrolabes (50 to 80 cm in diameter) were constructed with as great an accuracy as possible for calculating purposes. In Western Europe the planispheric astrolabe ceased to be made very suddenly after 1600, having been made redundant by new instruments and modes of calculation.

The astrolabe was introduced to Europe through Spain by the Islamic peoples who occupied the north coast of Africa and part of Spain in the tenth century. With the Christian reconquest of Spain, knowledge of the astrolabe penetrated medieval Europe during the eleventh century, and craft centres were established by the thirteenth in France, Germany, the Low Countries and England. The *Treatise on the Astrolabe*, written in 1391 by Geoffrey Chaucer, the great English literary figure, is not only an excellent introduction to the use of the instrument, but it is important as the first technical treatise written in the English native language.

Radius astronomicus or cross-staff

The purpose of the instruments used by astronomers in the Middle Ages was to measure for the stars the following celestial co-ordinates: altitudes and azimuths; declinations and right ascensions; celestial longitudes and latitudes. The *radius astronomicus* or cross-staff enjoyed a life of over 300 years, and was a late medieval invention. It was published in 1328 by the southern French astronomer and mathematician Rabbi Levi ben Gerson.

The cross-staff consists of a long wooden staff (whose cross-section is about 3×3 cm) with a perpendicular vane that moves over it. The staff is graduated trigonometrically so that angles can be measured by holding the staff to the eye and moving the vane until its ends are level with the points that are to be measured. Its main function was to measure directly angular distances between stars or planets, or the diameter of the sun or moon. Because of its simplicity, however, it was extremely versatile, and could also be used to measure the length of a comet's tail, and both celestial altitudes and terrestrial longitudes using the method of lunar distances. Its practical functions extended to surveying, where it was used to measure heights, lengths and distances, and to navigation. It soon became the sailor's preferred instrument for taking the altitude of the sun or stars.

The cross-staff was light, easy to construct and easy to dismantle and carry. To provide accuracy for the astronomer it could be as long as 450 cm, but was generally 90–180 cm in length for use by surveyors, and even smaller to take on board ship. During the sixteenth century, the surveyor's and seaman's staff was improved by such scholars as William Bourne (fl.1556–88) and John Dee (1527–1608), while Dee and Thomas Digges (c. 1546–95) also worked on the astronomer's staff. The improvements pioneered in England were gradually adopted by Continental makers.

Nocturnal

The nocturnal is a simple instrument for giving a rough indication of the time, perhaps to a quarter of an hour, during the night. Its use depends on being able to see the Pole Star and the Great Bear (Ursa major) or Plough, and on the fact that the stars appear to rotate about the Pole once in 24 hours (less 4 minutes each day). In appearance, the nocturnal looks like a small table-tennis bat, with a calendar scale engraved on it. Rotating over this is a disc marked with two periods of 12 hours. The hour positions are usually notched for counting by feel when used in the dark. Above this disc is a long rotating pointer, and the central rivet has a hole. To tell the time, the mark for 12 midnight is set to the date, the instrument is held upright, the Pole Star sighted through the rivet, and the pointer turned to be in line with the Guards of the Bear, that is the two prominent stars α and β in the constellation, which always line up with the Pole Star. The time is then shown by the pointer cutting across the hour disc.

The nocturnal was described in sixteenth-century texts, and examples exist dating from about 1500. It found a use on sea and on land into the seventeenth century, when the pocket watch made it redundant.

The reform of observatory instruments

Scientific knowledge progresses through increasing accuracy of measurement. The making of accurate angular measurements is a fundamental task of the astronomer, and from ancient times until the nineteenth century cataloguing of celestial positions was

79 Astronomical sextant made in 1600 at Prague by the Swiss instrument-maker Jost Bürgi (1572–1632). The whole instrument is made of iron, except for the brass limb (radius 112.2 cm) which carries the degree scale. It was used to measure the angular distance of any two objects in the sky in any position, provided they were not more than 60° apart. It was based on a design of Tycho Brahe, who died in Prague. The sextant was used during the years 1602–4 by Kepler and again on 20 January 1628 at Prague to observe the lunar eclipse. (National Technical Museum, Prague, 17–195)

his principal responsibility. Angular measure involves dividing a circle into 360° (or a quadrant into 90°), an inconvenient figure that would not have been chosen unless it had been accepted over many millennia. The techniques necessary to perform such divisions with exactitude occupied the best craftsmen for centuries. The quality of the instruments, and the skills in their proper use, enabled cosmology to rise above mere speculation.

There has, however, to be good reason for designing improvements, and one such was the problem of deep-sea navigation across the Atlantic to the New World. Navigation was now reliant on stellar astronomy through the use of printed tables. But the tables of the co-ordinates of some 1,000 stars remained, in effect, those produced by Ptolemy *c.* AD 150 (see p. 87), which had merely been adjusted for precession through the centuries. The Dane, Tycho Brahe (1546–1601), while a student at Leipzig, realised that the then current astronomical tables were defective. Tycho observed that a conjunction of Saturn and Jupiter in August 1563 varied by a month from the date expected from the tables. It was this observation (measured using a simple cross-staff) that Tycho regarded as the turning point in his career, and he abandoned the study of law for astronomy. Eventually, under the patronage of the Danish king, Tycho began to construct a large observatory in 1576 on the island of Hven in the Danish Sound, and conducted most of his systematic plotting of stellar positions during the 1580s using instruments to his own design made by the finest craftsmen in Germany (see pp. 207–10). Although all the instruments used on the island of Hven are lost, there remain two sextants constructed for Tycho now preserved in the National Technical Museum in Prague. Both were made in the year 1600, one by Erasmus Habermel, the other by Jost Bürgi (Fig. 79). These sextants were so ingenious they served as a model for subsequent decades.

Tycho Brahe was the first to break through the accuracy barrier in the measurement of angles. With his instruments and with his skill, he was able to record co-ordinates to an accuracy of better than 1' of arc. This surpassed the measurements made by the best Asian astronomers, notably Ulugh Beg (1394–1449) in Samarqand (see p. 164), whose accuracy was between 10' and 5' of arc. The stellar co-ordinates measured by Tycho enabled Johannes Kepler (1571–1630) to formulate his laws of planetary motion, which vindicated the Copernican view that the earth moved round the sun, and laid the groundwork for the celestial mechanics of Isaac Newton (1642–1727).

Navigational instruments
Astronomical navigation

While a ship is sailing along a coast, the determination of its position is no great problem. In the open sea, out of sight of land, a different sort of navigation has to be applied. The latitude (the angular distance from the Equator, 0°, to the Pole, 90°) can be measured by a quadrant or cross-staff by taking the altitude above the horizon of the Pole Star. The altitude of the sun at noon, when it crosses the meridian, can be taken by a quadrant or sea astrolabe, cross-staff or back-staff. Knowing the day of the month, the altitude, and hence the latitude, can be found from tables.

The longitude is more difficult. This is the angle around the globe, east to west. Longitude can be measured by a clock, because if you know the time at your port of departure, and can find out the local time by the sun, then the difference in time gives you the longitude, because 1 hour equals 15° of longitude. But clockwork mechanisms were not good enough to withstand the ship's motion and changes in temperature until after the middle of the eighteenth century when John Harrison perfected the marine chronometer. Before this, seamen had to rely on the compass, the log-line, and the traverse board to keep a rough tally of the progress of the journey.

When the Portuguese, during the early fifteenth century, began to explore the coast of West Africa, it was initially by coastal navigation. But with attempts to reach India, instrumental navigation became necessary. The Portuguese Prince Henry the Navigator (1394–1460) founded a training school at Sagres for officers. Here were brought mathematicians and astronomers and cartographers to found the nautical science of navigation, to devise instruments, and to draw up charts.

During the middle of the fifteenth century, Portuguese sailors used the plain altitude quadrant for angle measurements. The astronomical quadrant is known from classical times; for ship use they are small, about 30 cm in radius, and were engraved with a scale from 0° to 90°. With a pair of pinhole sights correctly aligned, the plumbline marked the angle. The Portuguese navigators knew that they could return to Lisbon by taking the winds out into the Atlantic until the latitude of Lisbon (38° 42') was reached, as shown by the altitude of the Pole Star. Then it remained to run east 'down the latitude'. Despite the invention of other instruments, the quadrant found favour among some sailors until the seventeenth century.

The sixteenth century dawned with maps, charts and terrestrial globes that depicted land masses unrecognisable in comparison with modern versions. These poor charts made navigation exceedingly hazardous, like riding a horse in a thick fog: ships were wrecked, men were killed, voyages were longer, and trade was diminished. Settlements made in America one year could not even be located the next because of the gross inaccuracies in plotting positions. The problem was that of finding the longitude sufficiently accurately, and the solution lay in astronomy with improved instrumentation. The technology of clocks was too primitive to be of assistance; accurate measurements of the moon in relation to stars were attempted in major observatories, but the resulting tables to carry on board were not sufficiently good, and the navigator and surveyor did not have satisfactory instruments for practical use. A great deal of effort was made, but a satisfactory advance came only with the new national observatories in Paris (1671) and Greenwich (1675).

An example of the rigorous search for more accurate navigation techniques is provided by Edward Wright. David Waters, in his masterly history of navigation, wrote of him: '... his book set the seal on the supremacy of the English in the theory and practice of the art of navigation at the end of the sixteenth century'. Wright was a Cambridge University mathematician, who became a fellow of his college, Caius. He was brought into the service of Queen Elizabeth as a result of the Armada, and he travelled on naval ships to gain practical experience. The result was *Certaine Errors in Navigation, arising either of the ordinarie erroneous making or using of the sea Chart, Compasses, Crosse staffe, and Table of declination of the Sunne, and fixed Starres detected and corrected*, published in 1599. Here Wright provided a most thorough mathematical treatment of errors in measurements and in the practices of seamen.

Mariner's astrolabe

For measuring the altitude of the sun near the meridian a new instrument was devised late in the fifteenth century. The mariner's astrolabe is a great simplification of the much earlier astronomer's planispheric astrolabe. It carries a degree scale around the rim, and it has an alidade or sighting rule, and there the similarity ends. The mariner's astrolabe is made of bronze or brass and is very heavy, especially at the bottom. The weight helps to keep it steady, and there are considerable portions cut away to reduce wind resistance, which would otherwise make its use harder or impossible. The mariner's astrolabe does no more than the quadrant: the sights are aligned and the angle read from the degree scale. The earliest recorded use of a mariner's astrolabe is during a voyage in 1481 by Diogo d'Azambuja down the coast of Africa. They were used by Portuguese, Spanish and Dutch navigators through the sixteenth century, their end coming during the 1640s.

Back-staff

For the seaman, the typical measurement made with the cross-staff was to find the ship's latitude by measuring the altitude of the Pole Star above the horizon. The sun's altitude could also be found, but this required the observer to face the sun, with the result that its light was in his eyes. To avoid this the back-staff was devised, and the cross-staff became known as the fore-staff.

The back-staff was invented in about 1594 by the English sea-captain John Davis. It was intended to be an improvement on the quadrant, mariner's astrolabe and cross-staff for finding the meridian altitude of the sun. The back-staff was also known as the Davis quadrant, and as the English quadrant by Continental seamen. The name quadrant arises because 90° can be measured although there is no 90° arc. The instrument looks like a large triangle with a 30° arc at one end, and a small 60° arc at the other. One-sided vanes with pinholes move over these arcs, and at the end opposite the large arc is a push-on vane with a slit through which the horizon can be viewed. By adjusting the vanes the sun's angle with the horizon can be found. The back-staff remained in use until John Hadley invented the octant in 1731.

Surveying

During the first half of the sixteenth century the scene was being set for the development of surveying into a profession, requiring increasing skill and accuracy, and also the use of instruments devised to make angular measurements, again basically addressing the same problems as those faced by astronomers and sailors. Surveying was to become a science alongside, and to a certain extent as a result of, the development of deep-sea navigation and the growing military skills of fortification and siege associated with the use of artillery. The accurate determination of position, both at sea and on land, became of increasing economic importance as ships left the security of coastal waters, and as new tracts of unexplored territory were discovered. As a result of these needs the art of globe and map-making grew rapidly in importance at a time when the necessary skills of engraving and printing were available, and most notably in the Low Countries. The engraver of copper-plate illustrations could also divide the circle on brass instruments for the use of the astronomer, seaman and surveyor. Another important requirement, both for instrument makers and for those who practised surveying and navigation, was mathematical knowledge. As with printing and engraving, mathematics, in particular in the practical applications, was developed on the Continent of Europe by such notable scholars as Martin Waldseemüller (1470–c.1518) in Lorraine, and Gemma Frisius at Louvain. During the mid sixteenth century, John Dee, a founder Fellow of Trinity College, Cambridge, spent three years studying in Paris, Brussels and Louvain, and on his return home he pioneered the provision of textbooks and translations of Continental practice in English. At much the same time, in 1543, the highly influential work by Robert Recorde, *The Ground of Artes Teaching the Worke and practise of Arithmetike*, was published, which, running through twenty-eight editions in the next 150 years, introduced pen reckoning and the use of Arabic numerals to a wide public. Before the new, angle-measuring techniques could be satisfactorily used, a prior requirement was a knowledge of arithmetic: Arabic had to replace Roman numerals. In the Exchequer Records Arabic numerals were first used at the end of the sixteenth century, but Roman did not disappear for another 60 years. Accounts were done by using ruled boards or cloths, and casting-counters: hence the phrase 'to cast one's accounts'. Recorde may be regarded as the founder of English mathematics, since his textbooks opened the way for self-education for the new class of technicians. He was also the founder of the English school of mathematical practitioners, among whom were later numbered many of the great London instrument makers.

What, then, were the techniques of the sixteenth-century surveyor? The basic requirements, then as now, comprise: measuring angles and the length of a baseline. After sighting prominent features with the angle-measuring instrument first at one and then at the other end of the baseline, and knowing its length, it is a simple matter to calculate the distance of a feature by the theorem of similar triangles. It is also possible to draw a plan or map to scale. Other requirements are to measure heights and depths and distance, to ascertain the orientation, and to achieve a level. When it came to opening up new lands in the Americas, the surveyor was obliged to fix his location by the stars, and thus his skills were close to those of the navigator.

Navigating and surveying are intensely practical activities, so that the development of instruments for their use was always conditioned by factors such as compactness, and speed and ease of use. The theory was established by the late sixteenth century and changed little over the next 300 years. It was the instruments that evolved, becoming more accurate and efficient, and better adapted to differing conditions.

The makers

Printers and copper-plate engravers were craftsmen from continental Europe who were to influence English practice. Robert Recorde's *The Ground of Artes* was printed by Reynor Wolfe, who left his native Drenthe in the north of the Netherlands to settle in London in 1533. Even 50 years later it was difficult to find an English printer who could set up a mathematical work correctly. Throughout the sixteenth century the English were indebted to the Continent, and especially to the Netherlands, north and south, for the skills of printing, of engraving book illustrations, maps and instruments, and for surveying techniques. All these activities are melded together in the work of Christopher Saxton (d.1596), who was obliged to farm out his engraving needs to several men to hasten the production of maps. His *An Atlas of England and Wales*, published at London in 1579, contains thirty-four maps, twenty-three of them bearing the engraver's name. There are seven different signatures, four of them by Flemings, and three by Englishmen. The English are: Augustine Ryther (four maps), Francis Scatter (two), and Nicholas Reynolds (one). Ryther was to become a leading instrument maker. The migration of skilled men increased during mid-century when the grip of Spanish religious persecution was tightening on the Netherlands, finally to result in the Revolt which broke out in 1564.

The instruments

The main materials used for making mathematical instruments were brass and wood, usually boxwood. Prior to the sixteenth century, brass that was needed for such uses as candlesticks, or memorials in churches, had been imported from the Continent, and was expensive. Now with the military threat from Catholic Europe and growing demand for brass and bronze, the Royal Charters for the Company of Mineral and Battery Works and the Company of Mines Royal for the production of brass and brass plate amongst other things, were granted in May 1568, thus allowing English manufacture for the first time. Humfrey Cole, of whom more will be said later, was closely involved in the setting up of the Mineral and Battery Works, which necessitated bringing in German craftsmen to train Englishmen in metalworking skills, and in prospecting for ores. Cole was engraver of dies for the Royal Mint, as well as the leading instrument-maker of the period.

It was from the Louvain area that came the man who can be regarded as the first to establish the scientific instrument-making trade in England. Thomas Gemini (*c.*1510–

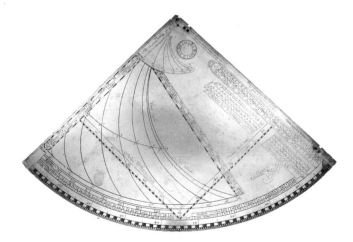

80 Horary quadrant made in London by Thomas Gemini; dated 1551. Radius 27 cm. Engraved 'Edwardus Rex' (for King Edward VI, who reigned from January 1547 to July 1553). The curved lines, calculated for the latitude of London, 51°34', are for time-telling in equal hours by the elevation of the sun in conjunction with a pair of sights and a plumb-line, now lost. Across the centre is a shadow square, for measuring elevations; comprehensive calendrical tables are distributed around the edges. (British Museum, MLA 1958,8–21,1)

62) came from a village near Liège, and as an engraver made his reputation with his plates for his own printing of the *Anatomy* of Andreas Vesalius, issued in 1545. This earned him an annuity of £10 from Henry VIII. At Blackfriars, he carried on the business of map-engraver and mathematical instrument-maker (Fig. 80). An astrolabe by him is in a Belgian museum bearing the arms of the Duke of Northumberland and of King Edward VI; it is dated 1552. In 1555 Gemini printed the *Prognostication* of Leonard Digges, and in the following year, Digges's *A Booke Named Tectonicon*. It was said there that the instruments could be obtained from Gemini. Another astrolabe was made for Queen Elizabeth I. This is dated 1559, and is engraved with the queen's name and the royal arms.

There is no doubt that Humfrey Cole (*c.*1530–91) was London's foremost mathematical instrument-maker of the sixteenth century. Cole was from the north of England, and was employed at the Royal Mint. He undertook to supply all the instruments described in the 1571 edition of Digges's *Pantometria*. Cole's masterpiece must surely be the large, two-foot diameter astrolabe, dated 1575, in the possession of the University of St Andrews, Scotland. This has several resemblances to the Gemini astrolabe made for Elizabeth I, both instruments having on the back a horizontal projection of the

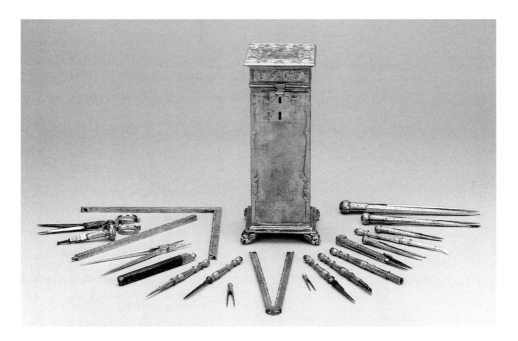

81 Case of architectural drawing instruments in gilt brass, made by Bartholomew Newsam of London in about 1570. Ht 19.3 cm. All four sides are engraved with classical figures, representing Peace, War, Poverty and Abundance. The instruments plug into shaped holes in a wooden block. There are twenty items (a few have been lost), which include 3:1 proportional dividers, adjustable dividers, compasses, beam compasses, folding rule, folding set-square, scissors, knives, pens and scribers. (British Museum, MLA 1912,2–8,1)

sphere derived from the planisphere of Gemma Frisius. Cole's production is both varied and extensive, judging by the twenty-six known instruments, and two engravings. It is clear, also, that he had an influence on subsequent makers, as would be expected. The existing instruments by Cole include: astronomical compendia, sundials, nocturnals, sectors, theodolites, armillary sphere and an astrolabe.

A key factor in the establishment of the English instrument-making trade was the way in which it could become grafted on to the existing guild structure. The City of London Guilds were medieval and, by the sixteenth century, what are known as the Twelve Great Livery Companies had emerged as leaders. To learn a craft and to practise it meant that an apprenticeship had to be served, and the arrangement properly recorded and approved by a City Company. Practising a new craft, mathematical instrument makers had to find a company as best they might; one way was to join a father's company, whichever it might be, under the patrimony arrangement. New companies were formed, of course: the Spectacle Makers in 1629, the Clockmakers in 1631. But the mathematical instrument-makers were captured to a great and surprising extent by one of the Twelve Great Livery Companies, the Grocers. Once a master-apprentice succession was established, the instrument-makers remained in the Company, and so a school was built up.

By 1600, the number of men trained as engravers had increased, and workshops had grown in number and size. Instruments became more complex and varied, with competition in producing new designs to catch attention and to do down a rival (Col. Pl. XVII). Earlier, craftsmen had worked in collaboration with the scholar-inventor; now they were to become capable of independent invention. Instruments for astronomy, time-telling, navigation, and surveying were created for economic reasons. There was, too, a market for wealthy men interested in scientific matters, and elaborately embellished instruments were provided for their delectation (Fig. 81). But this group is more obviously catered for by the products of the leading Continental makers rather than the English, whose instruments are austerely functional.

Bibliography

Bennett, J.A. 1987. *The Divided Circle: A History of Instruments for Astronomy, Navigation and Surveying.* Oxford: Phaidon Christie's.

Brown, J. 1979. *Mathematical Instrument-Makers in the Grocers' Company 1688–1830.* London: Science Museum.

Brown, J. 1979. Guild organization and the instrument-making trade, 1550–1830. *Annals of Science* 36, 1–34.

Brown, Lloyd A. 1949. *The Story of Maps.* Reprinted 1977. New York: Dover Publications.

Chapman, A. 1984. Tycho Brahe in China: the Jesuit mission to Peking and the iconography of European instrument-making processes. *Annals of Science* 41, 417–43.

Chapman, A. 1990. *Dividing the Circle: The Development of Critical Angular Measurement in*

Astronomy 1500–1850. 2nd edn, 1995. Chichester: John Wiley & Sons.

Cotter, Charles H. 1968. *A History of Nautical Astronomy.* London: Hollis & Carter.

Dekker, Elly 1993. Epact tables on instruments: their definition and use. *Annals of Science* 50, 303–24.

Dekker, Elly 1995. An unrecorded medieval astrolabe quadrant from *c.* 1300. *Annals of Science* 52, 1–47.

Gibbs, S., with Saliba, G. 1984. *Planispheric Astrolabes from the National Museum of American History.* Washington, D.C.

Gunther, R.T. 1932. *The Astrolabes of the World.* Reprinted 1976. London: Holland Press.

Horský, Z. and Škopová, O. 1968. *Astronomy Gnomonics: A Catalogue of Instruments of the 15th to the 19th Centuries in the Collections of the National Technical Museum, Prague.* Prague.

Karrow, R.W., Jr. 1993. *Mapmakers of the Sixteenth Century and their Maps.* Chicago: Newberry Library.

King, D. A. and Turner, G. L'E. 1994. The astrolabe presented by Regiomontanus to Cardinal Bessarion in 1462. *Nuncius: Annali di Storia della Scienza* 9, 165–206.

Krämer, K.E. 1980. *Mercator: Eine Biographie.* Duisburg.

Mackensen, L. von 1979. *Die erste Sternwarte Europas mit Ihren Instrumenten und Uhren: 400 Jahre Jost Bürgi in Kassel.* Munich: Callwey.

Maddison, F.R. 1969. *Medieval Scientific Instruments and the Development of Navigational Instruments in the XVth and XVIth Centuries.* Coimbra.

Mörzer Bruyns, W.F.J. 1994. *The Cross-Staff: History and Development of a Navigational Instrument.* Zutphen: Walburg Pers.

Pledge, H.T. 1939. *Science since 1500.* Reprinted 1966. London: Science Museum.

Poule, E. 1983. *Les Instruments astronomiques du Moyen Age.* Rev. edn. Paris.

Price, D.J. 1957. Precision instruments to 1500. In *A History of Technology,* ed. Charles Singer *et al,* vol. III, pp. 582–619. Oxford: Clarendon Press.

Ræder, H., Strömgren, E. and B. (eds) 1946. *Tycho Brahe's Description of his Instruments and Scientific Work as given in Astronomiae instauratae mechanica (Wandesburgi 1598).* Copenhagen: Munksgaard.

Richeson, A.W. 1966. *English Land Measuring to 1800: Instruments and Practice.* Cambridge, Mass.

Roche, J.J. 1981. The radius astronomicus in England. *Annals of Science* 38, 1–32.

Šima, Z. 1993. Prague sextants of Tycho Brahe. *Annals of Science* 50, 445–53.

Stimson, A. 1988. *The Mariner's Astrolabe: A Survey of Known, Surviving Sea Astrolabes.* Utrecht: HES.

Taylor, E.G.R. 1954. *The Mathematical Practitioners of Tudor and Stuart England.* Cambridge University Press.

Taylor, E.G.R. 1971. *The Haven-Finding Art: A History of Navigation from Odysseus to Captain Cook.* Rev. edn. London: Hollis & Carter.

Turner, A.J. 1987. *Early Scientific Instruments: Europe 1400–1800.* London: Sotheby's.

Turner, G.L'E. 1991. *Gli Strumenti: Storia della Scienza,* vol.I. Gen. ed. Paolo Galluzzi. Turin: Einaudi.

Turner, G.L'E. 1991. *Scientific Instruments and Experimental Philosophy 1550–1850.* Aldershot: Variorum.

Tyacke, S. (ed.) 1983. *English Map-Making 1500–1650: Historical Essays.* London: British Library.

Ward, F.A.B. 1981. *A Catalogue of European Scientific Instruments in the Department of Medieval and Later Antiquities of the British Museum.* London: British Museum Publications.

Waters, David W. 1958. *The Art of Navigation in England in Elizabethan and Early Stuart Times.* London: Hollis & Carter.

Whitfield, Peter 1995. *The Mapping of the Heavens.* London: British Library.

Zinner, E. 1956. *Deutsche und niederlandische astronomische Instrumente des 11. bis 18. Jahrhunderts.* 2nd edn 1967; reprinted 1979. Munich: Beck.

COLIN RONAN

Astronomy in China, Korea and Japan

Astronomy in both Korea and Japan was based on the astronomy of ancient China, which can be said to have dominated these East Asian cultures. This is not only because China was for long the dominant ancient civilisation in this region of the world, but also because the Chinese developed astronomy very early on in their history. It goes back to at least the second millennium BC.

Throughout most of its history, Chinese astronomy was based on a particular world view. This was the belief that the country's ruler was 'the Emperor of all under heaven'; he was a divine appointment and, in consequence, the heavens echoed his rule. If he headed a good and just administration, then the celestial bodies followed their appointed courses without deviation or the appearance of unexpected events. If his administration was unjust or defective in other ways, then this was indicated by the appearance of comets, the advent of new stars – what we today know as 'novae' and 'supernovae' – and a whole host of other unpredicted events in the sky.

In one sense, such a link was not unique. What we should now term 'portent astrology' was prevalent in at least some other early civilisations, for it is only to be expected when an absolute ruler and his people are both under the influence of gods in the heavens. Yet in China the concept was deeper and more sophisticated than this. The link was not between celestial deities and the emperor; it was a more holistic outlook. The earth, its emperor and the entire cosmos were all part of a gigantic organism. If one part were 'ill' or out-of-step, the whole being was affected. With this outlook it was logical to expect celestial events to presage disasters on earth, which might perhaps be avoided by administrative changes. Nevertheless, the Chinese did not themselves invent the personal astrology which uses horoscopes; that was a much later foreign importation, which seems to have arrived by way of India, probably about AD 220.

In China portent astrology can be said to have assisted astronomy. This was because it called for careful and regular observation of the skies. Moreover, the obvious importance of every portent demanded that results be noted down in detail, and this the Chinese assiduously did. The result is that China now possesses the longest unbroken run of astronomical records in the world, observations which are now proving of vital importance to modern astronomers whose research requires evidence of long-term celestial events (see pp. 329–41).

A contributory reason for the longevity of Chinese astronomical records is that, again in company with people in all early civilisations, the Chinese made attempts to foretell the future. Various methods of divination or 'techniques of destiny' were used,

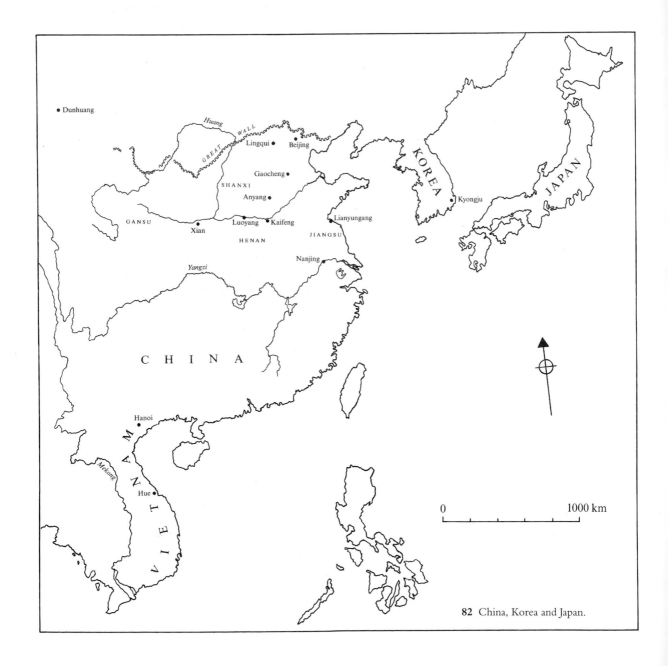

82 China, Korea and Japan.

but in the days of the Shang kingdom (1500 BC–1050 BC), one important system was scapulimancy, a technique that goes back to Neolithic times. It involved using ox and deer shoulder-blades or the shells of tortoises which, having been scraped clean of flesh, were dried and then polished. Holes were then drilled in them from the reverse side, and when a prognostication was required, a glowing brand was inserted into one of the holes. The resulting cracks in the bone or the tortoise shell were then examined by an expert diviner. But what is so significant about the Chinese use of this method, is that the prognostication, and later its results, were both recorded on the cracked material.

Though lost for many years, remnants of these 'oracle bones' were discovered in the last decades of the nineteenth century by farmers tilling their fields near Anyang in Henan province. Bought by a local drug-store as 'dragon bones' because they were reputed to have special medicinal purposes, they were dispensed until 1899 when Chinese administrators and scholars coming across them found that they contained very ancient writing, and recognised them for what they were. Now many thousands of these ancient bone pieces used for scapulimancy form a veritable treasure-house of information, some of it containing astronomical details, going right back to Shang times. Thus it is that Chinese observational records of the sky take us so very far into the past.

In ancient times in China the astronomical records were a closely guarded secret. The reason for this is obvious enough; if unusual celestial events presaged disasters and were pointers to maladministration, then the last thing a ruler wanted was for his people to know about such divine censure. In consequence, astronomers soon became an integral part of the administration of a country which developed the most elaborate bureaucracy in the ancient world. There was the equivalent of an Astronomer Royal heading the specialist staff, while others both in the administration and outside it were actively discouraged from astronomical study. In addition to astronomers, the Astronomer Royal had charge also of astrologers and meteorologists and *clepsydra* experts to tend to the ingenious water clocks (*clepsydras*) which the Chinese devised. These were all officers of the court, though even the Astronomer Royal was lower in the social scale than senior administrators.

Calendars in ancient China

Calendars were important in early China, as they were elsewhere, both to regulate religious observances and to organise civil life. However, as a consequence of the Chinese ruler's divine connection with the heavens, it became customary that after important changes of reign and always after a move to a new dynasty, a fresh calendar was drawn up. This made public that a new mandate and a new disposition of celestial influences had been granted; it was a tradition well established by Han times (206 BC to AD 220) and led, for instance, to some forty new calendars between early Han times and the advent of the Ming dynasty in 1368.

As happened in other early civilisations, the very first Chinese calendar was lunar. This was only to be expected as it is easy to see the regular changes in the moon's shape – the cycle of its phases – and to count the days between one new moon and the next. The seasons are, however, regulated by the sun, not the moon, and a solar calendar is more convenient for civil purposes. This requires more subtle observations over a far longer period. Nevertheless, oracle bone fragments show that by 1400 BC, in Shang times, the Chinese had adopted a lunar calendar with mean lunations (lunar months) of 29.5 days (modern value 29.53 days) by alternating months of 29 and 30 days. Most years contained 12 lunar months.

The two cycles, the mean lunar month and the year (of 365.2422 days), are incommensurable; one is not an exact multiple of the other. Some compromise has to be made, as was found in other countries besides China. Though it gradually became realised that a single valid calendar was not a very likely possibility, Chinese astronomers still sought one, and so developed their knowledge and understanding of astronomical science. Thus about the sixth century BC they came to recognise that there was a 19-year cycle after which the phases of the moon recur on the same day of the solar (or tropical) year. This period is known in the West as the Metonic cycle, having been discovered by Meton of Athens, about a century later (see pp. 46 and 70–71).

To keep generally in step with the solar year, an additional long or short month was appended to the lunar year when necessary. As time progressed, continued observation allowed knowledge of the moon's behaviour to become more accurate, and by the Tang dynasty (AD 618–906) a value very close to the modern one was applied to the calendar, giving not only an improved agreement between the appearance of the moon and the beginning of the month, but also leading to better prediction of lunar eclipses.

By the first century BC a new system of intercalation was adopted, based on the *er-shi-si jie-qi* or 'twenty-four fortnightly periods', each one corresponding to a 15° motion of the sun along the ecliptic (its apparent path in the sky). The actual period of each 'fortnight', 15.218 days, gives a year of 365.23 days. The year of this wonderfully descriptive calendar began on 5 February with 'The Beginning of Spring', moved on to 'The Rains' (20 February), 'The Awakening of Insects' [from hibernation] (7 March), the 'Spring Equinox' (22 March), 'Clear and Bright' (6 April), 'Grain Rain' (21 April), 'Beginning of Summer' (6 May), 'Lesser Fullness of Grain' (22 May), 'Grain in Ear' (7 June), 'Summer Solstice' (22 June), 'Lesser Heat' (8 July), 'Greater Heat' (24 July), 'Beginning of Autumn' (8 August), 'End of Heat' (24 August), 'White Dews' (8 September), 'Autumn Equinox' (24 September), 'Cold Dews' (9 October), 'Descent of Hoar Frost' (24 October), 'Beginning of Winter' (8 November), 'Lesser Snow' (23 November), 'Greater Snow' (7 December), 'Winter Solstice' (22 December), 'Lesser Cold' (6 January), and 'Greater Cold' (21 January).

As far as the lunar calendar was concerned, most months contained two fortnightly periods, but because the lunation is only 29.53 days, there is always a chance that a lunar month will fail to contain two full periods. An intercalary month was inserted when this happened, and therefore the Chinese considered this civil calendar as a luni–solar one. Nevertheless, lunar calendars were markedly improved over the centuries as ever more precise measurements were made of the moon's motion across the sky, a motion which we now know is complex due not only to the moon's elliptical orbit but also to gravitational effects between the earth, moon and sun. By the time the sixth century AD had dawned, the irregular apparent motion of the sun along the ecliptic was recognised, and in the seventh century the irregularities of both the sun and the moon were incorporated into the calendar. Improvements continued well into the thirteenth century.

As time went on the astronomers of China reached a better understanding of eclipses of both the sun and moon, and became able to predict their appearance, a very necessary matter for Chinese portent astrology. In the first century BC, during Han times, they recognised a period of 135 months during which eclipses of both kinds totalled 23. By the third century AD the astronomer Yang Wei was able to specify many details, even predicting approximate times during an eclipse of the sun when the edge of the moon would make first contact with the sun and when the eclipse would end. But all these predictions were based on observational evidence rather than any underlying theory of lunar and solar motions.

Foreign calendars were also introduced into China. During the Tang dynasty (AD 618–906) an Indian calendar arrived, and in the Yuan dynasty (AD 1279–1368) a Muslim calendar made its appearance. Both ran side by side with the indigenous calendars of the Chinese. Even when the highly developed Gregorian calendar was introduced from the West by Jesuit missionaries in the seventeenth century, though its admirable qualities were recognised, it did not officially supplant Chinese systems until 1912.

Besides lunar and solar calendars, the Chinese also operated a Day Count system. This worked independently of the sun and moon and was devised very early on. It combined the 'Ten Celestial Stems' or *gan* (a series of ten Chinese characters), and the 'Twelve Earthly Branches' or *zhi* (a series of twelve horary characters), dividing these into sixty units. Counting was done by way of *gan-zhi* combinations. For over 3,000 years no one tried to discard the system. From early times the Celestial Stems were used to define a 10-day period (*xun*), which functioned as a kind of week. The 7-day week was a later importation from outside and did not arrive at least until the Song dynasty (AD 960–1279).

The now well-known 12-year cycle, associating the years with various animals, may go back as far as the sixth century BC. It has no astronomical connections, being purely of archaeological and ethnological interest.

Closely connected with the calendar was the matter of timekeeping. At some early time in China it became accepted that the day began at midnight, as does ours now. The Mesopotamian method of beginning the day at sunset (see p. 45) was totally unknown to the ancient Chinese. As for the day itself, it was subdivided in two different ways. In the first place the day was divided into 12 double-hours, each of which equals 2 Western hours. Each of these double-hours was divided into halves, named 'beginnings' and 'mid-points'. Midnight and noon came at mid-points rather than at beginnings of a double-hour, so that the double-hours ran from what we should call 11 pm to 1 am, 1 am to 3 am, and so on. A second system subdivided the day into 100 equal intervals or *ke*, each of which corresponds therefore to 14 minutes and 24 seconds of Western time-reckoning, or roughly quarter of a (Western) hour. Each *ke* was itself divided into fractions called *fen*. The night from the end of dusk to the start of dawn was also divided into five night watches (*gong*), which varied in length from about 2.6 hours in winter to 1.7 hours in summer.

Charting the skies

The idea of charting the skies was not exclusive to China, though as it turned out the Chinese made some particularly notable advances in the technique. Indeed, the earliest Chinese 'star chart' goes back some 4,000 years to at least 2000 BC.

This early carving was discovered on a cliff at Jiangjunya about 9 km south-west of the city of Lianyungang in Jiangsu province, and lies on a sloping face of bare rock measuring about 22 m by 15 m. The carving contains many stars as well as human and animal heads. The heads appear alone, with no bodies, and seem very primitive, even in comparison with Neolithic art found elsewhere in the region. Nevertheless, some are adorned with head-dresses and these are similar to Neolithic depictions, one even on a pot sherd from Jiangjunya itself. Moreover, some other symbols echo decorations on Neolithic pottery vessels and can be interpreted from images on oracle bones. The human heads with head-dresses are to be found near the top of the carving and Chinese archaeologists suggest that perhaps these represent star spirits since they appear in the company of stars. Some human heads without head-dresses are taken to refer to sacrifice made on what may have been a giant altar stone. Then, near the lower part, there are what seem to be animal heads and crops, though possibly a circular symbol here may be depiction of an early concept of the universe itself.

There are two remarkable things about the carving which make it evident that it is an early star chart, even though it contains much symbolical material as well. First, there are discs which indicate the sun (Fig. 83), and these appear to have been given seasonal positions. Second, there is a region where the stone has been polished and where a number of bright stars as well as the moon appears, also in seasonal positions. This bright region (Fig. 84) is evidently a depiction of the Milky Way, that hazy band of starlight which crosses the whole heavens. That one can be sure of this is due not only to its position and its general appearance but also to the fact that the Milky Way shows gaps and divisions which are depicted on the carving.

The drawing up of actual charts of the sky rather than schematic pictures like the carving on the Jiangjunya cliff, means finding some way to depict positions as if one is

83 Carvings on the Jiangjunya cliff, depicting the sun in two different positions.

84 The Jiangjunya cliff carving of the Milky Way and the moon (depicted as a disc) in various positions.

drawing a map. This is made more difficult by the fact that the sky appears like a curved dome above one's head, whereas a chart is probably going to be drawn on a flat surface. Moreover, the dome does not stay still; it rotates once every day, approximately 4 minutes faster (on Western reckoning) than the full 24 hours taken by the sun. Because of this difference between the rotation period of the sky and the daily apparent rotation of the sun, the stars visible each night slowly change throughout the year, different groups appearing from season to season.

To those like the Chinese who dwell in the northern hemisphere of the earth, the pivot point about which this rotation occurs is known as the north celestial pole. Today this point lies very close to the star Polaris, or α Ursae Minoris.

In view of all this, if the skies are to be charted, it becomes necessary to divide up the heavens into areas which can readily be recognised and whose extent can be measured. Such divisions may take a variety of forms. Since, as will become evident later, Chinese astronomers as well as those of other civilisations, designated the positions of celestial bodies by measuring the angles between them, areas can be parcelled out reasonably precisely.

85 A European representation from the nineteenth century of the constellation of Orion, the Hunter. (Betelgeuse is here shown as Betelgue.) The characteristic seven brightest stars of the constellation can readily be seen at Orion's shoulders, sword-belt and legs. The Chinese recognised these stars, though sometimes they showed the constellation in two parts.

The Chinese carried out this process of division from very ancient times, refining it as time went by. One of their earliest schemes was that of the Five Palaces. One division – the Palace of Purple Tenuity – contained those stars close enough to the north celestial pole to render them visible above the horizon all night and every night. They never set, and are usually known as the circumpolar stars. Those comprising them vary in number with the geographical latitude of the observer. For an observer at the earth's North Pole (latitude 90°) they cover the entire collection of stars north of the 'celestial equator', but for those in lower latitudes the number becomes less, decreasing as the latitude decreases. An observer on the earth's equator (latitude 0°) will therefore see no circumpolar stars; all will rise and set every night. From the latitude of early capital cities such as Xian, Luoyang and Kaifeng (latitude 35° N) the area of circumpolar stars would be equivalent to those visible today from the islands of Crete and Cyprus. From Madrid and Ankara (latitude 40°), the circumpolar area would be a little larger and similar to that seen from Beijing.

The four other divisions were the Palace of the East, symbolised by the Azure Dragon; the Palace of the South, symbolised by the Vermilion Bird; the Palace of the West, symbolised by the White Tiger; finally, there was the Palace of the North, symbolised by the Dark Warrior (an entwined turtle and snake). There were other ways of selecting areas. The scheme of the Nine Fields was one. This also had a circumpolar region, accompanied however by eight sections not four. Such a division was connected to the eight hexagrams which appear in the divination book *Yi Jing* (*The Book of Changes*), which dates probably from the second century BC.

Rather different were the Jupiter stations, a twelvefold division of the celestial equator and, by analogy, the ecliptic along or near which not only the sun but also the planets appear to move. This division into twelve parts had nothing whatsoever to do with the Western scheme of the twelve signs of the zodiac, but was connected with the fact that Jupiter takes nearly 12 years to complete each orbit round the sun and, more specifically, with the *tai sui*, a 'shadow' planet invented so that it could imitate Jupiter's motion but in the opposite direction. The twelve 'Earthly Branches' which we have already met in discussing the Chinese day-count system, were also tied in with the Jupiter stations.

Of all the Chinese divisions of the sky, the best known in the West is probably the series of lunar mansions (or lunar lodges). This system, which is of considerable antiquity and goes back at least to the fifth century BC, divides the heavens into 28 unequal sectors. The use of 28 divisions is derived from the moon's apparent motion among the stars – the sidereal month of 27.32 days. The lunar mansions were important in portent astrology and thus the system was thought to be of particular significance.

Grouping the stars into recognisable patterns or constellations took place early on in China as it did in other ancient civilisations. However, the groupings are very different from those familiar in the Western tradition; indeed only two Western constellations are anywhere near the same. The first and most obvious is what we call Orion (Fig. 85), that wonderful group of stars so noticeable in the winter skies of the northern hemisphere. The Chinese call it *Shen*, which means 'a union of three' and thus indicates that what they noticed most of all were the three conspicuous stars of what we call 'the belt of Orion'; in this instance the name *Shen* is therefore sometimes translated as 'Triaster'. The other is that part of Ursa Major (which is now generally called the Great Bear) which contains seven bright stars and is now known in the West variously as The Plough or The Big Dipper. The Chinese refer to it as The Northern Bushel. In addition, some similarity is to be seen in about ten other constellations, but in essence the Chinese groupings are very different from those of the West. This is of considerable interest because it strongly suggests that the Chinese groupings grew up independently of the West's.

Mariners are now familiar enough with the fact that the two end stars of the bowl of the Big Dipper act as pointers to the Pole Star (Polaris), and thus to the north celestial pole. Yet this was not always so, for the celestial pole slowly changes its position. The reason this happens is that the earth's axis is slowly spinning in space. This spin, often likened to the way the axis of a top spins or precesses as the top slows down, means that each end of the axis traces out a circle in the sky. It takes 25,800 years for the celestial poles to move a full circle. The result is observable from earth because the observed point of rotation of the starry sky gradually changes from pointing in one direction to another. (Southern hemisphere observers experience a similar change for the south celestial pole.) In other words, while the axis is at present pointing close to Polaris for northern observers, in 13,000 years from now the north celestial pole will appear to be very close to the bright star Vega, currently some 50° away. The movement of the celestial poles causes a change of over 20 minutes between the length of the year measured by observations of the sun (the tropical year) and its length when measured by observing the stars (the sidereal year). However, the Chinese did not recognise this movement as a continuously occurring phenomenon, and called it merely *sui zha* or 'annual difference'. The astronomer Yu Xi was the first to mention *sui zha*, in AD 330, though in the West this 'precession of the equinoxes' had already been discovered by Hipparchus in the second century BC (see pp. 80–81).

The preparation of charts of the sky and of catalogues of star places requires some means of specifying positions of stars with reference to one another or, better still, to some reference grid. Since all astronomical observations of positions are made by measuring angles, the most convenient system, even today, is to consider the stars as fixed on the inside of a dome or a sphere. For the Chinese the basic reference system was based on the celestial pole and the celestial equator (Fig. 86). The celestial equator was divided into 365.25 degrees. This gave one co-ordinate axis: the other axis – at right angles to this – was the 'north polar distance', also measured in similar degrees. (Here, to avoid confusion, we shall continue to use the Western convention which divides a circle into 360°.) The north polar distance, where angles are measured from the north celestial pole (0°) to the celestial equator (90°), is equivalent to the modern

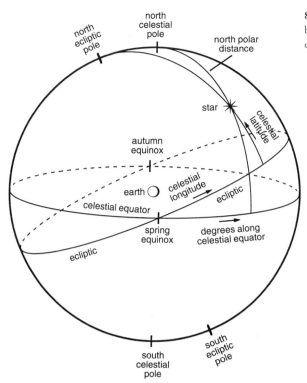

86 The celestial sphere, showing co-ordinates based both on the celestial equator and the ecliptic.

method used by astronomers, which measures distances north (0° to +90°) or south (0° to −90°) of the celestial equator in the opposite direction, i.e. from the celestial equator up to the North Pole (or down to the South), a co-ordinate known as 'declination'. Incidentally, in the West, the shift from measuring positions based on the ecliptic and the pole of the ecliptic (i.e. on celestial longitude and celestial latitude) to the equatorial system as used in China occurred only in the late sixteenth century.

The first catalogues of star positions appear to have been due to Shi Shen, Gan De and Wu Xian, three of the earliest notable astronomers of China, whose excellent work was carried out between 370 and 270 BC, that is, two centuries before Hipparchus. Together their lists enumerated 1,464 stars, grouped into 284 constellations. This means that their groupings were, and are, smaller than those used by Western astronomers whose constellations number only 88. For instance, our long meandering constellation Hydra with no less than sixteen stars bright enough to be readily seen, covers three lunar mansions together with eight other Chinese star groups. In AD 310 during the Western Jin dynasty (AD 265–316), this early work was collated by the Astronomer Royal Qian Luozhi, who had a bronze celestial globe cast and coloured the stars on it in red, black and white to distinguish the listings of the three astronomers.

The fact that positions of the stars could be catalogued and charted with some precision means that the lunar mansions could be precisely laid down. Each mansion was therefore specified by particular stars, so that the moon's entry into and passage out of a mansion could be readily observed. However, because of the way the mansions were specified, these stars frequently turn out to be rather dim ones.

Besides catalogues of stars, the Chinese also prepared charts of the sky. The earliest of these were probably produced in Han times. We know this from Han carvings and reliefs, for these show individual constellations or groups of stars (asterisms). This was done by depicting the stars as dots or small circles, which were connected by lines to delineate the constellation itself. This 'ball-and-link' convention is therefore old in China, though its adoption in the West seems not to have appeared until late in the nineteenth century.

The preparation of maps of the sky also involves the problem of depicting the curved surface of the sky – that is, of the celestial sphere – on a flat surface and, just as mapping the almost spherical surface of the earth, requires the use of some kind of 'map projection'. In both China and the West this goes back a long way when it comes to mapping the earth, but for star charts it seems first to have appeared in China during Han times in a chart, now lost, by Zhang Heng, the designer of the world's first seismoscope for detecting earthquakes. The oldest surviving Chinese star chart which depicts the whole of the visible sky is painted on paper and is now in the British Library. It comes from Dunhuang in Gansu province, and gives a flat representation of Qian Luozhi's three-coloured traditional chart on the celestial globe. Though it has no grid lines and so, in one sense, is imprecise, it shows over 1,350 stars, and is drawn up in thirteen sections; one is a planisphere – a flat map centred on the north pole – and twelve are flat maps centred on the celestial equator. The Dunhuang chart was probably prepared in the Tang dynasty, perhaps around AD 700 (Col. Pl. XVIII).

Many later star charts and planispheres still survive, for instance those of the Song dynasty by Su Song in AD 1094. These show co-ordinates as well as stars and were prepared for constructing an 'armillary clock', which is described on pp. 263–4. Su Song also developed a method of plotting the stars using two planispheres, one centred on the north celestial pole, and one on the south. Both became adopted by others, and put the Chinese in the forefront of those civilisations which charted the stars with ever increasing precision, for the West had little or nothing to compare with this until Renaissance times.

Observing instruments

The gnomon

Of all the instruments used in observing the heavens, certainly the simplest and, probably, the oldest is the gnomon. Basically, this is a stake stuck vertically in the ground, and was used by many civilisations in ancient times. Its original use was primarily to measure the sun's shadow at midday at different times of the year, and so assist in providing a solar calendar (Fig. 87).

87 A late Qing dynasty (AD 1644–1911) representation of the measurement of the sun's shadow at the summer solstice with a gnomon and a shadow template at the foot of the vertical rod. The measurement is being made by the legendary astronomer Xi Shu. (From *Historical Classic*, reproduced in Needham 1959, fig. 110)

Since the sun rises in the east and sets in the west, and pursues a curved path in between these points, midday is that time when it appears highest in the sky. At midday in summer the sun reaches its highest point in the sky above the horizon – the summer solstice – and its shadow will then be the shortest one to be recorded during the year. Conversely, at midwinter – at the winter solstice – the sun will appear at its lowest midday position above the horizon; then its shadow will reach its greatest length. In early times, and certainly in the seventh century BC, these determinations of shadow length were considered so important that the ruler himself attended the actual measurement. Later on, measurements were made at the equinoxes as well.

In the sixth century BC and then more extensively three centuries later, the Chinese standardised weights, measures, and many other practical details such as the width of roads. The gnomon was not forgotten and there came to be a standard size, namely one with a length of 8 *chi*, that is 2.4 m. Shadow lengths were measured using a *tu gui* or standard 'gnomon shadow template' made either of pottery, terracotta or jade. It was these standard instruments which were used in the most ambitious project, that of determining the terrestrial distance corresponding to an arc of the meridian. Such a determination of this north–south line was vital for precise calendar-making, because that entailed measuring accurately the latitude of those stations where observations were to be made. It was also significant in map-making and in determining the size of the earth. The most notable Chinese attempt to do this was made between AD 721 and 725 under the famous Buddhist astronomer and mathematician Yi Xing and the Chinese Astronomer Royal at the time, Nangong Yue.

No less than nine locations were selected, with latitudes ranging from 17.4° at Indrapura in the old kingdom of Champa (now near Hue in South Vietnam) up to the old city of Lingqui which lay close to the Great Wall in the north of Shanxi province at latitude 40° – the same latitude as present Beijing. This series of stations, though not strictly on a north–south line, covered the prodigious distance of just over 3,500 km. Simultaneous measurements of the shadows at summer and winter solstice were made at all stations, and these confirmed the previous value obtained three centuries before for the latitude of the city now known as Hanoi in North Vietnam. The main outcome of this remarkable feat was that the distance on the earth corresponding to 1° of latitude turned out to be almost 155 km, which is rather larger than today's value (111 km), but far more accurate than previous attempts. Indeed, the variation in shadow length with change of latitude was found to be four times the value previously accepted. Certainly no piece of field research like it was carried out anywhere else in the early Middle Ages.

Incidentally, in making his calculations of the results Yi Xing used 'tangent tables'. Though previously thought to have been a Muslim invention of the ninth century, it

88 Zhou Gong's Tower at Gaocheng, housing a 12-m gnomon.

now seems that the Chinese discovered the use of tangents and tabulated them at least a hundred years earlier.

In the tenth century AD Arab astronomers began to build huge stone measuring instruments in order to obtain greater precision in their measurements. This was because such instruments could have larger scale intervals engraved on them which could be read with far more accuracy than small ones. During the Yuan (Mongol) dynasty (AD 1279–1368) in China, which was established by Kublai Khan, use was made of foreigners, particularly Europeans and Arabs, in the administration. Thus it was that the Arab idea of using large instruments for astronomy came to be introduced. A legacy of this still stands at Gaocheng, some 80 km south-east of the old capital Luoyang, where in 1276 the Chinese astronomer Guo Shoujing constructed a giant gnomon with a height of some 12 m (Fig. 88). Built on a site used certainly since Han times for official solstice measurements, the structure is known as 'Zhou Gong's Tower for the measurement of the sun's shadow'. It consists of a brick tower shaped like a truncated pyramid, with sides measuring some 15 m at the bottom and 7.6 m at the bottom. Two staircases inside lead up to a small building at the top which has three rooms, the centre one giving a good view of the top of the gnomon and the shadow it cast. At ground level, and to the north of the tower, there protrudes a stone 'Sky Measuring Scale' with a length of over 36 m. This contains graduations and two long water channels to indicate that it lies flat and at right angles to the gnomon. The present structure is a Ming dynasty (1368–1644) renovation.

One innovation used in measurements from Zhou Gong's Tower was a 'Shadow Definer'. This consisted of a leaf of copper about 5 cm wide and 10 cm long, in the middle of which was a pinhole. This was pivoted in a square wooden framework, and allowed the shadow of a cross-bar at the top of the gnomon to be seen crossing a tiny image of the sun, thus assuring readings superior in accuracy to those using the shadow of the cross-bar only.

The armillary sphere

An armillary sphere is a device composed of a number of rings (Latin *armillae*) which depict the celestial sphere and its reference circles. How many rings are involved depends upon the number of 'great circles' – the celestial equator, the ecliptic and their meridians – which astronomers wish to depict. The Chinese, with their concentration on the celestial equator and its poles, always depicted these, though sometimes they also included the ecliptic and its poles, the presence of which made the armillary more complex (Fig. 89).

The early history of the armillary sphere in China is uncertain. Known as *hun yi*, the 'celestial sphere instrument', it is said that the astronomer Loxia Hong set up such a device in 104 BC. At all events, the Chinese certainly used the armillary to measure the positions of celestial bodies. The earliest Chinese armillaries would, probably, only have had rings to represent the celestial equator and a meridian. However, from the first

89 An armillary sphere constructed in bronze by the Jesuits in China and now at the Old Observatory in Beijing.

90 (*below*) A Ming Dynasty (AD 1368–1644) copy of Guo Shoujing's 'simplified instrument', preserved at the Purple Mountain Observatory of the Academia Sinica, near Nanjing.

century BC in Han times, developments occurred. Geng Shouchang introduced the first permanently fixed equatorial ring in 52 BC, while in AD 84 a ring to represent the ecliptic was fitted. Then, some 40 years later, in about AD 125, Zhang Heng added rings to represent the horizon and a meridian of altitude. After this, armillaries of both types – the simple and the more complex – continued to be made without change until in the thirteenth century AD a most significant development occurred.

To anyone who has watched the way the sun, moon and other celestial bodies move across the sky, the fact that they travel from east to west in curved paths will have become obvious. This is because the heavens appear to rotate about the celestial poles. For observers it means that any sighting instrument must be rotated not only from east to west but also upwards (or downwards) at the same time in order to keep a celestial body in view. However, in AD 1270 the Chinese astronomer Guo Shoujing devised a simple armillary consisting of only a single ring carrying a sighting bar, but mounted so that its axis of rotation lay parallel to the axis about which the heavens appear to rotate; as we now know that is really parallel to the axis about which the earth rotates. This 'equatorially mounted' armillary ring needed but one rotation – a movement from east to west – in order to follow the curved paths of the stars, and so made observing not only more convenient but permitted more precise observations to be made (Fig. 90).

Known in China as *jian yi* or 'the simplified instrument' because of its simple armillary ring, nevertheless it contained all kinds of refinements. For instance, there was also a vertical single ring armillary at the north (higher) end; this was to allow determination of the altitude and azimuth (angle round the horizon starting from north measured eastwards). The main instrument gave, of course, measured positions along the celestial equator and downwards from the north celestial pole (the north polar distance). Again, the cast circle at the top (north) end of the axis of the instrument carried a cross-piece inside it with a mask with a hole in it to allow the observer to determine the moment when the Pole Star reached its highest point. The circle at the lower end of the axis was used for measuring positions along the celestial equator, and ran on roller bearings. Also it had markers which showed the limiting positions of the lunar mansions. Finally, the whole instrument was mounted on a framework which contained channels filled with water. These allowed the base to be adjusted so that it was levelled by having the water at the same height in every channel. A Ming copy still survives in Nanjing.

The 'simplified instrument' was the precursor by many centuries of the equatorial mounting of astronomical instruments in the West. The first of these to be used was a small sighting instrument known as a *torquetum*, described first about 1284, but an equatorial mounting of the kind devised by Guo Shoujing did not appear until 1791 when it was used for an astronomical telescope made in England, whence its design became known as the 'English' mounting; whether it was based on a report of the Chinese instrument is not known.

With the Chinese belief that the aspects of the heavens and the emperor's administration were closely connected, as early as the second century AD it occurred to Zhang Heng that an armillary sphere rotated mechanically would show the heavens as they ought to be. If one checked against observations of rising or setting stars, say, then any discrepancies should readily be detectable. It is reported – though in a seventh century book – that in AD 132 he set up in a closed room such a device driven by a water-wheel activated by a jet of water, presumably from a constant head of water from a *clepsydra* or water-clock. The mechanism would, of course, have to drive the 'hidden' armillary very slowly, and how Zhang Heng achieved this is not known. That it could be solved is certain from the work of Yi Xing and a colleague Liang Lingzan in AD 723, for from this time onwards the mechanical rotation of armillaries for purposes of demonstration became regularly adopted. However, this development had another effect of great importance, for it led in AD 1090 to the construction in Kaifeng, then the capital, of a large public mechanical clock.

Time-keeping devices

Time-keeping has always been a province of the astronomer and the Chinese used a variety of methods to measure the passing of time. Sundials were one obvious development of the gnomon, and were not exclusive to China. Yet the Chinese used many designs, and certainly by Former Han times (206 BC–AD 24) such devices were used to check the time-keeping accuracy of water-clocks. By the time of the Southern Song dynasty (AD 1127–1279) they devised an ingenious equatorial sundial, the dial of which was a disc with a straight gnomon that protruded both above and below the disc. In summer, when the sun is north of the celestial equator, readings were taken from above the disc, while in winter when the sun is below the celestial equator, readings were made on the underside. Conveniently, its equal graduations gave equal hours.

As far as water-clocks or *clepsydra*s were concerned, these were not unique to China but were found in both Egypt and Mesopotamia. Like those two civilisations, to begin with the Chinese used an outflow clepsydra, in which water in a container flows out in a steady – or reasonably steady – stream. It appears to have been used at least as early as the seventh century BC, and was known either as *lou hu* ('drip vessel') or *ke lou* ('graduated leaker'). The Chinese also knew of another archaic device, a floating bowl with a hole in its bottom. Placed in a vessel of water it took a specific time to sink. This inflow clepsydra with a vertical indicator rod fixed inside it, came into wide use during Han times.

The problem with both types is that as the water level drops so the rate at which the water flows in or out changes. The chief way to overcome this difficulty was to use a series of water tanks, each feeding one below it. The more there were, the steadier the outflow, and sometimes six such tanks were used. But the Chinese noted two other sources of irregularity. One was the loss of water by evaporation, the other was that the water flow became sluggish at low temperatures and would cease altogether if the water froze. One way of overcoming the latter was to use mercury as the liquid and then allow it to flow into a vessel which was part of a balance; here the passing of time was indicated by increasing weight. From about AD 450 mercury was also used in this way for determining short intervals of time.

The great clock tower in Kaifeng mentioned above (Fig. 91) was constructed by Su Song, who not only introduced an elaborate design of compensating tanks for the flow of water to drive a water-wheel but also a device which was to be the heart of all later mechanical clocks all over the world. This was an escapement mechanism. The significance of this device is that it only permits the gears of the clock to move tooth by tooth with a regulating device keeping the same interval between the escapement of

91 A reconstruction, drawn by John Christiansen, of Su Song's astronomical clock-tower at Kaifeng. (From J. Needham, *Science and Civilisation in China*, vol. IV, pt 2, 1965, fig. 650)

each tooth. Later, in accurate Western clocks, the regulating device was a pendulum, but in Su Song's clock it was the time taken by flowing water to fill a small bucket at the end of each spoke on the escapement wheel. The important thing is that the principle for both is exactly the same. Certainly Su Song did not himself claim the invention of the principle; that seems to have been due to Yi Xing in AD 725. But what Su Song did was not only to cause the clock to drive a hidden armillary sphere inside the tower, but also to drive an observing armillary on a platform so that it too followed the apparent movement of celestial bodies across the sky. In addition time was indicated by the movement of 'jacks' or little statuettes. This was the word's first public mechanical clock, though its rotation of armillary spheres was not new. Nevertheless, it was not until the early nineteenth century that clock-driven telescopes became widely adopted by Western astronomers.

Cosmology

In ancient China we can distinguish three general views of cosmology. In no case were these derived from any desire to help in giving detailed descriptions of the movement of the sun, moon or planets. Chinese astronomers did consider these but not from a geometrical scheme of orbits, as came to be the hallmark of Western astronomy. In China such movements were considered from an arithmetical or what may be called an algebraic approach. They were concerned with recurrences, regularities and the determination of long periods after which relative positions of bodies with respect to each other were repeated. In brief, the Chinese astronomers took a different approach from their European counterparts, but this did not prevent them drawing up a picture of how the universe was laid out. Indeed, over the ages three different schemes were proposed.

The first of these was the *Gai Tian* layout, which may be translated as the 'heaven as a cover' theory. In this, the sky is regarded as a gently curved vault lying above and parallel to a gently curved earth beneath. Heaven rotates daily about a vertical axis through the centre – a position which corresponds to the north terrestrial pole. The distance between earth and sky was supposed to be some 80,000 *li* (46,000 km), a figure obtained by taking terrestrial measurements and applying them to the heavens. Though very ancient, it was a scientific theory, free from myths and human overtones.

The second cosmological idea was the *Hun Tian*, which is usually translated either as the 'celestial sphere', or as the 'enveloping sky' theory. The earliest propounder of the theory of whom we know is Loxia Hong, who was active between 140 and 104 BC. In about AD 117 in a book about the armillary sphere, Zhang Heng described this theory by an analogy: 'The heavens are like a hen's egg and as round as a crossbow bullet; the earth is like the yolk of an egg, and lies along in the centre. Heaven is large and earth is small.' He also claimed that there is water inside the lower part of the celestial sphere, that the heavens are supported by vapour and that the earth floats on those waters.

This was a concept which fitted in well with, or perhaps gave rise to, the armillary sphere.

Zhang Heng acknowledged that this celestial sphere theory was old, and later there were some criticisms of it, one being that it was difficult to imagine that the sun, being hot and fiery, could move through the waters below the horizon. Later, the belief in 'waters under the earth' fell into abeyance. Yet in essence, the *Hun Tian* theory, like its predecessor, arose from the problems which faced the practical astronomer.

There was a third cosmological theory in China, but this was relatively late, making its appearance only during the Han dynasty. Indeed, the earliest astronomer associated with the theory is Qi Meng who lived during the Later Han dynasty (AD 25–220) but he is said to have remembered what those before his time had taught concerning it. This astounding view of the universe conceived of infinite empty space with the sun, moon, planets and stars floating freely in it. All these bodies were believed to be condensed vapour and carried in their paths by a 'hard wind'. This last was possibly a concept derived from the blast of air generated by pumps during smelting. The idea of celestial bodies floating in space seemed to account for the behaviour of the moving celestial bodies which sometimes crossed the sky in one direction and then reversed their motion; the theory allowed them to do so because they were not rooted to anything or tied together in any way.

This *Xuan Ye* or 'infinite empty space' theory was a most enlightened view, but it was perhaps too far in advance of its time, for it seems to have found no favour with later Chinese astronomers. Such a fate also befell three other theories which appeared during the third and fourth centuries AD, but which are of little significance except that they show, clearly, that there was an active period of cosmological speculation in China during the first few centuries of our era.

Chinese astronomy, linked though it was to the holistic view of emperor, people and natural events, made notable and often unique contributions. It saw the introduction of effective calendars, of star charts, and of the equatorial system of celestial co-ordinates for specifying the positions of celestial bodies. The Chinese also possess long and continuous detailed astronomical records unequalled anywhere else in the world. In addition the Chinese introduced sophisticated astronomical instruments, including the first equatorial mounting, and were the pioneers of driving such instruments automatically using a clock to control their motions. Chinese astronomers were also the first to suggest an infinite universe.

Ancient astronomy in Korea

In Korea, the Chinese concept of the heavens displaying the activities of a divine ruler and his administration was accepted. As in China, the *Gai Tian* or 'heaven as a cover' theory of the universe held sway until the *Hun Tian* or 'enveloping sky' arrived after the Han period. The Koreans also adopted measurement of positions based on the equator

92 A Korean 10-won banknote illustrating the tower-like observatory built in AD 647 at Kyongju in the Silla kingdom. (British Museum)

and north celestial pole and divided the celestial sphere into five 'palaces', though besides the circumpolar regions, they recognised nine sections of the sky rather than four as in China. In addition Korean astronomers used the lunar mansions, Jupiter stations, and fortnightly periods of the Chinese and accepted that a circle possessed 365.25 degrees. They also adopted the Chinese calendar and their scheme of time-keeping.

In AD 647 a tower-like observatory was built at Kyongju in South Korea in the Silla kingdom. This appears on a modern Korean banknote (Fig. 92). It seems that instruments were mounted on the flat platform at the top, which has a north–south east–west alignment. The tower itself may also have acted as a gnomon. However, the acme of Korean observational astronomy was the re-equipment in the 1430s of the Royal Observatory undertaken by King Sejong (1418–50) who, besides being a great ruler, was also a patron of culture and scholarship and took a great interest in astronomy. Indeed, not only did the king initiate the refurbishing and provide the necessary finance, but also he personally oversaw the planning and execution of the work.

The new instruments were numerous and fell into a number of categories. For time-keeping there were various sundials – equatorial, horizontal and bowl-shaped – together with one designed for accurately determining the south point of the horizon. There were also various water-clocks, some portable, but two were elaborate *clepsydra*s which struck the hours, each housed in its own pavilion. One – the Striking Clepsydra – was housed in a two storey construction, on the upper floor of which three jacks in the form of immortals acted as announcers, sounding the double-hours by a bell, the night watches by a drum, and divisions of the latter by a gong. One storey below was a horizontal wheel with twelve immortals on iron rods and free to move up and down; each had a placard announcing one of the double-hours. The second – the Jade Clepsydra – fed mechanisms to indicate the rising and setting of the sun as it passed round a paper mountain some 1.7 m high, a jade immortal in the form of a girl who struck the hours, and four warriors as well as twelve other jacks to indicate the time. In addition, there was scenery to indicate the seasons.

Probably the most significant astronomical instruments were two versions of Guo Shoujing's 'simplified instrument', one larger than the other, and 'sun-and-stars time-determining instruments' derived from it for determining the time by observing the positions of stars near to the north celestial pole. In addition there was a measuring scale over 27 m long with a bronze (or brass) gnomon standing some 9 m high, and containing a shadow definer. This, clearly, was based on the Zhou Gong's tower with the 12-m gnomon near Luoyang. The Observatory also contained a rain-gauge and a Chou Foot Rule, and was therefore a repository of standards as well as an astronomical and meteorological centre.

Unfortunately, Sejong's Observatory was destroyed in the devastation which accompanied two unsuccessful invasions of Korea by the Japanese in 1592 and 1597, and though some items were later replaced, in the next century Western astronomy reached China and soon penetrated to Korea, with the result that indigenous developments were coloured or replaced by Western traditions.

Korea produced a wonderful late flowering of Chinese-based astronomy, yet fine though this was, it was essentially an heir of Chinese traditions.

Ancient astronomy in Japan

Astronomy did not become a science in Japan until long after contact with the cultures of China and Korea, and well after the invention of writing which was introduced from Korea in the third century AD. Indeed, technical matters were for long the concern of immigrant specialists, who were consulted when necessary. Only in AD 533 did the Japanese emperor request a Korean state to send professors of medicine, divination and calendar-making. Then about 70 years later a Korean priest Kwal-luk arrived and presented as a tribute books on astronomy, astrology, calendar-making and geography, as well as other subjects. Three or four pupils, of whom two were Korean immigrants, were selected to study under him.

Around these times, the reconsolidation of China into a gigantic centralised state inaugurated a period when Chinese political and cultural institutions were introduced into Japan, and Japanese were sent to the Chinese court. Thus it was that in AD 646 Japanese clans were federated under an emperor, and a bureaucratic state established. The outcome was that the Japanese became enthusiastic for Chinese culture, and so imitated Chinese practices in astronomy as well as in other fields.

By AD 628 the Japanese had already adopted the Chinese time-keeping system and had constructed *clepsydra*s, but now they went still further. In 675 an astronomical observatory began to function and then in 702 the Taiho civil code, which covered not only government administration but also education, was adopted. This ensured that astronomy, portent astrology and calendar-making became accepted subjects. Between 690 and 861 Chinese calendars were adopted and underwent four revisions. When a year later in Tang dynasty China there was another calendar revision, this was also adopted in Japan, yet following it no further revisions were made in spite of growing differences between calendar dates and observed events. This neglect may have been due to a gradual change in attitude in Japan, for astronomy, as part of the one-time enthusiasm for Chinese institutions, declined. Then, during the tenth until the first half of the sixteenth century, the old bureaucracy was replaced by a strictly hereditary court aristocracy, itself to be replaced later by an emergent warrior class. An almost inevitable consequence was that the office of 'Astronomer Royal' became purely hereditary; it followed a well established course of procedures, and there was no urge to develop the science.

Seemingly a somewhat humble subject brought in from Chinese culture, astronomy flourished in Japan while Chinese institutions were in favour, but lost its high esteem when attitudes changed. Unhappily, the Japanese themselves seem to have added little if anything to the science, although their recorded observations continue to be valuable.

Bibliography

Clark, D.H. and Stephenson, F.R. 1977. *The Historical Supernovae*. Oxford: Pergamon Press.

Cullen, C. 1982. An eighth century Chinese table of tangents. *Chinese Science* 5, 1–33.

Harley, J.B. and Woodward, D. (eds) 1994. *The History of Cartography*, vol. 2, pt 2: *Cartography in the Traditional East and Southeast Asian Societies* (with chapters on 'Chinese and Korean star maps and catalogs' (F.R. Stephenson) and 'Japanese celestial cartography before the Meiji Period' (Kazuhiko Miyajima). University of Chicago Press.

Nakayama, S. 1969. *A History of Japanese Astronomy*. Harvard University Press.

Needham, J. 1959. *Science and Civilisation in China*, vol. III, pp. 171–461. Cambridge University Press.

Needham, J., Lu, G-D., Combridge, J.H. and Major, J.S. 1986. *The Hall of Heavenly Records*. Cambridge University Press.

Ronan, C.A. 1981. *The Shorter Science and Civilisation in China*, vol. 2, pp. 67–221. Cambridge University Press.

Sivin, N. 1986. Chinese archaeoastronomy: between two worlds. In *World Archaeoastronomy*, ed. A. Aveni. 1989. Cambridge University Press.

Sivin, N. 1989. *Cosmos and Computation in Early Chinese Mathematical Astronomy*. Leiden: Brill.

Stephenson, F.R. and Clark, D.H. 1978. *Applications of Early Astronomical Records*. Bristol: Hilger.

ANTHONY F. AVENI

Astronomy in the Americas

The Americas are two isolated continents populated by races indigenous to Asia who migrated across the Bering land bridge more than 10,000 years ago. Since then the diverse cultures of these lands would evolve untouched by the civilised West until the time of Columbus. What a marvellous human-centred laboratory in which to explore important questions about the development of astronomy. Do people all over the world react in the same way to the visible impulses that rain down upon them from the starry skies? Was the world round for those people too? Did they also develop the idea of a sun-centred planetary system? Did the same sky events clocked and calibrated in the same way suggest identical calendars? Can we really expect New World astronomers to have anything in common with their Greek or Babylonian counterparts? Did they use the same observing techniques, keep similar notebooks, draw parallel conclusions? Would the astronomers' superiors, whoever they were, put celestial knowledge to the same use? Such questions about the astronomies of these sequestered cultures may be discussed and compared with information derived from other chapters of this book.

Our discoveries about astronomy and the human condition will reveal a paradox. We shall find on the one hand remarkably similar conceptualisations between the hemispheres east and west, especially when we review the evidence on the Mayan notions of Venus – so similar in fact that we will almost be compelled to inquire whether their astronomers once entered into trans-oceanic collusion with Babylonian skywatchers (see p. 276). But on the other hand our encounter with the Inca sun-watcher's pillars and Amazonian constellations will bring us face to face with astronomical perceptions and ideas indigenous to American soil so divergent from those of our European ancestors that we may even ask whether we are dealing with people living on the same planet.

In studying the astronomies of these diverse cultures we operate under a major handicap. Most of them transmitted their knowledge of the sky through non-literate means. They did not use stylus and tablet, quill and parchment – the traditional mode of conveying knowledge in the West. The exceptions are the Maya, who possessed writing and numerics akin to our own, and their distant relatives, the Aztecs and their

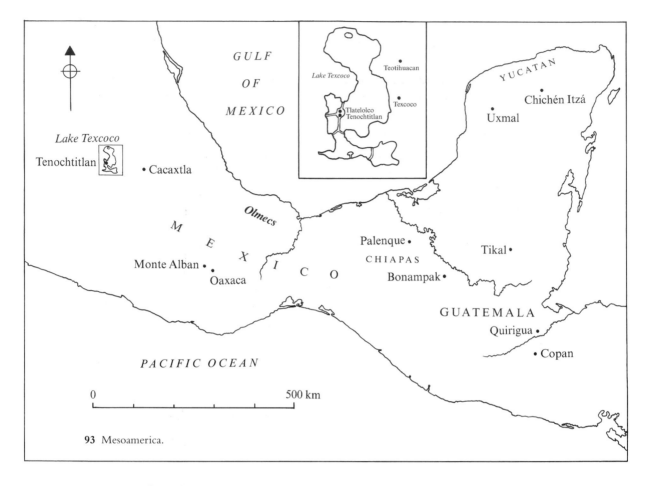

93 Mesoamerica.

ancestors of highland Mexico, who created pictographic books on deerskin. To make matters worse, Spanish historians who chronicled the conquest of the New World admit to having destroyed most of these documents because they understood them to be involved with the worship of the devil. Only four Mayan and about two dozen Central Mexican books have made it to the twentieth century intact. The only other evidence we possess for the study of astronomy comes from monumental sculpture, ceramic and architectural decoration; this can be combined with other unwritten testimony such as data concerning the arrangement and orientation of buildings to face sky phenomena that took place at the horizon. These data comprise the interdisciplinary field of archaeoastronomy. We must also include the histories written by the early Spanish prelate-scholars who often carefully reported details on native astronomical customs, ironically, so that they could abolish them. Finally, ethnoastronomy considers data collected by contemporary anthropologists that reveal a host of calendars, constellations and cosmologies that have survived in almost pristine form half a millennium after contact, especially in very remote areas.

We allot the largest space in this chapter to the cultures that are the most well researched and which exhibit astronomies that might contrast with Western astronomy in the most revealing light. We begin with the Maya who, with their elaborate hieroglyphs and mathematical dot and bar notation, are traditionally regarded as the paragon of scientific New World cultures. Perhaps this is because, at least on the surface, their astronomy seems more like our own – precise and predictive, logical and mathematical. We continue with the Aztecs and Incas, each stereotyped as bloodthirsty subjugators of people, yet demonstrably capable of scaling impressive intellectual heights, the former with their solar oriented temple and complex time cycles that interwove human destiny and divine history, the latter with a precise calendar embedded in sight lines connected with a system that radially partitioned the landscape of their capital city and incorporated an impressive mnemonic scheme of social organisation.

We conclude by reviewing the astronomies of a diverse array of societies organised in smaller units such as family and class. These include the Desana of Amazonia and certain native tribes of North America. These cultures offer us an array of media and methods for expressing a knowledge of the sky. They highlight for us some of the differences that exist between the astronomies of urbanised, highly stratified societies and those we tend to classify as simpler in their social organisation. We shall learn that from the hunter–gatherer on the slopes of the Sierra Madre to the highly specialised court astronomers of Mayan Copan, the sky offered a means of solving some of life's basic problems: how to regulate human activity, how to understand and worship the gods, how to know what it means to be a member of society. The texts, both written and unwritten, and the answers contained within them, bear testimony to human ingenuity in articulating ways of knowing the sky.

The Mayan pinnacle

Sometime around the sixth century BC in Mesoamerica, highly stratified societies began to express the measure of time quantitatively and in written form. Calendrical inscriptions carved in stone appear at the ruins of San José Mogote near the modern city of Oaxaca. Later (c. 300 BC) the inscriptions became more abundant at nearby Monte Alban. Meanwhile, along the Gulf coast, the Olmecs since the eighteenth century BC had been carving simple motifs that would later become recognisable as Mayan hieroglyphs. For reasons we perhaps will never know, these forms of expression coalesced to produce an abundance of carved inscriptions for which the Maya have become famous. Between AD 200 and 900 they produced thousands of astronomically related dates and intervals that served the purpose of framing their history.

The base of the Mayan counting system was twenty and its principal sub-unit was five, each no doubt arising from the use of the human body as a counting device. Dots, probably originating from making gestures with the tips of the fingers, stood for ones, and fives became bars, an abstraction of the extended hand with fingers held closely

94 Mayan calendrical glyphs in a monumental inscription, the Accession statement of Jaguar Paw Skull, ruler of Tikal (AD 488–537):

Introductory glyph (position A1)
Long count (positions B1, A2, B2, A3, B3): 9.2.13.0.0 = $(0 \times 1) + (0 \times 20) + (13 \times 360) + (2 \times 7200) + (9 \times 144,000) =$ 1,390,320 days after creation (he came to office)
Calendar round (positions A4, A7): 4 Ahau 13 Kayab
Lord of Night (position B4): the ninth lord
Phase of the moon (position A5): 17 days counted from new moon
Number of days in current month (position B6): 29
Additional glyphs give the other time cycles and titles. The ruler is portrayed on the reverse side.

together. Many of the early glyphs resemble these gestures. The vigesimal count was built upon a base of twenties to create a system of numeration by position that also contained an entry for zero; thus, ones, twenties, twenty times twenties, twenty times twenty times twenties, etc.

Mayan time was cyclic and repetitive. The day, expressed by the Mayan word *kin*, which also means 'sun' and 'time', was the smallest temporal unit recorded in the inscriptions, there being no expressions for subdivisions of the day. It is only when it came to counting the days that the Maya, like the Babylonians, changed the counting scheme. Thus, they identified the third position with 360 (18×20 rather than 20×20), perhaps as a way of handily reckoning the approximate length of the year. Thus they probably thought of the three lowest orders in a Mayan day count as years : 20-day 'months' : days, rather than the arithmetical sum of three quantities. By the beginning of our Christian era, they wrote inscriptions that reckoned the number of days lapsed since the most recent cycle of creation. This was expressed as a five-place number that we call the long count, an example of which is shown at the top of Fig. 94. Mayan mathematicians of the early Classical period fixed the great event, which we would transcribe as 0.0.0.0.0 on their temporal odometer to a date that corresponds in the

Christian calendar to 12 August 3114 BC. Though the reasons why they did it still elude us, this setting of a very long super-cycle of creation is relatively common in highly stratified societies (see Aveni 1989).

A shorter cycle, called the *tzol kin* (count of days), which the Maya still employ today, predated the long count by several centuries. Widespread throughout Mesoamerica yet unknown in the Old World, this 260-day round usually was expressed in the form of one of twenty day-names preceded by a numerical coefficient ranging from one to thirteen. Thirteen was the number of layers of heaven in Mayan cosmology, while the origin of twenty already is obvious. But other reasons may underlie its popularity. In contemporary Guatemala, there is solid evidence linking this basic time cycle to the human gestation period. This convenient round number approximates nine lunar synodic months (265.77 days) in the same way 360 approximates the length of the tropical year (365.2422 days). The period is also a close approximation to the mean interval of appearance of Venus as evening, or morning, star and it beats harmoniously in the ratio of $2:3$ with the eclipse half-year of 173.5 days, a fundamental cycle for reckoning eclipses. Moreover, the Maya paid close attention to each of these periods. Such proximities offer the advantage to the calendar keeper of assigning particular sets of dates in the *tzol kin* to eclipse or Venus warnings. Yet another astronomical facet of the 260-day cycle is that in southern Mesoamerica the sun at noon remains south of the zenith for approximately this period and north of it for the remainder of the year. Like all important constants of nature, this basic period of 13×20 probably became exalted in Mesoamerican calendars precisely when it was recognised to be commensurate with other perceived cycles of Mayan time.

Like the subtropical year period in our own calendar before the reform of Caesar (see p. 94) and in other calendars (see, for example, the Inca calendar on pp. 294–6), the *tzol kin* probably began as an early Mesoamerican agrarian device to keep time. Counting lunar months and body tallies that corresponded to active time – that is, the interval between burning off the old crops of the previous harvest and the completion of the harvest of the current cycle – would have constituted all the time that mattered for these people. The rest would go uncounted. An astronomic event such as a stellar heliacal rise would serve to reset the cycle to a seasonal beat.

By the first century AD we find inscriptions that refer to a second short cycle, the 365-day year or *Haab* of 18 months of 20 days plus 5 extra days. Placed alongside the 260-day period, a given date in the *Haab* does not roll around until 73×260 or 52×365 days have expired. This interval we call the calendar round, and since the Maya seem to have recognised that it ran about the length of a reasonably augmented human lifetime it may have been intended to express precisely that concept in a grander scheme of repetitive cosmic cycles. There is, incidentally, no solid evidence that the Maya ever cared about a leap year. Rather, it was the continuous flow of time they emphasised in their ideology; thus, they seem to have been content to let their time cycles roll on without a break, though they easily could calculate the anniversary of the tropical year within the *Haab*. For example, in one Mayan calendrical document they seem to have recorded the distance of the half-year mark from the summer solstice, and at different points in time they reversed the order of representative pictures that refer to solstice events.

Part of our fascination with the Maya can be attributed to the fact that they (or at least the élite) were literate in the most popular definition of that skill; that is, the Classical Maya possessed a visible language that consisted of letters (syllables actually) and a grammar, and one of the products of that literacy was the book. Descriptions by the first outsiders to see them indicated that Mayan books distinctly resembled those developed in the Western tradition from Egyptian papyrus to Dead Sea Scrolls to Medieval Codex – in fact, today we still inappropriately call the four Mayan books that survive 'codices'. A sixteenth century bishop of Yucatan, Diego de Landa, tells of the books and their users:

> They provided priests for the towns when they were needed, examining them in the sciences and ceremonies, and committed to them the duties of their office, and provided them with books and set them forth ... Their books were written on a large sheet doubled in folds ... and they wrote on both sides in columns following the order of the folds ... The sciences they taught were ... computation of the year's months and days ... the fateful days and seasons, their methods of divination and their prophecies ... and how to read and write with the letters and characters ... and drawings which illustrate the meaning of the writings.
>
> Tozzer 1941, pp. 27–8

Astronomers, along with the mathematicians and scribes who worked with them, were members of a courtly class (Fig. 95). Indeed, a recent élite burial at Copan proved to contain the remains of a scribe. These specialists were required to know with meticulous accuracy every celestial cycle that might conceivably guide human destiny. The control over these time periods by the Maya lords, their superiors, enabled them to schedule their own affairs as well as those of the state. As in China, such court professionals advised the ruler about the appropriate time to make decisions. When should we enter into armed conflict, set a date for a royal marriage, or an ascent to the throne? When ought we to conduct a ritual, or make an offering to pay the debt to the gods for their assistance in producing a good crop or a healthy newborn child? We can imagine royal appointees passing from town to town, their codices tucked under their arms, ready and willing to advise the local rulers on such issues.

Alas, no Mayan astronomers' notebooks survive, but we can study the contents of the books that escaped the Hispanic conflagration. And we can reasonably deduce what sorts of events they watched, even the methodology they employed to time the periods. The format of Mayan astronomical tables in the codices incorporates numbers, words and pictures – a stark contrast to the monotonous strictly tabular Babylonian texts (see

95 Mayan calendar scribes depicted on a ceramic vessel. The one on the extreme left and the third from the right may be teachers. One of them has pen in hand as he leans over a text in conversation with one of his students. Mayan dot and bar numerals issue from his mouth. (From Justin Kerr, *The Maya Vase Book*, Kerr Associates, New York, 1989, p. 67)

Fig. 19). But make no mistake, the content is the same: they are omen laden. Unlike our wall calendar, which simply appends events to a continuous sequence of running numbers (Modern America's 1993 February calendar, for example, places St Valentine's Day on the 14th, President's Day on the 15th and a full moon on the 6th), Mayan calendars stress a sequence that consists of linking one event (usually a picture) with the next and the next, etc., through a sequence of tabulated intervals. Unlike the monumental inscriptions, these tables were not simply historical records but true ephemerides, for they pin astronomical events to real time both past and future (often hundreds of years apart) and they offer mechanisms for predicting when the events will take place.

In the Dresden Codex we find, for example, a table for predicting solar and lunar eclipses, an ephemeris for Mars and another for Venus; the Paris Codex contains a table that charts the place of the moon in the Mayan zodiac, a menagerie of very diverse animals not unlike the zodiac of Old World Egypt. An exquisite example of the pinnacle of Mayan astronomical precision, the Venus table is among the most fascinating of all Mayan documents. It projects to us a class of specialist that became as carried away with rigour and detail as our own Babylonian, Greek or Renaissance astronomers. We reproduce in Fig. 96 and Col. Pls XIX–XX a pair of pages of this twelfth-century codex from north Yucatan. One of these is part of a Venus table that runs through 65 synodic cycles of the sky's brightest planet and encompasses a total of 146 *tzol kin* (104 *Haab*). A correction table that precedes the main text can be used to adjust predictions of when Venus will rise heliacally, thus rendering the scheme accurate to one day in five centuries. But religious constraints also bore on the calendar. Also, thanks to the correction table, one who knew how to use the table could re-enter it precisely on the day name of Venus, which is 1 Ahau in the 260-day count. The table also integrates the *Haab* date of Venus's first appearances in perfect step with the Venus cycle, thanks to the natural harmonic fit between the synodic period of the planet (584 days) and the approximate length of the seasons (365 days) – a perfect 5 : 8 ratio. In effect, one measure of the Mayan control of Venus time lay in the recognition that once the planet made a heliacal rise, it would do so again almost exactly on the same date eight solar years later.

As evidenced in the red coloured numbers at the bottom of each page, the 584-day synodic period is fractured into four sub-intervals in an unusual way: 250 days are assigned to the time Venus spends as evening star, while 236 are accorded to the morning star apparition. In between 90 and 8 days are set equal to the two disappearance periods. This is remarkably similar to the way the Babylonians segmented the Venus period (see p. 42). Now it turns out that only one of these interval assignments comes close to reality. This is the last – the one immediately preceding each of the five Venus pictures, which show the Venus god hurling spears that likely represent his daggers of light that bring omens of (usually) ill fortune to the theatre of human affairs. Again, the choice of the too long 90 days as well as the unequal 236 and 250-day periods (the last two in reality ought to be equal at approximately 263) has an underlying, if foreign, logic. It is likely that the Maya wanted to reckon the appearance of each of the Venusian stations with observable points in the lunar phase cycle. By so choosing these intervals, they thus guaranteed, for example, that Venus's first station (appearance) would correspond to the last date on which the morning star would still be visible when the moon was in the same phase it exhibited as when Venus made its heliacal rise. This tendency to lock changes in Venus appearances into points in the lunar phase cycle may be an indication that the old Mayan calendar once had a lunar rather than a solar base line.

It is probably no coincidence that the scribes (or their superiors?) chose to recopy an eclipse warning table next to the Venus Table in the Dresden Codex (see Fig. 96). Each interval across the bottom line of this eight-page table translates into a time packet of lunar synodic intervals – repeated clumpings of six lunar synodic months (178 days) followed by one set of five (148 days). Each bunch of five moons is followed by a picture, the whole table encompassing 405 full moons as well as 46 *tzol kin*, spanning more than three decades. If there were any doubt that eclipses were involved, some of these illustrations depict half-light, half-dark discs with lunar crescents opposing the *kin* glyph, symbol of the sun. Dragons devouring the sun and a dead lunar goddess hanging by her hair from a segmented sky-serpent also appear in the pictorial portion of the table.

The Dresden Eclipse Table seems to have contained mechanisms for warning which full moons might be eclipsed and which new moons might eclipse the sun. It must have taken all of a century or more, i.e. several generations of perceptive astronomical observing, for specialists in skywatching to work through to a conclusion which their Chinese and Babylonian counterparts also had arrived at – that once a lunar or solar eclipse occurs, it is not possible to have another (of the same kind) until six, or more rarely five, lunar months pass.

Of the long-range 405-moon cycle in the Dresden Codex, this much is certain:

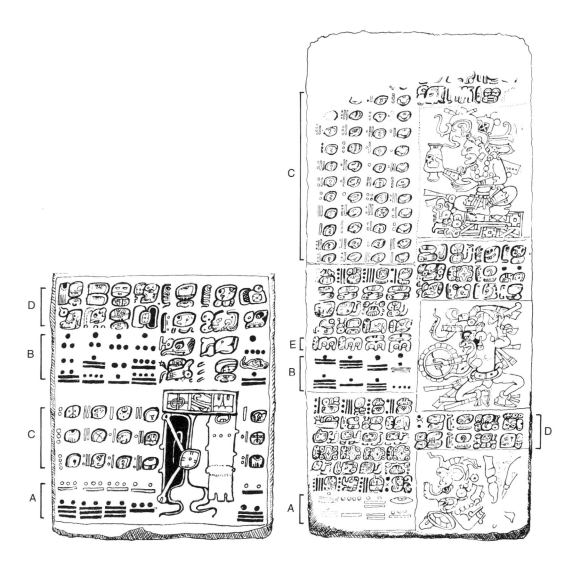

96 Pages of two astronomical warning tables from the Mayan Dresden Codex (see also Col. Pls XIX, XX). On the left is an eclipse table, and on the right a table marking the times of appearance and disappearance of Venus on its synodic round. The format is the same. Position A marks intervals between events: groups of 177 or 148 days between eclipses, or the different station intervals for Venus. Each entry is clearly readable in dot–bar base-20 notation. Position B gives the cumulative number of days obtained by adding successive intervals to previous totals as the reader moves from left to right. Position C gives possible coefficients and day names in the *tzol kin* for each event. The large pictures portray the event: for the eclipse a sun symbol positioned within a half-light, half-dark field hanging from a sky band; for Venus the heliacal rise signalled by the Venus god flinging darts of light at victims whom he impales below. Position D gives the omens and E the Venus hieroglyphs, which are also prominent in astronomically aligned architecture. (From J. Antonio Villacorta and Carlos A. Villacorta, *Codices Mayas Reproducidos y Desarrolados*, Tipográfica Nacional, Guatemala City, 1977, pp. 118 and 102)

first, it was used to gain control of astronomical time; and second, it was a time cycle derived from the observation of eclipsed and uneclipsed moons of the past, which could be used as data to generate a model for anticipating when eclipses might occur in the future.

But why would forerunners of the Dresden Codex and their copyists settle on 405 as the basic choice in which to cast their picture–interval–picture eclipse format? Part of the answer is that ritual time, the 260-day side of the coin of all of Mesoamerican temporal reckoning, was as important to them as the time meted out by the stars. In fact, most of the tables in the codices are 260-day almanacs. So, too, the Venus Table is thoroughly imbued with principles of commensuration. The Maya were fond of discovering and rediscovering that two or more calendrical cycles could be equated to fit precisely into larger cycles by separately multiplying each one by different small whole numbers. Thus $8 \times 365 = 5 \times 584 = 2920$ relates the seasonal year to the Venus cycle in the five-page Venus Table. And $104 \times 365 = 65 \times 584 = 146 \times 260 = 37,960$ in a full run of the table ties both of them to the *tzol kin*. Likewise in the Eclipse Table $405 \times 29.53 = 46 \times 260$. In one very unusual case, a long count number, which marks one of the Venus stations since the most recent creation, permutes a host of cycles: $5256 \times 260 = 3744 \times 365 = 2340 \times 584 = 1752 \times 3 \times 260 = 36 \times 37,960 = 1,366,560$. It has been called the 'super number' of the Mayan codices.

These mathematical operations resemble our habit of resetting a clock to noon or midnight by making its hands point straight up when the sun crosses the meridian. Magical multipliers possess the practical value that they place the observed celestial periods on track with one another as well as with other religious cycles; for example, they can restore Mars and Venus back to the same time of appearance in the sky or they can realign eclipses with the seasons. In the case of the super number, the Maya tried to fix a time in history more than three millennia after creation, when Venus would repeat its first appearance as morning star, and moreover they reset the correct day-name in the 260-day ritual count to fit with that appearance. Great cross-referencers, the Classical Maya became so carried away with finding and fitting together commensurate periods that they even concocted schemes in which they could predict the occurrences of phenomena associated with two different celestial bodies, both within the same calendrical table. For example, the heliacal rising of Venus following inferior conjunction, as anticipated by each of the tabulated dates in the Dresden Venus Table, always occurs after the beginning but before the termination of the next lunar month following a lunar eclipse that was visible in Yucatan. Indeed, Mayan astronomers forced Venus to march to the celestial beat of a lunar drummer, as opposed to the more familiar rhythm that dominates our own calendar. But anyone who pays attention to the calendar of, say, Islam, which still operates with a lunar base, will not find such a timekeeping format so unusual. Thus there is a very good reason for placing the Eclipse and Venus Tables next to one another in the codex.

We suggested earlier that there is no evidence for a Mesoamerican leap year. Yet this seems strangely at odds with the mechanism the Maya used to reset the canonic Venus to track the real luminary the astronomers surely must have followed across the sky. Indeed, the corrections Mayan astronomers employed to update their Venus predictions seem not so different from the way Caesar in 45 BC (see p. 94) and Pope Gregory XIII in 1582 reformed our own Roman calendar. To put the too short seasonal calendar of 365 days on course with the solar (tropical) year (365.2422 days), they called a halt to the calendar wheel, sliced a segment out of it to restore the marking of the equinox date to fit the time of year when days and nights really were equal, and then introduced a foolproof plan that would be employed thereafter to indicate far into the future when to make periodic minor adjustments to reset the canon. Caesar's scheme was clever enough to last fifteen centuries before misalignment of the calendar with reality became embarrassingly obvious. Gregory's concoction, with minor alterations that came well after his time, has become so effective that it exceeds the degree to which we can reckon the variable period of revolution of the earth about the sun! In passing, it is worth remembering that religious reasons (setting the date of the Easter festival) rather than scientific ones motivated the desire to change.

The priests who set the Mayan Venus Table into real time were engaged in similar activities. They assessed single (4-day) and double (8-day) corrections to the tabulated Venus round every 104 years, thereby making up for the 0.08 day shortfall between the average observed synodic period of the planet (583.92 days) and the whole number of 584 days that appears in the table. The 'aberrant' numbers on the introductory page of the Venus table prove it. Though there is still disagreement about exactly in what order these near-whole, large multiples of the observed Venus period were employed and for how long they were effective, they surely were part of the grand design for crafting a sophisticated script that told with staggering precision what Venus did in the sky.

Like eclipse cycles and Venus appearances, Mars also possessed a numerical link with other known cycles. The Mars Table, which precedes the Venus Table in the Dresden Codex, exploits the unusual coincidence between a triple *tzol kin* (3×260 days) and the Mars synodic period (780 days) – just the sort of coincidence that would titillate a commensurate-minded Mayan skywatcher. The planet is represented by a series of pictures of a long-nosed beast shown descending to varying degrees from a segmented sky serpent. If the key event anticipated in the Venus Table is the first morning appearance of the planet, that in the Mars Table appears to be the first perceptible backward motion and subsequent forward movement of the planet on its retrograde loop – a celestial swerve that baffled even Kepler.

Make no mistake about it, though the astronomy in the Mayan codices impresses us with its logic, rigour and precision, it was essentially a scientific means to a practical and religious end – sustenance and correct behaviour in the service of the gods. The glyphs between the numbers typify the astrological ends of Mayan date prediction: 'Woe to the turtle [drought?]; woe to the warrior and pregnant female; his sacrifice, his divine punishment is set; evil excessive sun; the misery of the maize seed.' Clearly, here was a universe that did not operate apart from human concern. That high mathematics and celestial prediction are constrained by ritual and divinatory considerations as opposed to being free to operate on their own as they did in the later West should give us cause to admire rather than criticise the Maya for what they were able to achieve. Furthermore, that they acquired their data with practically no technology, ought to place them in our awe.

The only inkling we have of a Mesoamerican observational technology are place-name pictographs from central Mexican codices that depict men holding or peering over what could be a sighting device that consists of a pair of crossed-sticks. Often these characters (were they astronomers?) are depicted in the recesses of temples, which gives rise to the notion that such edifices might have been preferentially aligned toward events that took place at the local horizon. Such a sighting method would be useful for empirically establishing solar, lunar or planetary periodicities by simply witnessing when and from what direction a rising or setting object resumed its earlier position along the axis of the doorway. Such a scheme is supported by at least one statement (see p. 287) in the post-conquest chronicles as well as by numerous investigations of oddly skewed misshapen buildings and buildings with narrow slots or windows that do seem to align astronomically (Aveni 1980).

Perhaps the most famous example of an astronomically oriented building is the Caracol of Chichén Itzá, Yucatan, a Mayan building with Toltec embellishments of the ninth to tenth century. This cylindrical edifice rests on a skewed double-decker platform (Fig. 97). A concentric circular maze of entryways leads to a solid core pierced by a spiral passageway that leads to a once sealed 2×2 m square chamber in the upper storey. The Caracol (or snail) receives its name from this coiled passageway, first excavated by nineteenth century archaeologists. At the top one confronts the remains of three long horizontal shafts, two of them being so narrow that they encompass only a few degrees of the flat horizon. As one of the Caracol's excavators once remarked: these could only have been made to look through. But what to see?

Long thought to be a military watchtower, the Caracol was first subjected in the 1930s to rigorous astronomical theorising. Alignments measured along diagonals through the shafts (e.g. inside left to outside right jamb) matched the equinox sunset and lunar extremes. More recent studies, which added measurements taken on the lower platforms, suggest more reasonable match-ups with the extreme positions of the planet Venus at the horizon. Given the elaborate detail connected to Mayan Venus-watching that appears in the Dresden Codex, it would not be surprising, especially in a place like north Yucatan with its featureless horizon, to discover architecture that incorporates sighting directions to a planet that disappears as evening star in the west. Such sightings could serve as a means of anticipating how long the planet would be obscured from view

97 The Caracol of Chichén Itzá, an astronomically orientated Mayan building as it may originally have appeared in the tenth century. Horizontal shafts in the turret and perpendicular lines to the skewed base align with Venus extremes. (From Proskouriakoff 1946)

before it would reappear in the east as evening star. But evidence of this kind from archaeoastronomy can be very circumstantial and must be evaluated in the light of other evidence about the practice of astronomy, not to mention other information about the specific event with which the alignments coincide. The case of the Caracol may be further aided by the existence on a nearby building of a zodiacal frieze resembling that in one of the codices. Adjacent to it are several rain-god masks adorned with Venus symbols – the same ones that appear on pages of the Venus Table. Therefore, this area of Chichén is rife with astronomical symbolism.

Though the Caracol and other circular towers just like it are often called Mayan 'observatories', that term alone does not give a complete description of how the architecture might have functioned. Astronomically oriented buildings such as the House of the Governor at Uxmal or the Temple of Venus at Copan – both aligned to Venus extremes and both exhibiting Venus-related iconography in their decorative elements – were probably integral facets of ceremonial complexes that served the same function as Romanesque and Gothic cathedrals or mosques that flourished in Christendom and Islam. However, we must realise that as New World religion developed in tropical climes, the open exterior space of the natural environment took the place of the vast enclosed shell of a Chartres or Mesquita (Córdoba). Anyone who walks the plaza that fronts the buildings of Palenque's Group of the Cross, for example, should appreciate that he/she treads once sacred ground, where people gathered not just to pay debt to the gods but also to witness their very appearance as part of a cosmic hierophany. Let us examine one particularly well-documented event that gives an idea of the importance of Mayan cosmic ritual and its connection with practical astronomy.

One of the most spectacular of all sky events recorded in the Mayan monumental inscriptions has been attributed to Chan Bahlum, of Palenque. It happened on 20 July AD 690, the date called 2 Cib 14 Mol in their 52-year calendar round. (The first half of this date is the position in the 260-day cycle while the second half, the 14th day of the Month of Mol, gives the position in the 365-day year.) This date is inscribed in several places on the stuccoed plaques at Palenque's ruins. These texts imply that the three major sky gods responsible for the most recent cyclic creation of the world have reassembled in the sky over the city where the chosen people lived. Now, during the early summer of 690 Mars, Jupiter and Saturn had been dancing about together in the night sky. Would their close conjunction reaffirm the descent of Chan Bahlum's dynasty through the bloodline of the gods themselves?

Thanks to the court astronomers it all came at a most appropriate time, for the young Chan Bahlum had big shoes to fill. Icons that accompany the glyphs suggest that the king conducted a public rite in which he let blood from his member with a sting ray spine and offered it to the sky god-ancestors to seal the kinship bond. What better way for the newly invested ruler to legitimise the continuation of the rule of his famous father Shield Pacal than by staging a rare celestial spectacle in the plaza fronting the temples where the words of creation are written? How, then, shall we view the motive for this public exhibition of astronomical phenomena connected with major events in the life of the Mayan city and its rule? Perhaps as a way of justifying his wealth, status and power, a Mayan king seems to be telling the world through the medium of public architecture that he was indeed capable of seeing into the future. Only then could he and his subjects prepare for the practical as well as ritual activities that directly manifested his unique dialogue with the gods of nature. Here was a philosophy the Mayan citizenry, high as well as low class, must have taken quite seriously.

Monumental inscriptions reinforce the ties between the Mayan polity and the sky already established in the architecture of the ceremonial centre. Figure 94, which shows but a single Mayan inscription on a stela – a 'stone tree of time' as they would have called it – gives us pause to realise just how far Mayan chronologists were willing to go in order to elaborate their histories in the pristine framework of astronomically based cycles. This monument, along with half a dozen others, which once resided in a publicly accessible area of the ruins of Tikal, documents the life history of Jaguar Paw Skull and his wife. The long count that opens the inscription pins down the location of the accession event in the framework of the current creation cycle (AD 488 in our calendar) while the calendar round localises the event in the 52-year cycle that contains the all important omen bearing days of the *tzol kin*. But this is not enough to anchor Jaguar Paw Skull in cosmic history. There follows hieroglyphic inscriptions that tell the phase of the moon, the number of the month in a cycle of 6 months (recall how the Maya used 6-month periods to aid in the prediction of eclipses) and the number of days in that month (either 29 or 30). Another glyph names the ruling lord of the night (one of nine rotating nocturnal guardians); other glyphs may refer to the constellation of the zodiac in which the moon appeared and still others depict cycles whose meaning yet eludes us. Transformed to real time, a significant number of the dates on these inscriptions match specific astronomical events. The flip side of the stela portrays the ruler himself fully garbed in his coronation costume.

Certain Mayan rulers seem to have adopted their own patron planets the way modern sports fans follow a favourite team. For example, Chan Bahlum developed a fondness for Jupiter. Inscriptions that record events in his life (e.g. his heir designation ceremony, accession to rule and apotheosis) seem quite deliberately attached to the movements of that planet, especially to the two stationary points of its retrograde loop. Events heralded in the inscriptions belonging to Yax Pac, or First-Sun-At-Horizon, a king of Copan, favour evening over morning star Venus appearances, which he employed to schedule raids on nearby cities like Quirigua. His grandfather, 18 Rabbit, who had built Copan's Temple of Venus with its slotted west-facing window, had done the same two generations earlier.

The recently deciphered inscriptions of Yax Pac's father, Smoke Monkey, yield Venus-related dates that mirror image those of his father's, 18 Rabbit. Smoke Monkey seems to have hitched his affairs to Venus, but not as evening star. Instead he timed the dedication of his palace to occur two Venus rounds after that of his predecessor, but he based it on a morning rather than an evening star event. He also closed off the west-facing window in his father's temple. Why would a son revert to the celestial predilections of his grandfather? There may be in every offspring an impulse to be different from his parents. In the serious world of juxtaposing royal Mayan history, religion and astronomy, young Smoke Monkey may have sought his measure of independence by selecting certain aspects of a celestial body that represented particular powers of the gods of nature.

Venus was as widespread a patron of Mayan warfare as was Mars in the Old World. Deep in the rain forest that straddles the border of the Chiapas State of Mexico and the country of Guatemala, an astral mural decorates a wall of the Great Palace in the ruins of Bonampak. It shows a dramatic battle scene executed in bright colours. The painting is 4.6 m long, 2.7 m high and contains more than 100 figures clad in battle regalia. At the centre, bigger than life size, stands the early eighth-century king of Bonampak, Chaan-Muan, attired in a jaguar skin. Below him, on the steps of his palace, his tortured captives appear, looking as if they had just been dragged in from the battlefield. Tired and weary, shamed and humiliated, they shed drops of blood from their hands at the king's feet as they prepare to endure their punishment. Will they be beheaded, or spared and returned to their home city? Above this scene of supplication are painted four depictions of constellations, among them a turtle, which has been identified with our belt of Orion (see Fig. 85), and a pair of peccaries, which may correspond to another star group (Gemini) not very far away in the sky. Certain of these signs are similar to the zodiacal band that appears on a page of one of the Mayan codices and on the frieze of a building near the Caracol at Chichén Itzá. On the back of the turtle and arrayed all about the other constellations are Venus hieroglyphs, the very same ones that appear in the Dresden Venus Table.

But there is something even more bizarre about these events. Practically all the dates on the stela fronting the building can be pinpointed in the November to mid-February part of the calendar. Furthermore, heliacal rising and setting events in the Venus cycle match very closely with the dates recorded on the monuments. Some of them may even be related to Venus's greatest apparent distance from the sun in the sky (greatest elongation) – when it stands highest in the sky at dawn or dusk, or more precisely, the first perceptible falling motion from its celestial high point. Was the turning point or pivot in the planetary cycle as well as its appearance or disappearance a signal that called the warriors to battle?

Does this 'star wars' scene describe reality or is it just propaganda? If it happened, was the attack actually timed by the appearance of Venus in the sky? If so, could not the enemy, presumably as wise in the ways of astronomy as Bonampak's astronomers, have cracked the secret 'star-war' code? It could scarcely have come as a surprise, for all raids were conducted during specifically designated war seasons. This had to have been a time when crops necessary to feed an army of foot soldiers would be standing in the fields. Also, war would take place during a season when intensive agricultural labour was not required.

Clearly, we have much to learn about Mayan warfare and its relation to the cosmos. Nonetheless, Venus war imagery tied specifically to the Central Mexican god Tlaloc (god of rain and fertility) has been traced back to 1,000 years before the Maya. Some of the iconographic signs that later developed into hieroglyphic Venus symbols appear at Teotihuacan, though they are not accompanied by any recognisable date-recording system. Once established, the Venus cult may have spread from the Maya back toward the west to Cacaxtla in the central Mexican highlands, where we find mural paintings filled with fantastic creatures, many of them adorned with Venus-related iconography. There is even a special waiting room in which the lords may have kept their victims before sacrificing them to the deity.

As we have learned, the Maya exhibited an embedded fatalism that strove to recover from sky events, observed and recorded in the past, certain repeatable patterns that could be projected into the future. For them, such patterns constituted realisable proof of the long-held Mayan belief that the future was contained in the past – that the unfolding of events over time's near and distant horizon actually could be foretold by looking with introspection over one's historical shoulder. Such a conception of the cosmos seems more temporally based than that of ancient Greece or Egypt. A Mayan astronomer would not likely have been interested in whether the earth or sun lay at the centre of the universe. Space was not so important to them as it is to us.

The Maya developed an astronomy in the service of politics and religion, yet nowhere else in the New World but among the Maya do we witness the mathematisation of nature's time cycles carried to the borders of obsession. Looking at both codex and stela, one acquires the feeling that the numerical and written contents of Mayan public monuments could not have been comprehensible to an ordinary citizen. This raises the question of how the peasant in the field comprehended the meaning of this rather exotic astronomy. Did priests recite a litany in front of Yax Pac's stela? Just why did the Maya become so carried away and could esoteric isolation of these élites have contributed to their downfall? We may never know.

Aztec city and cosmos

The Aztecs were relative newcomers to the Valley of Mexico at the time of the Spanish contact. They controlled the basin of Lake Texcoco from 1325 until 1519, when Cortés entered their capital of Tenochtitlan on the site of modern Mexico City. By then their empire had expanded to both the Gulf and Pacific coasts. Bedazzled by the well-described practice of Aztec human sacrifice, many modern writers misunderstood their ideology. Aztecs offered the blood of captives to the sun god Tonatiuh to keep the heavens in motion, lest eternal darkness should befall them. So it was pictured in their holy texts. Politically we can think of such a policy as a mechanism of control and domination exerted by the Aztec rulers over their newly conquered tributaries all around the lake basin. But from a religious perspective we might better regard these sacrificial acts as mandatory rites of renewal of the life of the Aztec city and its dedication to military conquest. They saw the subjugation of their tributaries in no less a righteous vein than the Spanish conquerors who dominated them.

The shape of Aztec time is manifest in their codices. The Aztec historical record is decidedly cyclic and it is pictured with the events of nature, among them eclipses, smoking stars (comets and/or meteors), perhaps even the zodiacal light. These books tell the reader that the sky is very much a part of civil life; thus pictorials are juxtaposed alongside events of social importance, such as deaths, conquests and accessions, all connected to year bearer dates. The year bearer is the name and number of the day in the ubiquitous 260-day *tonalpohualli* that corresponds to the initiation of the New Year. The Aztecs eschewed the long count and kept their count of days only in the calendar round. As there are 20 day-names and 365 days in the year (*xihuitl*), only 4 such names can be assigned to the first day of the year. These would consist of every fifth name in the list. Given thirteen coefficients that precede the names, the number appended to

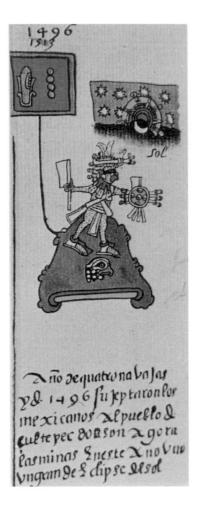

98 Astronomical events punctuate historical time in the conquest-period Aztec Codex Telleriano Remesis. In this frame a Spanish gloss helps us to pinpoint a total eclipse of the sun visible at the Aztec capital of Tenochtitlan in 1496, which was tied to the conquest of neighbouring Tlatelolco. (From Lord Kingsborough, *Antiquities of Mexico*, vol. 3, 1831)

the day-name of the year bearer would advance by one every year. Therefore the year bearer in Fig. 98, from a document produced shortly after the Spanish contact but undoubtedly partially copied from an earlier time, reads: 4 Flint (Tecpatl). The next one would read 5 House (Calli), followed by 6 Rabbit (Tochtli), 7 Acatl (Reed), 8 Flint, etc.

As one can readily see, it is difficult to pin down specific events in the Aztec record because only the year is specified. But in some cases celestial events are identifiable without question. For example, thanks to the aid of a Spanish commentator who inked in the Christian years along the top of the text (after an earlier scribe had mistakenly done so), we can be sure that the picture linked to the date in Fig. 98 refers to a total eclipse of the sun visible from Mexico City late on the afternoon of 8 August 1496. The scribe has even correctly pictured the sun, still partially eclipsed, setting over the high mountains that ring the western periphery of the Aztec capital. Incidentally, a total eclipse also took place in the year the Aztecs claim to have founded their city (1325). It may also be significant that the start of that year coincided with the March equinox. By the time of contact the *xihuitl* commenced in our early February. Does this mean the lords of Tenochtitlan, like their counterparts at Palenque, rewrote their history after the fact? Investigators have suggested that like-in-kind events, such as conquests, pestilence, even bad weather, exhibit patterned recurrences based on multiples of 52-year periods in the Aztec chronicles.

The future is contained in the past and it is interwoven with the cosmos. This principle is exhibited on a larger scale in artefacts such as the Aztec sun stone, a celebrated carved monument dug up near the centre of Mexico City in the seventeenth century. There the sun god with flint-knife tongue is surrounded by calendrical dates that adorn pictures of the four previous cataclysmic destructions of the world, by water, wind, etc. We live in the fifth world surrounded by the days of the present calendar epoch that ring the outside of the stone and it is our responsibility to keep time's wheel in motion forever by paying our blood debt to the sun. Much has been written about the many other monuments strategically placed about the capital in order to reinforce the connection between the flow of time and the need of military conquest in order to supply the blood of sacrifice. But our objective is to understand the empirical basis for the Aztec way of comprehending time.

Registering the appearance of celestial bodies in the doorways of specialised buildings, as we have seen, may have been widespread throughout ancient Mesoamerica. The conquerors had heard about it when they entered the great cities. For example, an informant told the Spanish friar Toribio Benavente (Motolinía) that a certain festival took place when the sun stood in the middle of the great temple, and because it was a little crooked, Moctezuma needed to pull the temple down and straighten it.

The temple in question, called by the Spaniards the Templo Mayor, is the most famous of all the ancient Aztec buildings in Mexico City. It was thoroughly re-excavated in the early 1980s after electrical workers engaged in installing one of the city's new subway lines accidentally broke into an offertory cache of jades, decorative shells, skulls and flint knives placed there 500 years before. Subsequently, Mexican archaeologists excavated seven nearly identical buildings, one inside the other and spanning two centuries, exposing each facade as they went, so that the alignments now could be measured. The resulting east of north by south of east orientation is just what it would have had to be to permit the rising equinox sun to fall into the notch between the twin temples that once surmounted the flat-topped 40-m high pyramid. When the sun arrived there, Spanish chroniclers tell us, a royal observer situated in the plaza fronting the bottom of the stairs carefully watched it. Like a town crier he would signal the time to begin the ritual of sacrifice that attended that particular month of the year. Like the Maya, the Aztec realigned the seasonal year with the unbroken count of cyclic time not by inserting leap days but by making observations of the stars. In fact Father Bernardino de Sahagún writes in the sixteenth century that the precise sighting of the Pleiades determined not just when to start the year, but also when to begin a 52-year

287

calendar round. It was then that the ceremony of the Binding of the Years took place. It began when the Pleiades, which Sahagún depicts in a drawing in his *History of the Things of New Spain*, crossed the overhead position at midnight (about mid-November). When the time approached, the priests ascended the Hill of the Star to watch the movement of the Pleiades with great anxiety:

> And when they saw that they had now passed the zenith, they knew that the movements of the heavens had not ceased and that the end of the world was not then, but that they would have another 52 years, assured that the world would not come to an end. Sahagún 1957, p. 143.

While Sahagún specifically identifies the Pleiades in his drawing it is unfortunate that he does not specify names of other star groups that also appear close by. One of them could be the same Scorpion identified in the Mayan zodiac, while the meaning of some of the other pictures, for example certain ones with sun and moon faces joined, seem fairly obvious. In the volume of his history dedicated to astrology, Sahagún tells us something that is reminiscent of a Mayan custom (see p. 282). He says that Venus was associated with a blood sacrifice; warriors spiked their arms and legs with cactus thorns, then flipped their blood at the newly arrived morning star.

Consolidating the calendars – creating Aztec standard time so to speak – must have offered numerous problems for the imperial heads of state, for chronicles from neighbouring cities indicate that each polity had its own set of year bearers. As we shall see, similar problems were shared by the Inca rulers whose empire extended over 30° of latitude. This resulted in very different sky orientations at the northern and southern

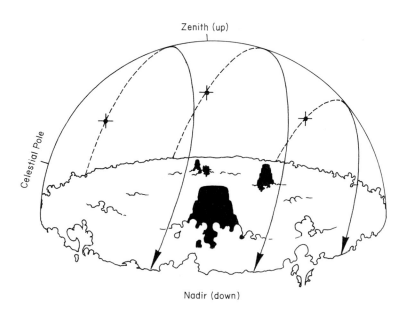

99 The zenith–horizon astronomical system is especially effective in the tropics. In Tikal, latitude 17°, the celestial pole lies low in the sky and all sky objects remain at relatively fixed azimuths as they rise and move straight up and over the sky. By contrast, in temperate latitudes objects circle at low angles to the horizon round a point high in the sky. (From Aveni 1980)

100 Teotihuacan, city of the gods (looking south). Its east-west axis once aligned with the Pleiades via a pair of pecked cross petroglyphs (like the one shown in the inset) which may also have served as calendrical counting devices. Note the count of 20 along each axis.

limits of the state. There, too, we shall learn that zenithal observations were very important, as we might anticipate among cultures of the tropics. Figure 99 explains.

Aztec predilections towards astronomy, particularly the idea of creating their city in the image of the celestial deities, can be traced to a sacred centre, which predated their culture by more than a millennium. Aztecs who visited Teotihuacan during Europe's Middle Ages (it was built before the time of Christ and already abandoned for several centuries) said that it was the birthplace of the gods. But this city was not built helter-skelter, the way the medieval cities of Europe evolved. Archaeological evidence suggests that it was carefully pre-planned and that, like Mecca or Jerusalem, it was as much a holy place as it was a centre of commerce. Along its Street of the Dead, in the stucco floor of a building located not too far from the great Pyramid of the Sun (names assigned well after the conquest), we find eroded yet visible evidence attesting to the precise arrangement fixed by its architects (Fig. 100). More than 2,000 years ago, they had surveyed and laid out the 130 square km ceremonial centre, the periphery of which would come to house more than 100,000 people. The evidence consists of a design made up of pecked marks in the shape of a double circle centred on a cross. It matches almost exactly another design carved on a rock outcrop 2.4 km to the west of the sun pyramid (Fig. 100, inset). In the 1960s archaeologists discovered that a line between the pair of bench-marks is almost exactly parallel to the east–west street of the ancient capital. A third marker lies high on a mountain to the north that overlooks the city, and a fourth on the south may have marked out another significant geographic direction. It is likely that these bench-marks were put there to establish that the city was an integral part of the cosmos. Only those directions deemed valid by divine will should be reflected in the domain the gods had once established here on earth – the home of the rulers of Teotihuacan who descended from them and the Aztec polity that claimed its own inheritance in the great city.

If you stood over the marker on the Street of the Dead 2,000 years ago and cast your eyes over the petroglyph on the western horizon at the correct time of year, you would have seen the Pleiades setting. The alignment of the Pleiades at Teotihuacan probably had a twofold significance. First, the star group passed directly overhead in the latitude of Teotihuacan, thus signalling the fifth cardinal direction. And second, when the Pleiades rose heliacally, after having been lost in the light of a glaring sun for approximately 40 days, it was the very day the sun passed the zenith.

Here then, was a highly visible, convenient stellar timing mechanism to signal the start of the new year, known 1,500 years before the Aztecs and likely passed onto them. The Pleiades, being both prominent and in the right place at the right time, became the celestial timer of choice to highly innovative astronomers who helped both shape the cosmological ideas and keep the calendar of Teotihuacan. But the ancient Teotihuacan orientation calendar was more natural because of yet another coincidence. Its tally system easily accommodated the body tally, for the city lay at the precise location where exactly twice the number of fingers and toes (i.e. 40 days) marked the period between sunrise on the cosmic axis between the bench-marks and the day of the observed solar zenith passage.

This magical orientation, approximately $15\frac{1}{2}°$ to the east of north and west of south, runs against the natural contours of the local topography. It was copied all over Mesoamerica for generations after the decline of Teotihuacan. Indeed, dozens of petro-glyphs resembling the Teotihuacan pecked-circles-centred-on-cross designs have been discovered by archaeologists at sites ranging from the far north of present-day Mexico to the remote southerly Mayan ruins of Guatemala. An exquisite pair of petroglyphs even marks the Tropic of Cancer, where the noonday sun arrives at the zenith on only one day of the year (the June solstice). While we cannot assign a significance to every one of these petroglyphs in alignment schemes and while there are other petroglyphs at Teotihuacan that cannot be incorporated into its axial orientation, the quadripartite symbol nonetheless remains an icon that demonstrates that the ancient city of the gods was thought to harbour deep and lasting cosmic roots.

Just how far back can we trace this habit of event-scheduling by the stars? Archaic clues about Mesoamerican skywatching emanate from the pictograms and petroglyphs that appear on rock outcrops that surround nascent ceremonial centres in North-west Mexico. Tally marks that make up the months and years were made by hunter-gatherers who spent but a handful of moons in one location, depending upon their remembrance of whether the deer were running or when the berries beckoned in full bloom upon the bush. Like the relatively small kin-related tribes who roamed prehistoric England's Salisbury Plain, once these itinerant people began to band together in larger groups, and especially once they became sedentary agriculturists, they honed their simple arithmetic tallying into more carefully crafted calendrical computations. There is good evidence that the pecked cross-circle design descended from such tallying schemes. In most cases the number of holes along the axes is equal to 20 and a few examples total 260, give or take a few elements.

Just how would such day-counting schemes have functioned in practice? The archaeological record is silent on such a question, but field anthropology may give us a possible answer. A nineteenth-century Ojibwa woman from Canada recorded in her diary:

> My father kept count of the days on a stick. He had a stick long enough to last a year and he always began a new stick in the fall. He cut a big notch for the first day of a new moon and a small notch for each of the other days. I will begin my story at the time he began counting a new stick.
>
> A. Marshack in Aveni (ed.) 1989, 313.

Then she goes on to describe how her mother began storing goods like maple sugar and rice for the winter during the first moon phase interval.

The more we examine the surviving evidence on Mesoamerica's first astronomers, the more we realise that theirs was a single continuous culture, not the disparate wandering, disconnected tribes portrayed in much of the older literature. If the Teotihuacanos emerge as the New World's ancient Greeks, perhaps we can think of the Maya as the symbol of cultural Renaissance and the Aztecs as the last attempt in this part of the world to organise skywatching in the setting of a great imperial state, before much of it was destroyed by European invaders seeking gold and precious stones, not to mention fresh souls for Christian salvation.

Astronomy and Inca social order

'They had observatories with windows' to catch the first and last rays of the sun at horizon. Their observations were employed to know when to shear the llamas, sheep and alpacas and when to sow and harvest crops, such events occurring at different seasons of the year. When their state astrologers watched the changing aspect of the sun in the local environment they said that the sun 'sits in his chair one day and rules from that principal degree [of the December solstice]. Then he sits in another chair where he rests and rules from that degree [of the other solstice].' From one seat to the other 'he moves each day without resting'. During the solstices he rests for more than a day in his chair, when at those days the motion of the sun from day to day on the horizon becomes imperceptible.

These statements, made by the half-Spanish, half-native chronicler Felipe Guamán Poma are brief but they establish both a purpose and a methodology for Inca observational astronomy. His words imply that the Inca dynasty, which ruled the Andes for little more than a century prior to Spanish contact, had developed a sophisticated system of positional astronomy based upon accurate horizon observations. As in Mesoamerica their astronomy was built into the city they inhabited. Said an earlier Spanish visitor: while most towns in the Indies 'lack design, order, or polity to command them, Cuzco has distinction to a degree that those who founded it must have been people of great wealth'.

Of all the reasons we can give for the success of the Inca empire, perhaps the most profound is the strict order and the high degree of organisation that was built into every component of it. Cuzco is situated in latitude $13\frac{1}{2}°$S at the junction of two rivers in a 3,200-m high mountain valley in the central Andes. They may have called it *Tahuantinsuyu*, which means 'the four quarters of the universe', because of its basic quadripartite plan. The city is segmented into halves called upper (*Hanan*) and lower (*Hurin*) Cuzco. Each half in turn is split into sectors, or *suyus*. Now, none of these four regions occupies a 90° segment of a circle, for the principal rationale for dividing the city this way had more to do with the watershed environment and kinship rank rather than considerations of pure geometry.

Four major roads departed Cuzco, one from each corner of the central square; these served as dividing lines among the *suyus*. Theoretically they extended to the remotest domains of the Inca empire – as far as Quito to the north and central Chile to the south. *Suyus* were ranked, as were the hierarchically organised kinship groups who lived within them, the organising principle of the moiety division being whether they were located up-river (higher ranking/*Hanan*) or down-river (lower ranking/*Hurin*).

But what do kinship and geography have to do with the stars? The *ceque* system, a giant cosmogram and mnemonic map built into Cuzco's natural and built topography, explains the connection. It served to unify Inca ideas about religion, social organisation, calendar, astronomy and hydrology. The sixteenth century Spanish chronicler Bernabe Cobo has left us a thorough and detailed description of this unique arrangement. *Ceques* were imaginary radial lines grouped like spokes on a wheel according to their location within each of the four *suyus*. The wheel's hub was the Coricancha (mistakenly called by the Spaniards the Temple of the Sun, Fig. 101). Covered over today by the Church of Santo Domingo, this structure was the most important temple of ancestor worship. Cobo lists nine *ceques* associated with each of the *suyus*: the north-east (centred around

101 Coricancha, temple of the ancestors of the Incas of Peru and centre of the radial *ceque* system. As is evident in this sixteenth-century sketch of the building, astronomical phenomena played a major role in organising the state. At the top lies the Southern Cross, then, descending on either side, the sun and moon, morning and evening star, Pleiades; even a rainbow is visible. From Santa Cruz Pachacuti Yamqui (Aveni 1980)

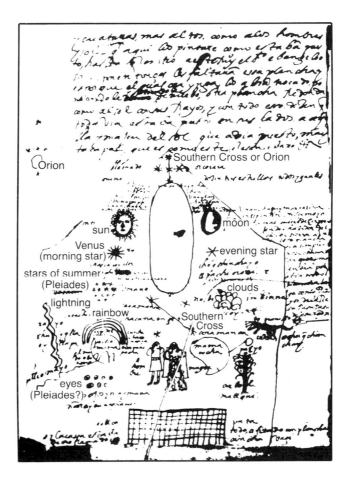

the district named Chinchaysuyu), south-east (Antisuyu), and south-west (Collasuyu), while 14 *ceques* were associated with the north-west (Cuntisuyu) quadrant, thus making a total 9 + 9 + 9 + 14, or 41 radial lines. These were worshipped and cared for in rotation by various kin groups.

According to Cobo's description, each *ceque* was traceable by its line of *huacas* going outward from the Coricancha across the landscape. At the *huacas* worshippers communed with the gods who controlled the cosmic forces. They believed the *huacas* to be openings in the body of *Pachamama* or Mother Earth and there they left her offerings. The locations of these *huacas* must have been rather important, for Cobo goes through considerable trouble to locate and describe each one in detail. There were 328 in all; some were built temples, others intricately-carved rock formations, bends in rivers, fields, natural springs called *puquios*, hills – even impermanent objects such as trees. And some were astronomical foresights involved with the establishment of a calendar. For example:

Chinchaysuyu – ceque 6, huaca 9: 'A hill called Quiangalla that is on the road to Yucay where there were two monuments or pillars that they had for signs and when the sun arrived there it was the beginning of summer'

Cobo, 1653, p. 172.

and

Cuntisuyu – ceque 13, huaca 3: 'Chinchincalla is a large hill where there were two monuments at which, when the sun arrived, it was time to sow'.

Cobo, 1653, p. 185.

Unlike the Maya, whose written record serves us well in delineating their astronomy, the Incas have left us no writing in the traditional sense, save for the *quipu* or knotted string device they may have employed to record quantitative information; however, *quipus* have yet to be deciphered. In order to reconstruct the astronomical arguments given by the chroniclers, one must proceed by attempting to locate the *huacas* in the field and to map out the *ceque* system. It must be remembered, however, that the chroniclers who inform us were holy men, scarcely educated in the ways of the astronomy of the Renaissance West, much less in Inca cosmology.

Today, investigators disagree on the precise placement of *huacas* and the degree of straightness of the *ceques*, but all agree that the Incas had developed a horizon-based solar calendar and set it into operation in the remotest regions of their empire. One chronicler says that they erected pillars (Fig. 102) in Quito just like the ones in Cuzco. Another gives details on four very important little pillars that were visible on a hill overlooking the city from the west:

When the sun passed the first pillar they prepared themselves for planting in the higher altitudes, as ripening takes longer.

When the sun entered the space between the two pillars in the middle it became the general time to plant in Cuzco; this was always the month of August.

[Down in the main plaza was] a pillar of well worked stone about one estado [1.8 m] high, called the *Ushnu*, from which they viewed it.

What can we make of this curious statement? Evidently, the northernmost pillar served as a warning device that the planting season in the Cuzco valley was approaching. People who cultivated crops at higher altitudes, where growth occurred at a slower pace, therefore would be allowed sufficient additional time to sow their seeds before planting commenced in the valley. The logic of this Inca space–time calendar is not as simple as it might appear at first glance. If we take the chronicler at his word, the *ceque* containing the pillars could not itself have been a sight line, nor could the centre of the *ceque* system have been the *ushnu*, from where the observations were made. Rather, the observer in

102 *Mojonadores* were stone masons or tower builders of ancient Cuzco. Pillars like the one shown in the background once adorned the skyline of the Inca capital, where they marked the solar course. (From F. Guamán Poma de Ayala, *El Primer Nueva Crónica y Buen Gobierno* (1584–1614), Travaux et Mémoires de l'Institut d'Ethnologie, 23, University of Paris, 1936)

charge of the calendar must have stood somewhere in the present-day main plaza of Cuzco, the Plaza de Armas, a few hundred metres from the Coricancha, to watch the sun set.

Horizontal space linked to vertical time. Cleverly, the Incas used the horizontal array of sun pillars to mark out the sun's 'chairs' or 'degrees' that fit with the vertically time-structured planting seasons. Moreover, the general time when the sun passed the pillars was the middle of August, when the sun lies opposite in the year to the date when it crosses the overhead position. Recall how important the zenith looms in other tropical systems of astronomy discussed in this chapter (cf. Fig. 99). Apparently the Incas had discovered that important agricultural dates coincided with one of the prominent, conveniently visible celestial phenomena in the environment of Cuzco. This may have led them to regard the act of planting and the passing of the sun underneath the world as like-in-kind events. At this time of year, an Inca legend says, the earth mother 'opens up'. *Pachamama* is then at her peak of fertility and can be penetrated both by the tiller (with his plough) and the sun (with his rays). The sight line across the landscape that connects the rising and setting sun on the days of overhead passage and its reciprocal date may have been an Inca way of expressing in horizontal space the vertical structure of the ecology of the Andes.

Like Teotihuacan's pecked crosses, Cuzco's *ceque* map was not just an astronomical directional scheme. It was also a seasonal calendar, each *huaca* representing a day in the year, and clusters of *ceques* signifying the lunar months. The number of *ceque* lines (41) doubled may have served as a count of three sidereal lunar months, the shortest period that returns the moon to the same constellation at the same time of night. Furthermore, the number of *huacas* in the *ceque* system, 328, is a good approximation to a year comprised of 12 such months ($12 \times 27\frac{1}{3} = 328$). And the disappearance period of the Pleiades from the sky corresponds approximately to the 37-day difference between the seasonal year of 365 days and the year of 328 days counted by the *huacas*.

The association of planting, irrigation and the Pleiades probably developed when the Incas recognised that the absence of this prominent star group (called *Collca* or the storehouse for the harvest) coincided with the time between the end of the harvest and the beginning of the next planting season. It became left-over time or uncounted time in the annual calendar, like our 12 days of Christmas or the 5 bad luck days that terminate the Mayan year. There is evidence that the Incas used an alignment of the walls of the Coricancha itself to correlate the place where the Pleiades rose with the June solstice timing of their major festival of *Inti Raymi*. The terminal *huaca* of the *ceque* that indicates this general alignment is called *Susunmarca*, one of the Inca names for the Pleiades. In fifteenth-century Cuzco, the Pleiades returned to view just before the solstice and the first full moon after their reappearance marked the festival.

Not unlike the way we vary the number of days in our months or change to daylight saving time, for convenience the Incas adjusted their month periods slightly in order to force them to fit into significant real events in the agricultural cycle, such as the time of ploughing, planting, the appearance of water and harvesting. There was much concern that certain festal and ritual activities be celebrated during a period commenced by a named full moon – like our antiquated habit of beginning the harvest during the month initiated by the appearance of the Harvest Moon or of opening the hunting season with the Hunter's Moon. The 12 months of the lunar synodic calendar, which had not yet been satisfactorily correlated with the sidereal months, were further subdivided and parcelled among the population of Cuzco and its vicinity in such a way that each social group was assigned the responsibility to tend to the ceremonies and sacrifices that were performed at the *huacas* that were associated with their particular month of the year in the calendar-counting scheme. Here was astronomy at the service of good government. Like the Aztecs' mandate to fuel the sun in order to keep it on course, the families that made up Cuzco's populace had a role in feeding Mother Earth and thus in articulating the Andean pathway of time.

Without the chronicler's statements, perhaps we would not have expected an imperial state like that of the Incas to have taken its astronomy so seriously. But the sky was very much a part of their concept of order. For Cuzco was a map of itself – a map that unified, bounded and subdivided natural and social space and time. The natural order that the Incas and their subjects perceived in the skyscape helped to structure the cityscape. The whole system worked when each social class performed its assigned function of making offerings to their gods in the proper place along its *ceque* line at the correct time in the calendar. In this confrontation between nature and culture in a harsh and variable agricultural environment, the *ceque* system, like the Magna Carta or the Ten Commandments, became a code to live by – a rigorous scheme designed by Inca royalty to prescribe proper human action. The core of the idea was based upon residence and kinship in a radial, fourfold geographic framework. Like the flow of water, astronomy was a part of that order, but it was inseparable from all the other components. If we attempt to pull it out of context it will disintegrate.

Other great indigenous cities of the Americas also used the pristine order evident in the sky to establish social order here on earth. For example, the same general kind of sky symmetry we find in Cuzco is apparent at Cahokia, located near where the Mississippi and Missouri Rivers join. Built 1,000 years ago, Cahokia was truly a significant economic and political centre of great proportions; it controlled the distribution of maize and exotic trade items over a very wide area. Though no chroniclers ever wrote about it and no indigenous writing system survives it, its axis of orientation is cardinal, and mound alignments imply that the sun was a major object of attention. By following and marking out the annual solar path along the horizon, rulers of this economic hub could regulate the seasonal flow of goods and services and schedule the holidays and their accompanying solar rituals that would take place when the local populace and the tributaries of the state turned out in the plaza in front of the great Monk's mound. In a similar vein, farther down the Mississippi, French missionaries witnessed the Natchez tribes assemble in early colonial times to worship the rising sun believed to be incarnated as their chiefs, whom they called 'Great Suns'.

North America's Hopi were also ardent sun watchers. Hopi orientation bears no relation to north or south; rather it is the points on the horizon which mark the places of sunrise and sunset at the solstices that matter. Here ethnoastronomy comes to the rescue, for this conclusion was reached by a pioneer anthropologist who visited Tewa (New Mexico) village in 1893. The map accompanying explorer Alexander Stephen's letter suggests that these descendants of the Pueblo-Anasazi of northern Arizona/New Mexico also seem to have been attempting to tie their centre to the world periphery by making a calendrical instrument out of the surrounding environment. Some Sun Priests even eagerly ventured out along the rim of the hill to the sun's path to help clear his way.

The Hopi marked the solstices, which the elders referred to as 'houses', where the sun stops in his travels along the horizon. At these places along the high mesa the priests erected small shrines. There a Sun Priest in charge of the calendar would deposit prayer sticks, an offering to welcome the sun and to encourage him along on his celestial journey. Some of these shrines have special openings that allow shafts of sunlight to penetrate particular directions, thus serving as another way to mark the appropriate time. Sometimes the Sun Priest would gesture to the sun, whirling a shield, on which was painted a sun design, to imitate the sun's turning motion, hastening away any malevolent spirits who might impede the great luminary.

To determine the locations of shrines, which really functioned as solar foresights like Cuzco's sun pillars, the priest would need to situate himself at an appropriate sun-watching station (a backsight like the *ushnu*). These were simple stone piles giving convenient access to the expanse of the eastern horizon along which the sun would be seen to rise. One of these stations described by archaeologists around the turn of the century consisted of a flat stone with a sun face carved on top and each of the four quarters marked on its sides. The calendar specialist would sit on it and carefully sight the nearest distinct peak or valley where the sun would make its last perceptible slow-down prior to winter or summer standstill. Such a sighting would help to anticipate the actual solstice by a few days so that the people could then have time to prepare for the ceremony. The Soyal Ceremony at winter solstice, for example, lasts 9 days and its announcement is made 4 days in advance by the Sun Priest and the Soyal Chief, both elders of the same clan. Modern studies prove that timing by the sun-watching scheme has been remarkably consistent from year to year, varying scarcely by a day or two from a seasonal mean date.

Why was it so important to know exactly where to intercept the sun god? So he could accept the people's offering; he was, after all, the one god who controlled the growth of crops and who knew the future because he alone travelled in the underworld. The Hopi astronomer was thus vested with the important responsibility of drawing a bead on the solar deity. Failure in his task could draw serious criticism. As one Hopi clan member gesticulated concerning an errant Sun Chief: '*He* was the one responsible for all that bad weather we had last winter by being late in tracking the sun.'

Native American astronomy then and now

We have focused attention in this chapter on the high cultures of the Americas – the Maya, Aztecs and Incas. Where possible we have attempted to trace the roots of their calendars and observational systems. We exalt these cultures to the status of those in the Classical world by comparing their architecture, sculpture, art and writing with those of the Greeks, Babylonians and Egyptians, from whom we claim our Western astronomical heritage. Indeed, we even dare to assign the Mayan culture its own 'classical' period. Given the paucity of the material record and the foreignness of the New World taxonomy of nature, we are at a loss to claim anywhere near the degree of knowledge of ancient American astronomy we believe we possess for the Old World. On the other hand, these continents offer us one distinct advantage in our studies. There are less nucleated, more simply stratified cultures still extant and still relatively isolated from the domination of the West, whose astronomical customs and beliefs can still be retrieved. Data from these living cultures may offer us inroads into the development of the more complex systems of which we have evidence in the great ancient cities of Copan, Tenochtitlan and Cuzco. Therefore, to close this chapter we attempt, where we can, to tie past to present by turning our attention to some of the more revealing evidence from the realm of the ethnologist concerning the practice of astronomy among contemporary indigenous American societies.

Present day Quechua-speaking people of South America still chart constellations that date all the way back to Inca times, if not earlier. For example, the Pleiades are still called *collca*, and α and β Centauri, which are among the few bright stars that may figure in alignments tied to the *huacas* of Cuzco's *ceque* system, represent the 'Eyes of the Llama'. These are part of a parade of dark cloud constellations that, along with star-to-star constellations like our own, comprise the Milky Way, which is so much more prominent in the southern than in the northern hemisphere. Other animals in the zodiac-like array across the firmament include fox, partridge, toad and a great anaconda. In fact, the Milky Way is still a functioning environmental calendar – just like ancient Cuzco's sun pillars. Thus the interval during which many of these sky creatures are visible corresponds to periods when their terrestrial counterparts are active. To give an example, the partridge, called *yutu*, makes its heliacal rise early in September and disappears from view in mid-April. Anthropologists are told that the beginning of this period is just the time that farmers needed to guard their crops against these birds. Likewise, real toads re-emerge from the earth just about the time the toad constellation first clears the horizon, bringing the rain with him. The same was true in the ancient past: when the chronicler Polo de Ondegardo wrote that 'all animals and birds on the earth had their likeness in the sky in whose responsibility was their procreation and augmentation', one has to wonder whether he realised that he was dealing with a highly organised astronomical system of tracking environmental time. More than a memory assist, these star groups held the power within them to bring good or bad fortune to the hard-bitten farmer of the high Andes.

The idea that the sky is a blueprint for ideas and social action also flourishes on the other side of the Andes in the rain forest of the Amazon and the Orinoco. The Desana of Colombia are one of the few thorough case studies in ethnoastronomy. They divide the year into two rainy and two dry seasons and reckon the central point of their calendar by the place where the shaman's staff will cast no shadow when held upright. Like a horizontal lid, a celestial hexagonal template consisting of the bright stars Procyon, Pollux, Capella, Canopus, Achernar, and one of the stars of our constellation Eridanus (τ), overlies the earth at sunrise/sunset, just as the sun is positioned at the equinoxes. At this time, when heavenly symmetry is in force, a vertical shaft of sunlight is said to fall on a mirror-like lake below, thus fertilising the earth. Furthermore, the original tribes

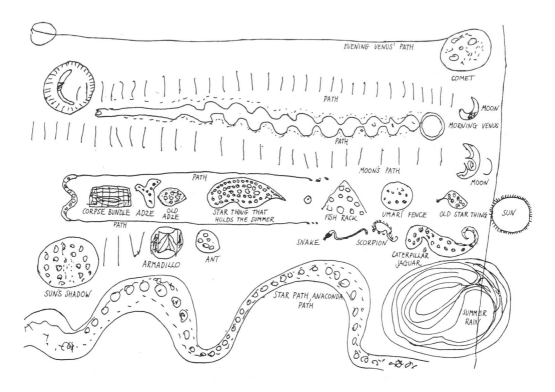

103 The heavenly bodies as drawn by a Barasana (Colombia) shaman (*above*), shown alongside a map of the Barasana zodiac (*below*) into which some of these celestial denizens are placed. (S. Hugh Jones, in Aveni and Urton 1982)

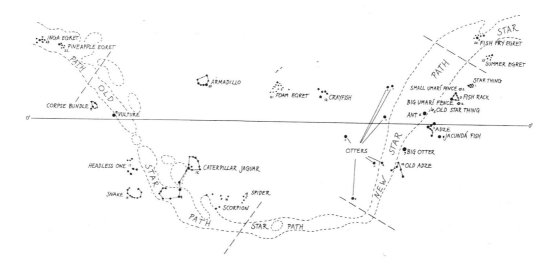

were said to be six in number and they still organise themselves socially in a hexagonal model.

The canopy crystal also served as an architectural model for the Desana longhouses, which are built on a six-sided plan. Informants say that each vertex of the hexagon consists of a house-post that can be identified with one of the basic support stars. The bisector of the hexagon on earth is a ridge-pole identified with the Pleiades, which rise in this area today just after spring equinox, thus signalling the beginning of the main fruiting season. To commemorate the event, piles of palm fruits identified with the Pleiades are heaped at the centre of the house.

Sky mirrors life, and life progresses like the seasons. So, the Desana mark out youth, maturity and old age on their sky hexagon and bind it to their terrestrial six-sided domicile. Expressed in a dance, a symbolic journey around the longhouse typifies the cyclic journey of both men and women through life. Each of the stellar vertices represents a significant marking post along life's road. Men, for example, move clockwise through Capella (naming) to Pollux (initiation) to Sirius (marriage). Women travel counter-clockwise, but only until they arrive at Sirius; then they turn about and join their husbands. When all return to their starting point, the Pleiades and Aldebaran, they are reborn, precisely on the equinox line.

We rediscover this principle of the cosmic house again and again across the Americas, from the Pawnee lodge of the Midwest United States, which had a smoke-hole through which observations of certain groups of stars used in storytelling could be seen, to the bell-shaped quarters of the Warao of Venezuela's Orinoco delta, with its zenith pole marked out with a yearly calendar calibrated by following the ascension of the solar image at noon. If your home incorporates a design for life and your calendar regulates activity, it is easy to understand why cities such as Cuzco and Tenochtitlan would be imbued with similar cosmic imagery. The difference between house and city is only one of size and social complexity.

Conclusion

Let us think back to the astronomical questions we raised at the outset of this chapter. Given what we have learned, what strikes us as most significant about the astronomies of ancient America? Whether ancient or contemporary, but perhaps by analogy reflective of ancient customs among less hierarchically organised societies, all of the sky observations seem to have been acquired with either low technology or no technology. These societies used neither wheel nor gear, and few of them employed metals. Their mental devices diverge from those of Western astronomy as well. We have mentioned neither fractions nor Euclidean geometry. Only the Maya developed numeration by position and the concept of zero, and only they seem to have taken the trouble to calculate in written form astronomical periodicities and predictions that reached millennia backward into the past and forged centuries forward into the future. No indigenous Americans raised questions about the rotundity of the earth nor did any speculate whether the sun or any distant celestial body might lie at the centre of the universe. We must remember that the spatial view of the universe – our concept of orbits, maps, deep space – is one of the gifts of the Greeks. Such concepts are culture-bound and we dare not anticipate that indigenous Americans would have had any interest in such matters. But all of this does not mean that native Americans did not philosophise or theorise about the world around them. Their speculations were basically human centred. Theirs was not a mechanistic universe that operated as an entity apart from human consciousness. The Western scientific credo – to mathematically express and empirically test verifiable natural laws that describe the way the world works – would have been anathema to their way of thinking, though the Mayan ideology admittedly skirts that of the West in such affairs.

The absence in these cultures of all the hallmarks of Western science along with no hint of any idea of progress in comprehending nature leaves us with a timely lesson. We realise the uniqueness – not the superiority – of our own science. This is the lasting contribution we, as members of Western-based scientific culture, acquire from studying Native American astronomy. We discover exotic concepts. Instead of a polar-equatorial co-ordinate system of location, we find instead a horizon-based directional scheme and a zenith solar-based temporal one. In place of a map of the solar system, we confront the architecture of the domicile and the city skyscape. Rather than digging up their leap years we encounter continuously running calendars with huge segments of uncounted, inactive time – calendars still deeply embedded in native folk ways. Their time emerges as every day activity itself rather than as a clock or metronome that beats out its inanimate pulse in total ignorance of the human psyche. One overriding similarity among all native American astronomical cycles is that they seem to serve as ways of understanding and expressing the cycle of human subsistence. Yet, even though every culture ultimately must run on its stomach, these sky timings are not solely concerned with practical matters, for the stars also were used to create colourful metaphors that express meaning in the deepest religious and philosophical sense. Indigenous American astronomy was ingrained towards the deepest thoughts in the minds of those who watched the sky. The firmament raised questions about concepts of the real world, about where the forces of nature originate, about who controls them, and where the human condition fits into the world view.

Bibliography

Aveni, A. 1980. *Skywatchers of Ancient Mexico*. Austin: University of Texas Press.

Aveni, A. 1981. Tropical Archaeoastronomy. *Science* 213, 161–71.

Aveni, A. (ed.) 1982. *Archaeoastronomy in the New World*. Cambridge University Press.

Aveni, A. 1989. *Empires of Time*. New York: Basic Books.

Aveni, A. (ed.) 1989. *World Archaeoastronomy*. Cambridge University Press.

Aveni, A. (ed.) 1992. *The Sky in Mayan Literature*. Oxford University Press.

Aveni, A. and Urton, G. (eds) 1982. *Ethnoastronomy and Archaeoastronomy in the American Tropics*. Proceedings of the New York Academy of Science, vol. 385.

Cobo, B. 1965 [1653]. *Historia del Nuevo Mondo*. Madrid: Bibliotheca de Autores Españoles, vols 91–2.

Lounsbury, F. 1978. Maya numeration, computation, and calendrical astronomy. In *Dictionary of Scientific Biography*, ed. G.C. Gillispie, vol. 15, suppl. 1, pp. 759–818. New York: Charles Scribner's Sons.

Proskouriakoff, T. 1946. *An Album of Maya Architecture*. Norman: University of Oklahoma Press.

Sahagún, B. de 1957 [1585?]. *Florentine Codex: General History of the Things of New Spain*, book 7. Trans. and ed. C. Dibble and A. Anderson. Santa Fe: School of American Research and Ogden: University of Utah Press.

Stephen, A. 1936 [1893]. Quoted in *Hopi Journal*, ed. E.C. Parsons. Columbia University Contributions to Anthropology, no. 23.

Tedlock, B. 1982. *Time and the Highland Maya*. Albuquerque: University of New Mexico Press.

Thompson, J.E.S. 1972. *A Commentary on the Dresden Codex*, Philadelphia: American Philosophical Society.

Tozzer, A. 1941. *Landa's Relación de las Cosas de Yucatan*. Papers of the Peabody Museum of American Archaeology and Ethnology, vol. 18. Harvard.

Urton, G. 1982. *At the Crossroads of the Earth and the Sky*. Austin: University of Texas Press.

Williamson, R. 1984. *Living the Sky*. Boston: Houghton-Mifflin.

Williamson, R. and Farrer, C. 1992. *Earth and Sky: Visions of the Cosmos in Native American Folklore*. Albuquerque: University of New Mexico Press.

Zuidema, R.T. 1977. The Inca Calendar. In *Native American Astronomy*, ed. A. Aveni, pp. 719–59. Austin: University of Texas Press.

BRIAN WARNER

Traditional Astronomical Knowledge in Africa

Throughout Africa, as on other continents, there was extensive traditional appropriation of celestial objects for mythological purposes. Commonly, their involvement was in religious rites, legends and folk tales in which human personal and social relationships were reflected symbolically in celestial relationships (Zahan 1979); or aspects of human life, such as life and death, found representations in the cyclical nature of astronomical events; or celestial bodies were endowed with male/female, positive/negative characteristics that are distillates of human character.

Most of these uses of the heavens in African belief systems lie outside the brief of the present chapter, which concentrates on the practical uses to which knowledge of the sky was put. Some basic cosmological beliefs will, however, be touched upon; and use will be made of the fact that the incorporation of celestial objects and events into traditional beliefs proves a widespread awareness of both the reliability and the mutability of the sky.

Outside of Egypt there are no indigenous African written records prior to the last century. Furthermore, there are few records of traditional beliefs made by visitors prior to 1800; the earliest, perhaps, is the record by the Romans that Africans were good astrologers (Westmarck 1926). Therefore, studies must rely heavily on nineteenth and early twentieth-century collections. Inevitably these are African beliefs filtered through (mostly) European minds, made at times when some external influence may already have taken place. The more recent collections are particularly prone to contamination – not only of non-African origin (European and Eastern), but because improved communications can lead to borrowing from neighbouring areas, with consequent blurring of traditions (Werner 1925).

A much-discussed example is given by the apparently traditional central belief of the Dogon tribe of the Republic of Mali (in former Western Sudan) that Sirius has a small, very dense companion in orbit around it. This was reported by anthropologists (Griaule and Dieterlen 1950, 1965) from information collected in the 1940s. Furthermore, the Dogon are purported to have had a heliocentric theory of the solar system

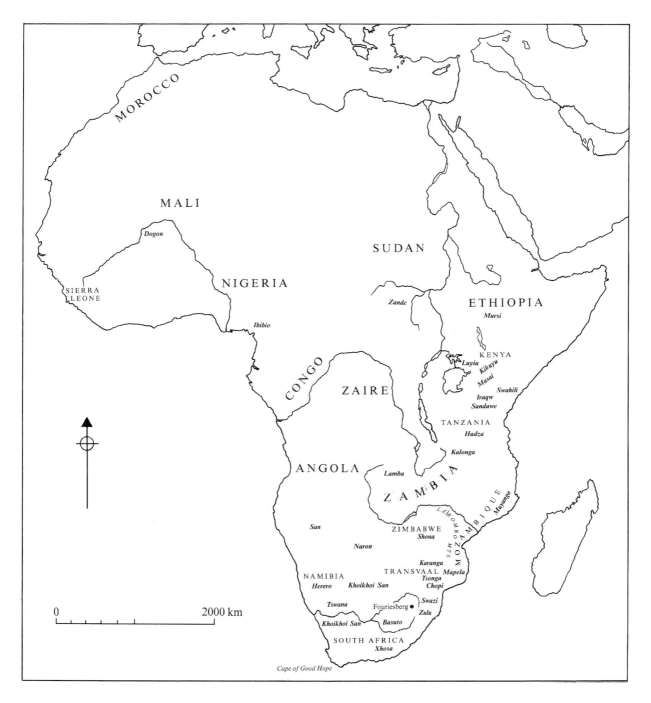

104 Africa (location of peoples based on information from D.H. Price, *Atlas of World Cultures*, Sage Publications, Newbury Park, Ca., 1989).

and to possess knowledge of the satellites of Jupiter and the rings of Saturn (Temple 1975). This specialised knowledge is most reasonably attributed to earlier twentieth-century contact with European culture, the rapid assimilation into central dogma having arisen from the dominant importance of twins and twin-ness in Dogon beliefs (Pesch and Pesch 1977). Other examples of rapid cultural transfer are given by Brecher (1979) in his discussion (and dismissal) of the possibility of such independent knowledge by the Dogon.

It is probably safe to conclude that any African tradition that does not resemble that of any other known contact group is indigenous. But the reverse is not necessarily true: for example, the heliacal rising or setting of the Pleiades was used as an indicator for crop planting throughout Africa; so too were they used for agricultural purposes in, among others, Borneo, Mexico, Asia and Polynesia (Bryant 1949).

Not all diffusion of ideas into Africa should be seen as recent European: trading down the east coast from the Middle East was common from about 2,000 years ago, and for many centuries before that Indonesians sailed to and settled on the east coast, introducing food plants and musical instruments and music scales that can be traced across to West Africa (Murdock 1959). Furthermore, it would be presumptuous to believe that diffusion was a one-way process.

There is a major division of indigenous African peoples into hunter–gatherers and pastoralists. Of the former only the San (Bushmen) of Southern Africa and the Sandawe and Hadza of Tanzania can be examined for traditional beliefs – another group, the pygmies, have entirely lost their original languages and much of their lore derives from the group whose language they now use (Murdock 1959).

The influence of Islam across the northern parts of Africa mixed or overlaid traditional African beliefs (Westmarck 1926); these are not considered here.

Keeping track of time

The sun, moon and seasons provide natural, though incommensurate, clocks for counting time. Traditionally in Africa there was almost universal adoption of a 13-month year, each month, often with a descriptive name (usually based on agricultural activities), starting at new moon, with adjustments every few years to keep the months and years in constant relationship. This is such a ubiquitous system, in Africa and elsewhere (Hadingham 1983), that any example of a 12-month year should probably be viewed as the result of cultural transfer.

The realisation that the lunar months and season had slipped out of phase was a time of extended discussion among elders, often leading to a renaming of the previous few months. Among the Mursi of South-western Ethiopia comparison and correction of the name of the month is a topic of constant debate, leading to both removal and addition of months, and making it impossible to maintain any chronology of events (Turton and Ruggles 1978).

This is one example of why it should not be presumed that an African group has the same concept of time as we do. There is well-documented evidence throughout the world, and in Africa in particular, that although time may be partitioned by use of time indicators (new moons, hot seasons), it is not necessarily seen as a continuous, one-dimensional sequence. The Iraqw of Tanzania, for example, use days, months and years as evinced by celestial cycles, but do not use them to generate a chronology against which other events may be compared (Thornton 1980). This lack of recognition that small units of time accumulate to add up to the next largest unit is not uncommon among the earliest people (Nilsson 1920; Pocock 1967).

The lunar phases furnish an easily visible subdivision of the month and were sometimes named descriptively (Bryant 1949), but the use of a 'week' of seven days as a quarter of a lunar month was unknown in Africa. The year would initially have been defined by the cyclical variations in weather, plant growth, etc. This gave natural subdivisions into seasons, of which there were commonly four, corresponding closely to the spring, summer, autumn and winter of European culture. Exceptions occurred, examples being the three seasons of the Basuto (Beyer 1919) and Swazi (but who recognised 'spring' as a distinct early part of summer: Marwick 1966), effectively five seasons of the Zulu, caused by dividing summer into two distinct parts (Krige 1936), and the two seasons of the Swahili (wet and very wet: Bryant 1949). An example of extreme indifference to the passage of time is that of the Kikuyu of Kenya, who are reported to have had two seasons and no knowledge or concern of the number of days in a month or the number of months in a year (Cagnolo 1933).

Many African people used bright stars or, most commonly, the Pleiades cluster to define the year with precision. As Africa extends from above the Tropic of Cancer to below the Tropic of Capricorn, the positions of stars have a different relationship to the seasons in the northern and southern halves of the continent. Among the pastoralist groups of Southern Africa the first visibility (in August or early September) of the Pleiades in morning twilight heralded the imminence of the planting season, and usually denoted the beginning of the year. The names given to the Pleiades by the various groups (Hammond Tooke 1886; Soga 1932; von Sicard 1966; Breutz 1969), all of which mean, in effect, 'the ploughing cluster', show the widespread diffusion of astronomical knowledge: *Isilimela* (Zulu and Xhosa), *selemela* (Sotho-Tswana), *shirimela* (Tsonga), *tshilimela* (Venda), *kilimia* (Swahili: East Africa and Zanzibar), *chirimera* (Karanga: Zimbabwe), *kelemera* (Nyabungu: Ruanda), *lemila* (Nyasa). Furthermore, von Sicard (1966) suggests that the word (and, by implication, perhaps the use of the Pleiades as a calendrical sign) was introduced from the East African coast by traders from the Middle East, where the Pleiades cluster is *kimah* in Hebrew. Such trade occurred from at least as early as the first century AD (Inskeep 1978). Breutz (1969) also suspects that some astronomical mythology was introduced this way from ancient Semitic and Ethiopian sources.

North of the equator the Pleiades were also used for instigating crop planting

(Frazer 1925): for the Masai and Kikuyu (Kenya and Tanzania) the Pleiades must be setting in the evening, similarly in Sierra Leone.

In contrast, the Khoikhoi (Hottentots) who, with the San, belong to an entirely different language group, called the Pleiades *!Khunuseti*, but also use them to forecast the start of the rainy season (Hahn 1881).

Diffusion (or at least acceptance) of ideas was not, however, complete. Neighbouring tribes could maintain quite separate mythologies. The Luyia of Kenya and the Lamba of Zaire used Venus as a herald of dawn but made no other use of the night sky, for ritual or practical purposes (Wagner 1954; Doke 1928).

The Mapela tribe of the Northern Transvaal used the first observed rising of Canopus (in May) to readjust the monthly and annual calendars (Jackson 1969) and it was used by the Tswana to denote the beginning of winter (Breutz 1969). In the Ibo-speaking region of Nigeria a 13-month year was used but the year began at different months for different tribes (Thomas 1913).

Reports of observations of the solstices – by noting the (approximate) dates when the sun sets and rises at its most northerly or southerly places, as judged by distant landmarks (e.g. 'a mountain or a forest': Callaway 1970) – come predominantly from one region of Southern Africa: the Swazi (Kuper 1947), the Basuto (Norton 1909; Beyer 1919), the Zulu (Krige 1936) and the Sotho (Jackson 1969).

In regard to distinguishing years from each other, the absence of a sophisticated counting system resulted in years being enumerated from especially memorable events, usually battles, deaths of chiefs, droughts, etc., but also great comets (e.g. 1843, 1848, 1884 by the Xhosa: Soga 1932) and probably eclipses (deduced from the existence of traditional words for these phenomena: e.g. McLaren 1929). Although only one or two hundred years of such time-markers remained in the memories of African peoples when this information was collected by linguists, no doubt the method extended back into antiquity.

The seasons were equally of importance to hunter–gatherers, but the lunar subdivisions employed by pastoralists were less necessary. This is evident in the large collections of San folklore made by Bleek and Lloyd (1911; and manuscript material in the archives of the University of Cape Town), much of which concerns astronomical mythology (Schapera 1930). Among the various groups of San, who ranged from the southern Cape to Angola, some adopted three annual seasons and others used four. There was considerable independence in the various groups' practical use of celestial objects: the Naron in Angola recognised the Pleiades as seasonal indicators and named a few prominent constellations, whereas other north-western San took no interest in the stars. The Cape San had names for many constellations and stars and knew their seasons of rising and setting. The San in general used the moon's phases to estimate short lengths of time, but did not know how many lunar months make up a season or a year.

During the night the passage of time was visualised by the steady change in orientation of the Milky Way (and, perhaps, in that of the Magellanic Clouds). An example is given by the Tswana who say that the Milky Way 'turns the sun towards the east'. Arcturus is said to 'pull the night across' (Clegg 1986). The Tswana also note the changing sky during the year and say that the Southern Cross 'moves with the year' and that it moves along the Milky Way (Breutz 1969).

Division of the day was usually effected by reference to the altitude of the sun – or its effects on the quality of the light or the intensity of its heat (e.g. Beyer 1919; Junod 1927). Although Shapera (1930) found that the San studied by him did not attach descriptive names to the times of the day, Silberbauer (1981) records many such names, including pre-dawn and post-sunset periods when light was recognised to be dominated by the zodiacal light.

The sun

For most of Africa the nature of the sun cannot be divorced from the over all concept of the universe, which is entirely geocentric, or, more embracingly, egocentric (Zahan 1979). A widespread concept is that the sky is a solid (e.g. blue rock), concave vault resting on the earth, upon which the sun moves (Samuelson 1929; Breutz 1969). At night the sun returns to the east, either under the earth (e.g. the Luyia of Kenya: Wagner 1954) or above the solid sky (e.g. Zulus: Krige 1936) producing stars through holes in the rock (e.g. Tswana: Clegg 1986). (This is still commonly believed, to judge by the example of the present author's Xhosa occasional gardener.) But for some (Basuto: Beyer 1919) each day has its own sun, the previous one having merely disappeared. Those dwelling on the east coast may believe that the sun rises from the sea (e.g. Tsonga: Junod 1927; Zulus: Callaway 1970). For some southern Nigerians, the vault of the sky was not thought to be very high, for they estimated the sun and the moon to be as large as a small table-top or large plate (Talbot 1925).

As an example of mythology that makes a lesser attempt to interpret observation in terms of terrestrial structure, the San believe that the sun was once a man from whose armpits light was emitted. Having kept the illumination selfishly to himself, some children of the ancient race threw him into the sky, since when he shines over the entire earth (Hammond Tooke 1886).

The rising and setting directions of the sun provide two of the cardinal points, which were sufficient for many peoples (Doke 1928; Breutz 1969; Zahan 1979). The varying path of the sun from solstice to solstice provided a celestial meaning to 'high' and 'low'; the north–south direction, when used at all, derived more from local topography than from any sense of orthogonality to the east–west direction.

The moon

Arising from its steady change of appearance during the month, and the regularity and short-term predictability of its monthly disappearance, the moon is particularly steeped in mythology (extensively described by Breutz 1969). New moon, starting the new month, was usually a day of festivity or rest, taboo for performing many common tasks but an opportune time to be purified in special rites. A form of moon worship, or reverence, was common in Africa. The Khoikhoi identified the moon with 'the lord of light and life' and became particularly anxious at a time of lunar eclipse (Hahn 1881). Dampier (1968), visiting the Cape of Good Hope on a voyage around the world in 1691, described their dancing and singing all night at times of new and full moon, as also did Kolbe (1731).

The Kikuyu believed that the moon was suspended from the nearest star to it, and that there would be different moons visible in different countries (Cagnolo 1933). An example of transfer of beliefs is given by the name *inyanga*, the Xhosa and Zulu name for moon, which is believed to be of Khoikhoi origin (Hammond Tooke 1886). Where modern people see the 'man in the moon', many Africans traditionally saw a man or woman carrying a bundle of sticks (Junod 1927).

Stars and constellations

The brightest stars and a few of the most conspicuous constellations are recognised and named throughout almost all of Africa. In the north, Orion and Ursa Major are used for calendrical purposes; in the south, Canopus and the Southern Cross are similarly employed (Zahan 1979). These are in addition to the Pleiades mentioned earlier. There are exceptions: the Tsonga recognise only the Pleiades and have one word that encompasses all stars (Junod 1927).

In most cases the agricultural tribes named only the individual bright stars that had some practical use. Thus the Basuto had names for Canopus and Achernar (Beyer 1919), and the Zulu used *Inqonqoli* for a star (presumably any suitable star) that rose an hour before the morning star (Venus) and *Inkwenkwezi* for a star preceding the appearance of the Pleiades, but neither had a name for Sirius as an individual star. The Basuto incorporated Sirius into a constellation called *Magakgala* which also included Rigel, Betelgeuse and Procyon (Beyer 1919: Hoffman 1925). The Zulu, however, had a word *Intsanta* for 'the star that is specially bright and scintillates numerous rays' which may have applied only to Sirius or may have been used for any bright star (Samuelson 1929).

The constellation of Orion is normally not distinguished as such, but the three stars of the 'belt' are known among eastern and southern Africans as the three pigs (Norton 1909; Beyer 1919; Samuelson 1929; von Sicard 1966), or, in the Congo, a dog, a huntsman and a palm rat (Hammond Tooke 1886), or simply as 'three and three' in Nigeria (Thomas 1913), or as three dogs in the 'belt' chasing three pigs in the 'sword' (Clegg 1986). The belt of Orion plays a significant role in the layout of the Dogon granary (Griaule and Dieterlen 1965).

The Ibibio of Nigeria knew Aldebaran (with the Pleiades) as 'the fowl mother of chicks' and called the stars, picturesquely, 'sand of the moon' (Talbot 1925).

Among the southern Africans the Southern Cross (Crucis) was grouped with the Pointers (α and β Centauri) to produce a constellation known as Giraffes, one male and one female (Beyer 1919; Breutz 1969; Clegg 1986). The Basutos also grouped Castor and Pollux into a constellation *Maselathako* (Beyer 1919). The Magellanic Clouds were named by some tribes: e.g. *Mazhara* and *Maguta* (Famine and Plenty) by the Karanga (von Sicard 1966), *se-thlako sa naka* and *se-thlako sa se-nakane* (Canopus's and Achernar's shields respectively) by the Basutos (Beyer 1919), and *Kgoro* and *Tlala* (related to predicting drought or rainfall according to their appearance) by the Tswana (Clegg 1986, who reports the same folklore for the Shona and Kalonga).

The Milky Way was widely recognised as dividing the sky into two parts; it was incorporated into mythology as the track of an imaginary beast, or of ancestors.

The San and some of the Khoikhoi representatives of the hunter–gatherer life style read different meanings into the stars than those of their agriculturalist neighbours (Hahn 1881; Hammond Tooke 1886; Bleek and Lloyd 1911; Norton 1909; Dornan 1921; Schapera 1930; Starr 1990). As pointed out by Hammond Tooke, the list of stars named by the San is longer than that used by Homer in his epic poems. In San mythology the stars are animals or people of earlier times; the names given to them usually represent animals that are abundantly available at the seasons when the stars are prominent. Few constellations are recognised as such, but neighbouring bright stars often share similar names. Thus α Centauri and β Centauri are male lions, α, β and γ Crucis are lionesses, Aldebaran is a male gnu, Betelgeuse is a female gnu, Procyon is a male eland and Castor and Pollux are his wives, the Magellanic Clouds are steenboks, the sword and belt of Orion are male and female tortoises respectively.

Canopus and Sirius played a different role. The former was known as the 'ant's egg star' because it was prominent during the time of year when this gastronomic delicacy was abundant. The times of year when these stars first appeared were times for special ceremonies involving rapidly moving burning sticks pointed at them, apparently imitating the prominent twinkling of these stars when near the horizon. Regulus was known as 'firewood finisher', because it set when the fire was finished (Silberbauer 1981).

In one eastern Namibian San group the Pleiades, Canopus and Capella were related as a constellation and the stars of Orion were recognised as a constellation about which a celestial hunting legend was related (Marshall 1975). The Milky Way was generally pictured by the San as 'night's backbone' (Silberbauer 1981) and in mythology was created by a girl of an early race who threw a handful of wood ashes into the sky. The same girl threw bits of an edible root into the sky, the old (red) pieces creating red stars and the young (white) pieces creating white stars (Bleek and Lloyd 1911). This is one of the few records of recognition of star colours; furthermore, the San recorded that the stars 'become white when the sun comes out'. The Zulus perhaps recognised that bright stars are reddened near the horizon and record having seen stars (probably Venus) during the day (Callaway 1970).

Venus

As the most prominent object in the night sky, for protracted periods before dawn or after sunset, Venus was recognised throughout Africa and played a prominent part in many mythologies. It often had names meaning 'evening' star, 'the drawer of days to their close', or 'forerunner of the sun', but there were more poetic names in use: 'evening fugitive', 'the watching one', 'the peeper into pots' (at the evening meal), 'the moon's wife'. Most Africans did not realise that the morning and evening appearances were the same object, and therefore gave it different names. A few groups, however, had made the connection (e.g. Karanga: von Sicard 1966; Khoikhoi: Schapera 1930; the Xhosa, where the morning appearance is called *Ikwezi lokusa* and the evening *Ikwezi*: Soga 1932). An extensive discussion of the appearances of Venus in African mythology is given by von Sicard (1966).

Other planets

Although there was awareness that certain bright stars were not always visible, only rarely does the name for planet imply motion relative to the fixed stars. An example is the Herero (of Angola) name which means 'to rove about' (Hammond Tooke 1886). The Zulu were aware that 'there are stars which travel, and which die like the moon' (Callaway 1970).

Surprisingly, Mercury was noted by a few groups, in particular the Basuto and Khoikhoi, both of whom only recognised it in the morning sky and used it to indicate imminent sunrise or time to milk the cows (Hoffman 1925; Schapera 1930).

Saturn was known to the Xhosa as *Ucanzibe*, which was a name also used to denote the equivalent of the month of April (Soga 1932). Jupiter was commonly known by the Basuto as *Matosabosiu*, 'the drawer up of night' or as *Tosa*, 'the shining knob by which the sky is rotated' (Norton 1909), and by the Khoikhoi was sometimes confused with Venus (Shapera 1930). Another Basuto name for Jupiter is *Moliana*, which is a diminutive form of 'to delay' and may indicate that the retrograde motion had been detected. The San refer to the 'Dawn's Heart' star, which is a central element in their mythology, representing a mythical male character, and nearby bright stars would receive the title 'daughters of Dawn's Heart'. Bleek (1875) identified 'Dawn's Heart' with Jupiter, and later writers have followed his lead (e.g. Norton 1909, Bleek and Lloyd 1911, Beyer 1919, Breutz 1969). However, there is no explanation given by Bleek for his identification, and the latest scholarship suggests that 'Dawn's Heart' is more probably Venus (K. Snedegar, private communication).

Meteors

Traditional interpretations of the meaning of meteors arise from every part of Africa, which proves that they were universally observed. A major review of the subject has been given by Lagercrantz (1964). Meteors were almost always regarded as ill omens, signifying the death of a chief (usually in the direction that the meteor was seen), war, disease or famine. This is the situation from the extreme north to at least as far south as the Zulu people of South Africa. There are some interesting and curiously distributed exceptions: the San of northern Namibia, the Masai of Kenya/Tanzania and a few groups in Morocco considered meteors to be favourable omens, foretelling good rains. For the Tswana a meteor foreshadowed great earthly events and a 'booming shooting star' (apparently signifying one that falls nearby) indicated a good season ahead (Clegg 1986). The Zande of Zaire evidently paid particular attention to the night sky, for their language includes a word ('heavenly smoke') for a meteor shower.

On any dark African night meteors can commonly be seen at a rate of many per hour, so we must deduce that either only exceptionally bright meteors (fireballs) were noted, or the majority of the peoples of Africa were prodigiously pessimistic.

Comets

As with meteors, comets were usually considered to portend disaster (Lagercrantz 1964). This parallel with medieval European belief extends even to some of the words used to denote a comet, which include 'tail star', 'hair star', 'star with long feathers' and 'star of dust'. The first of these is used throughout Africa. Again there are more optimistic viewpoints in a few areas, most notably the San of northern Namibia. The Masai, and some neighbouring tribes, consider comets to be the embodiment of important gods; but in general comets are not awarded more than portent status. Bright comets are sufficiently frequent that it is not too difficult to relate their appearance to the death of a chief or a famine within a year or two. Nevertheless, when used as year markers by the Xhosas it is the comet rather than its effect that is remembered (Soga 1932).

Novae

There is one indirect piece of evidence that one tribe at least, the Tswana, may have been aware of novae (or supernovae). An elder of the tribe reported: 'When the big stars have come out, small ones are afraid of their light and then they move away and burn themselves up [i.e. as meteors]. A star of which it is unknown when it appears is either a sign of sound health or of great famine.' (Breutz 1969).

Physical representations of celestial objects

Little has been published on African artefacts or pictorial representations of celestial objects. Earthy (1924) noted that among the Chopi and some other tribes of Mozambique women frequently incised circular marks in their foreheads or elsewhere as denoting the moon; one clan, the Muyanga, whose name derives from the heat of the sun, incised their backs in the form of rays from the rising sun. A line on the forearm, representing a meteor, was also in use. Women of some southern Sudanese tribes ornamented their lower lips with small polished quartz pieces which they considered to be stars that had fallen to earth (Lagercrantz 1964).

Although there are star-like images (sets of radiating lines) used as decorative motifs on a variety of African artefacts (e.g. the ethnographic collections in the South African Museum have wooden bowls and walking sticks, collected in the early twentieth century), unless supported by oral tradition these can not be certainly attributed to intentional astronomical depictions. More plausible are star-like and crescent-shaped incisions in rock pavements in parts of Namibia (Dowson 1992).

Zahan (1979) claims that the zigzag decorative motif common on African utensils and house facades is a two-dimensional representation of the path made by the sun in its annual oscillation in declination. This motif would appear, however, to spread over a wider range of peoples than those aware of such motions.

It might be thought that the most rewarding area to seek pictorial representations of celestial bodies would be among the numerous San rock paintings of Southern Africa, some of which date back more than 25,000 years (Wendt 1976). However, the majority of these are now realised to be depictions of trance-related experiences (Lewis–Williams 1981; Lewis–Williams and Dowson 1989), so they are unlikely to contain astronomical topics unless the latter are incorporated into the beliefs associated with trance. Just such a proposal has been made by Thackeray (1988; see also Thackeray and Knox–Shaw 1992) who points out that meteors and comets could well have been thought by the San to be analogous to the flashes and streaks of light perceived on entering trance. This

105 Cave painting of a comet or fireball, Fouriesberg, South Africa. (From Woodhouse 1986b)

106 Bone tally stick, possibly used as a lunar calendar. From Border Cave, Lebombo Mountains, South Africa. (From Beaumont 1973)

could explain the appearance of comets or fireballs as components of some San trance paintings and other paintings in Zambia (Woodhouse 1986a,b,c). One of these is represented in Fig. 105.

Bones or pieces of wood with notches cut in them are commonly found in Stone Age deposits. Claims have been made that at least some of these tally sticks served a calendric purpose, but this remains a controversial subject (Marshack 1989). That the Khoikhoi, after contact with Europeans, constructed weekly calendars by drilling holes in wood is attested by Schapera (1930), who states that these had seven holes through which a leather string could be threaded daily. The examples in the South African Museum have eight holes, the first being used to anchor the string.

With this example in mind, it is suggestive that a tally stick bearing twenty-nine notches (and made from the fibula of a baboon) has been excavated from Border Cave in South Africa (Beaumont 1973: and see Fig. 106). Although fractured at one end, it is thought not to be missing any of the original notches. It is possible that this device served as a lunar calendar – the twenty-nine notches furnishing twenty-eight spaces between them, or the first notch serving to attach a string. Fragments of other tally sticks are commonly found; this is the only complete example and it is dated reliably to 38,000 BP. Other fragments date perhaps as far back as 100,000 BP.

For completeness, it should be mentioned that the carved wooden platter, found in a cave near the ruins of Great Zimbabwe and supposed by Schlichter (1899) and Keane (1901) to be thousands of years old and to depict signs of the Zodiac, is at most a few hundred years old and is not now considered to represent the sky (Webb 1952).

Acknowledgements

I am indebted to the following persons for assistance in locating many of the original sources on which this chapter is based: E. Lastovica (South African Astronomical Observatory, Cape Town), R. Papini (Local History Museum, Durban), G.M. Morcom (East London Museum), M.M. Hirst (Kaffrarian Museum, King William's Town) P. Beaumont (McGregor Museum, Kimberley), M.G. Whisson (Rhodes University), P. Davison, G. Klinghardt and E. Willis (South African Museum, Cape Town), and J.F. Thackeray (Transvaal Museum, Pretoria). I also thank J. Parkington (University of Cape Town) and G. Klinghardt for advice.

Bibliography

Beaumont, P. B. 1973. Border Cave – a Progress Report. *South African Journal of Science* LXIX, 41.

Beyer, G. 1919. Suto astronomy, *South African Journal of Science* XVI, 206.

Bleek, W. H. I. 1875. *A Brief Account of Bushman Folk-Lore and Other Texts.* Cape Town: J. C. Juta.

Bleek, W. H. I. and Lloyd, L.C. 1911. *Specimens of Bushman Folklore.* London: Allen.

Brecher, K. 1979. Sirius enigmas. In *Astronomy of the Ancients*, eds K. Brecher and M. Feirtag, p. 91. Cambridge, Mass: MIT Press.

Breutz, P. L. 1969. Sotho-Tswana celestial concepts. *Ethnological Publications*, no. 52, 199. Pretoria: Department of Bantu Adminstration and Development, Government Printer.

Bryant, A. T. 1949. *The Zulu People.* Pietermaritzburg: Shuter & Shooter.

Cagnolo, C. 1933. *The Akikuyu.* Nyeri: Mission Printing School.

Callaway, H. 1970. *The Religious System of the Amazulu.* Cape Town: Struik.

Clegg, A. 1986. Some aspects of Tswana cosmology. *Botswana Notes & Records* XVIII, 33.

Dampier, W. 1968. *A New Voyage Around the World.* Reprint. New York: Dover Publications.

Doke, C. M. 1928. Lamba ideas of cosmogony. *South African Geographical Journal* XI, 18.

Dornan, S. S. 1921. The heavenly bodies in South African mythology. *South African Journal of Science* XVIII, 430–37.

Dowson, T. E. 1992. *Rock Engravings of Southern Africa.* Johannesburg: Witwatersrand University Press.

Earthy, E. D. 1924. On the significance of the body markings of some natives of Portuguese East Africa. *South African Journal of Science* XXI, 573.

Frazer, J. G. 1925. *The Golden Bough.* London: Macmillan.

Griaule, M. and Dieterlen, G. 1950. Un système Soudanais de Sirius. *Journal de la Société des Africanistes* XX, 273.

Griaule, M. and Dieterlen, G. 1965. *Le Reynard Pâle*, p. 468. Paris: Institut d'Ethnologie.

Hadingham, E. 1983. *Early Man and the Cosmos.* London: Heinemann.

Hahn, T. 1881. *Tsuni-Goam: The Supreme Being of Koi-Koi.* London: Trübner.

Hammond-Tooke, W. 1886. The star lore of the South African natives. *Transactions of the South African Philosophical Society* V, 304.

Hoffman, C. 1925. The natives and astronomy. *The South African Nation* 7 (Nov.), 14.

Inskeep, R. R. 1978. *The Peopling of Southern Africa.* Cape Town: David Philip.

Jackson, A. O. 1969. The Langa Ndebele Calendar and annual agricultural ceremonies. *Ethnological Publications*, no. 52, 233. Pretoria: Department of Bantu Administration and Development, Government Printer.

Junod, H. A. 1927. *The Life of a South African Tribe.* London: Macmillan.

Keane, A. H. 1901. *The Gold of Ophir.* London: Edward Stanford.

Kolbe, P. 1731. *The Present State of the Cape of Good Hope.* London: W. Innys.

Krige, E. J. 1936. *The Social System of the Zulus.* Pietermaritzburg: Shuter & Shooter.

Kuper, H. 1947. *An African Aristocracy – Rank among the Swazi.* Oxford University Press.

Lagercrantz, S. 1964. Traditional beliefs in Africa concerning meteors, comets and shooting stars. In *Festschrift für Ad. E. Jensen.* Munich: Klaus Renner Verlag.

Lewis-Williams, J. D. 1981. *Believing and Seeing: Symbolic Meanings in Southern San Rock Paintings.* London: Academic Press.

Lewis-Williams, J. D. and Dowson, T. 1989. *Rock Engravings of Southern Africa.* Johannesburg: Southern Book Publishers.

Marshack, A. 1989. On wishful thinking and lunar 'calendars'. *Current Anthropology* XXX, 491.

Marshall, L. 1975. Two Jũ/wã constellations, *Botswana Notes and Records* VII, 153.

Marwick, B. A. 1966. *The Swazi*. London: Frank Cass & Co.

McLaren, J. 1929. *A Grammar of the Kaffir Language*. London: Longmans, Green.

Murdock, G. P. 1959. *Africa: Its People and their Culture History*. New York: McGraw-Hill.

Nilsson, M. P. 1920. *Primitive Time Reckoning*. Lund (Sweden).

Norton, Rev. Father 1909. Native star names. *South African Journal of Science* IX, 306.

Pesch, P. and Pesch, R. 1977. The Dogon and Sirius. *Observatory* XCVII, 26.

Pocock, D. F. 1967. The anthropology of time reckoning. In *Myth and Cosmos*, ed. J. Middleton, p. 303. New York: The Natural History Press.

Samuelson, R. C. A. 1929. *Long, Long Ago*. Durban: Knox Painting & Publishing Co.

Schapera, I. 1930. *The Koisan Peoples of South Africa*. London: Routledge & Kegan Paul.

Schlichter, 1899. Travels and researches in Rhodesia. *Geographical Journal* XIII, 376.

Silberbauer, G. B. 1981. *Hunter and Habitat in the Central Kalahari Desert*. Cambridge University Press.

Soga, J. H. 1932. *The Ama-Xhosa: Life and Customs*. Lovedale Press.

Starr, E. M. 1990. Sub-Saharan African astronomical mythology. *The Planetarian* XIX, 8.

Talbot, P. Amaury 1925. *The Peoples of Southern Nigeria*, vol. III. Oxford University Press.

Temple, R.K.G. 1975. Response to appeal by W. H. McCrea concerning Sirius. *Observatory* XCV, 52.

Thackeray, J. F. 1988. Comets, meteors and trance: Were these conceptually associated in Southern African pre-history? *Monthly Notes of the Astronomical Society of Southern Africa* XLVII, 49.

Thackeray, J. F. and Knox-Shaw, P. 1992. Astronomical and entoptic phenomena, *Monthly Notes of the Astronomical Society of Southern Africa* LI, 6.

Thomas, N. W. 1913. *Ibo-Speaking Peoples of Nigeria*, part I. London: Harrison & Sons.

Thornton, R. J. 1980. *Space, Time and Culture among the Iraqw of Tanzania*. London: Academic Press.

Turton, D. and Ruggles, C. 1978. Agreeing to disagree: the measurement of duration in a Southwestern Ethiopian community. *Current Anthropology* XIX, 585.

Von Sicard, H. 1966. Karanga Stars. *NADA* IX, 42.

Wagner, G. 1954. The Abaluyia of Kavirondo (Kenya). In *African Worlds*, ed. D. Forde. Oxford University Press.

Webb, E. J. 1952. *The Names of the Stars*. London: Nisbet.

Wendt, W. E. 1976. 'Art Mobilier' from the Apollo 11 Cave, South West Africa: Africa's oldest dated works of art. *South African Archaeological Bulletin* XXXI, 5.

Werner, A. 1925. *Mythology of All Races: African*. Boston: Archaeological Institute of America, Marshall Jones Co.

Westmarck, E. 1926. *Ritual and Belief in Morocco*. London: Macmillan.

Woodhouse, H. C. 1986a. Bushmen painting of comets? *Monthly Notes of the Astronomical Society of Southern Africa* XLV, 33.

Woodhouse, H. C. 1986b. Halley's Comet viewed from Southern Africa. *South African Journal of Science* LXXXII, 132.

Woodhouse, H.C. 1986c. Comets in the rock art of Southern Africa. *Papers of the Archaeological Society of New Mexico* XII, 55.

Zahan, D. 1979. *The Religion, Spirituality and Thought of Traditional Africa*. University of Chicago Press.

WAYNE ORCHISTON

Australian Aboriginal, Polynesian and Maori Astronomy

The vast expanse of the Pacific Ocean encompasses a number of different geographical island groupings. One of these is Polynesia, which together with Australia forms the focus of this chapter. While the Australian Aborigines first settled that ancient island continent between 60,000 and 100,000 years ago, the scattered islands of Polynesia were one of the last regions of the globe to acquire human colonists. Given their contrasting settlement histories and cultural and biological dissimilarities, it is not surprising that the indigenous astronomical systems of Australia and Polynesia were also very different.

Australia

Australia is an enormous continent, stretching some 3,500 km from north to south and 4,200 km from east to west (Fig. 107). It encompasses almost 7.7 million square km and is much larger than Europe and comparable in size to the United States. With its latitudinal range from tropical north to temperate south and luxuriant coastal landscapes contrasting markedly with the arid interior, it is a continent of geographical diversity, and this had a profound ecological and hence cultural effect on its indigenous Aboriginal occupants. Given the marked regional variations that occurred in Aboriginal culture, we are justified in assuming that widely ranging indigenous astronomical systems must have existed throughout the continent. But since very few explorers or anthropologists have paid any attention to Aboriginal astronomy, this must remain a supposition. All we can do here is summarise what is known, for those few tribal areas where documentation does occur, and try to identify some common elements.

The Aborigines had no special instruments or devices for their celestial observations, yet their knowledge of the southern sky '... was probably the most precise possible for a people dependent on the naked eye' (Haynes 1992). They had names for the sun, moon, naked-eye planets, comets and meteors, many of the brighter stars and some fainter ones, the Milky Way, the Coal Sack, and the Magellanic Clouds. Brightness was not always important – sometimes it was colour that mattered. For instance, the Aranda tribes of Central Australia recognised white, blue, yellow and red stars. Antares was described as *tataka indora*, or 'very red', while some stars in the Hyades were *tataka* and others were *tjilkera* (white).

The Aborigines also recognised various star groups, but the interesting point here was that only certain stars were included while others – sometimes very conspicuous ones – were ignored. In other words, pattern was sometimes more important than magnitude. Thus, the Aborigines of Groote Eylandt in the Gulf of Carpentaria recognised *Unwala* (the Crab), a group of five relatively faint naked eye stars in Hydra, yet they ignored the two adjacent, first-magnitude stars, Procyon and Regulus.

And as would be expected given the different cultural perspectives involved, even when the Aborigines did recognise the same constellations as we do, their descriptions were totally different. For instance, to the Boorong tribe from the Mallee region of Victoria, Corona Borealis was a boomerang, Corvus a kangaroo, Delphinus a great fish, and Coma Berenices a tree with three main branches. MacPherson (1881) noted that

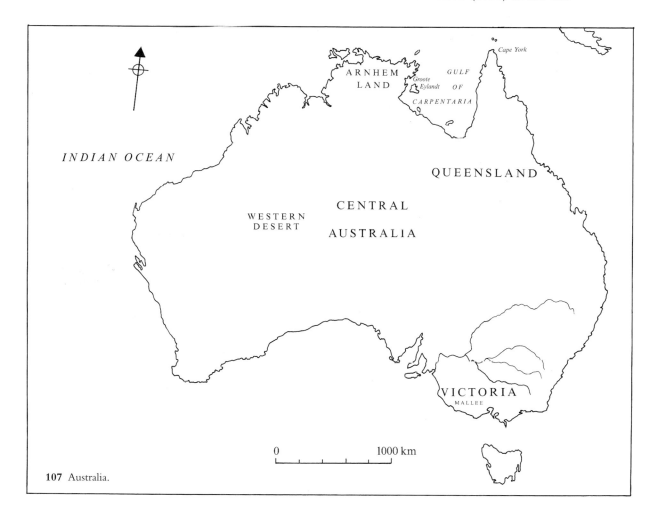

107 Australia.

these Aborigines showed '. . . evidences of imagination in tracing resemblances between objects on the earth and the outline formed by certain stars'.

The annual movements of the stars (as the earth followed its path around the sun) were also recognised by the Aborigines, and they used this knowledge to develop a complex seasonal calendar. This was vital to a hunter–gatherer people dependent upon nature for survival. In the tropical far north, the appearance of Arcturus in the eastern sky at sunrise indicated that it was time to harvest the spike-rush (which was made into baskets and fish-traps), while to the south, among the Boorong, Arcturus coincided with that time of year when pupae of the wood-ant formed a significant dietary component. Meanwhile, Vega's presence in October marked a time when the eggs of the Mallee hen were searched out and eaten. Throughout Australia, then, the Aboriginal tribes recognised a number of different stars that served as vital seasonal indicators and regulated the food-quest or other equally important ecological activities.

Because the Aborigines did not understand the true nature of the various astronomical objects and phenomena they observed, they developed various myths which served them by way of explanation. In addition, these myths often provided a link between humans and nature and an on-going reminder of appropriate modes of behaviour in different cultural situations. The myths were passed down from generation to generation in song, dance and mime, and through images preserved as rock engravings or paintings, or on bark. In a non-literate society, these methods, and word-of-mouth, were the only means whereby astronomical knowledge could be preserved.

To the Aborigines, the sun was seen as a woman who by day carried a fire from east to west across the sky. At the start and end of each journey she decorated herself with a powder made from ground red ochre, in the process spilling some and colouring the adjacent clouds and sky. At night she travelled underground back to her eastern homeland, and readied herself for the next day's journey.

The moon was always a male, and generally was linked to fertility. In some areas of Australia, Aboriginal girls who gazed at the moon stood the chance of getting pregnant! Various 'logical' explanations were also advanced to explain the moon's phases. Eclipses of the sun occurred when the sun woman attempted to mate with the moon man – one would have to assume that she was usually unsuccessful, given the scarcity of total solar eclipses.

While pan-continental basic interpretations of the sun and moon seem to have existed, this was certainly not the case for other celestial objects. To illustrate this we will look at the different ways Aborigines throughout Australia conceptualised the Pleiades or 'Seven Sisters', a distinct cluster of naked-eye stars in Taurus, a constellation ncighbouring Orion.

To the Boorong of Victoria the Pleiades were seen as a group of young females (*Larnankurrk*), playing to a corroboree party of young men (*Kulkunbulla*) making up Orion's belt and sword. The Pleiades and Orion were identified as summer constellations.

Among the Pitjandjara of the Western Desert the Pleiades were known as *Kung-karungkara*. Originally a group of women who lived on earth, they kept a pack of dingos as protection from a man named *Njiru* who was intent on ravishing them. After one of the women was raped and died the others fled into the sky becoming the Pleiades. *Njiru* pursued them, and now forms the stars of Orion's belt. To the Pitjandjara, the appearance of *Kungkarungkara* in the eastern sky in autumn had an extra significance, for it marked the start of the dingo-breeding season. Fertility ceremonies were then mandatory.

The Ngulugwongga of Arnhem Land had a somewhat similar account. According to this tribe a man named *Pingal* lived on earth with his wife, *Abobi*, and seven daughters. Eventually *Abobi* left *Pingal*, changed herself into a star, and went to live in the sky. As his daughters grew up, *Pingal* lusted after them. One day when they were out collecting food he began raping the oldest daughter, but *Abobi* saw this. She threw her daughters a rope, and they climbed up into the sky, becoming the Pleiades. *Pingal* also went to live in the sky, but as the moon. His daughters now stay close together, for they are still afraid of their father.

On the coast of Arnhem Land, the Pleiades–Orion region had a different association again. Among the Yirrkalla there was a legend about a group of fishermen in one canoe and their wives in another, voyaging from the east. En route the men caught a turtle and the women two large fish. As they neared shore a storm blew up, the canoes capsized, and all of the inhabitants drowned. The fishermen and their canoe now form the constellation of Orion, and the women and their canoe, the Pleiades. The turtle and two large fish form groupings of stars nearby, in the Milky Way. The appearance of Orion and the Pleiades throughout the wet season served as a warning to the living of the potential danger of undertaking sea voyages at this time of year!

A variant of this interpretation occurred on nearby Groote Eylandt. There three fishermen known as *Burumburum-runja* comprise the three stars of Orion's belt, while their wives, *Wutarinja*, make up the Pleiades. Variations on this theme are found in other parts of Arnhem Land (Fig. 108).

For upwards of 1,000 years Indonesian fisherman voyaged to the northern coast of Australia each year and spent many months living there among the Aborigines and interacting with them. Given this situation, it would be interesting to see whether any of the astronomical myths found in Arnhem Land can be traced back to Indonesia. Similarly, it is not known if the Aborigines of Cape York and the northern Queensland coast adopted any elements of Melanesian astronomy introduced by Papuans from New Guinea, with whom they maintained close contacts for many centuries.

Polynesia

Far to the east and north-east of Australia lie the many scattered islands of Polynesia, within a great expanse of the Pacific Ocean bounded by Easter Island, the Hawaiian Islands, and New Zealand (Fig. 109). In stark contrast to the environmental opulence of

108 Bark painting from Milingimbi in Arnhem Land, showing a large canoe containing Orion's belt (the three stars to the left), the Pleiades (in the centre) and the Hyades (the fish within the canoe). The fish swimming in the water are stars in the Milky Way. (From Mountford 1956)

the volcanic 'high islands' were the depauperated atolls. With few attractive resources, and hardly standing above sea level, they bore the brunt of the region's cyclones. Further south was subcontinental New Zealand, ranging from subtropical to subantarctic in climate, and with a floral and faunal régime largely unlike anything met elsewhere in Polynesia.

Despite this geographical pot pourri, the Polynesians from the different island groups were culturally and linguistically related, deriving as they did from a common ancestral stock, and this is reflected in their astronomical practices and beliefs. But, as in the case of Aboriginal Australia, little has been written on this subject, the only works of any moment being Best's two little Dominion Museum Monographs, *The Astronomical Knowledge of the Maori* and *The Maori Division of Time*, first published in the 1920s.

Among the Maori of New Zealand, the study of heavenly bodies (*whanau marama*) was the domain of specialists known as *tohunga kokorangi*, who would spend long lonely evenings contemplating the stars. Like the Aborigines they had to rely solely on the naked eye for their studies, yet some *tohunga kokorangi* had such phenomenal eyesight that they are reputed to have seen the four Galilean moons of Jupiter and many more than seven stars in the Pleiades.

These Maori astronomers, and their island Polynesian cousins, identified and named the sun, moon, naked-eye planets, brighter stars, Milky Way, Coal Sack, both Magellanic Clouds, and even the zodiacal light. There were names for comets and meteors, and for the different phases of the moon. To the Maori, stars, in general, were *whetu* (in other parts of Polynesia, *fetu*, *hetu* and *etu*), and planets *whetu ao*.

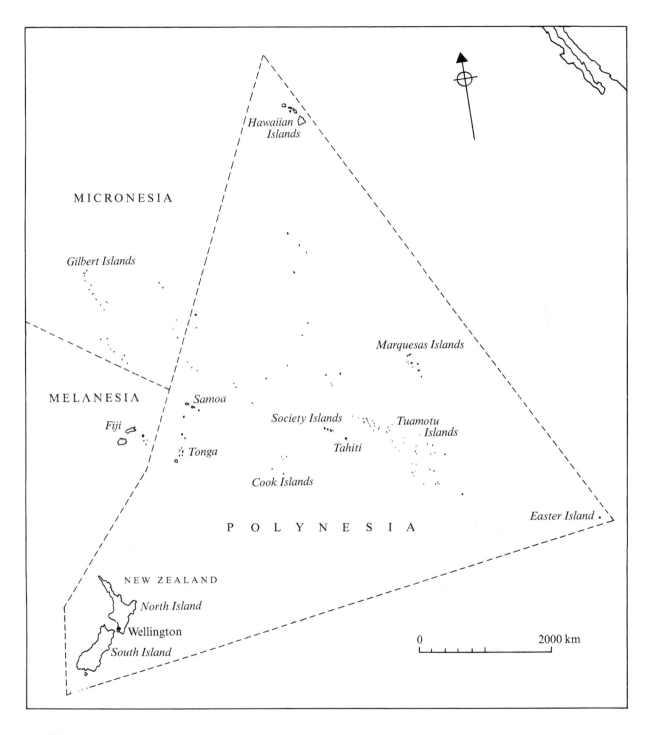

109 Polynesia.

Often, different terms were used throughout New Zealand to describe the same astronomical object, a reflection of the distinct culture areas that typified the country at the time of initial European settlement. For instance, Best reports the following names for the Milky Way:

Te Ika a Maui	*Te Mangoroa*
Te Ika-matua a Tangaroa	*Mangoroiata*
Te Ika-o-te-rangi	*Te Tuahiwi-nui-o-rangi*
Te Ikaroa	*Tuahiwi o Rangi-nui*
Te Ikaroa-o-te-rangi	*Whiti-kaupeka*
Te Ika-whenua-o-te-rangi	

while it was known as *Mokoroa-i-ata* in the Cook Islands. Among the Maori, at least seven different names were used for the bright star Canopus and five for the Southern Cross.

The Maori and other Polynesians recognised various constellations and assigned them names. Some, like the Southern Cross and Leo, corresponded to our own constellations, while others (for example, Orion's belt, the Hyades, the Pleiades, and the tail of the scorpion in Scorpio) represented only portions of these.

Seasons were also recognised, based on the presence of diagnostic stars or groups of stars, and were used to regulate horticultural, fishing, and other ecological activities. In New Zealand, for example, the appearance of the Pleiades heralded the birding season, when the catch was preserved in fat for consumption during the lean winter months. The Pleiades, together with Rigel and Orion's belt, were also associated with the planting of the *kumara* (a Maori dietary staple somewhat similar in form to the potato), while the appearance of the star Vega indicated that the time to harvest the crop had arrived. The annual appearance of certain stars, and the associated ceremonies, certainly evoked considerable sentiment.

Another major use to which the Polynesians put their astronomical knowledge was in voyaging. After the Vikings, they were arguably the world's most accomplished mariners. Two-way voyaging was endemic throughout the central Polynesian region, and celestial objects were habitually used for navigation. These skills were also enjoyed by the Micronesians, and in 1931 Grimble published a detailed account of the ways in which the Gilbert Islanders used their astronomical knowledge for voyaging purposes. He remarks that in the Gilberts there was no separate word for 'astronomer'; if you wanted an expert on the stars you asked for a *tiaborau*, or navigator!

Rather than merely recognising seasons (and the number varied according to locality), the Polynesians developed their own calendar systems, based mainly on movements of the moon. However, Best (1959) describes these systems, at best, as 'crude and incomplete'. The Maori and island Polynesian year generally contained 12 months, though 13 have been reliably reported in both areas. Among the Tuhoe tribe of New

Zealand, who lived in a mountainous somewhat inhospitable inland region of the North Island, 12 months were recognised; their names, and English translations of the associated Maori descriptions, are given below (after Best 1959, 19).

1 *Piripi* All things on earth come together owing to the cold; likewise man.
2 *Hongonui* Man is now extremely cold, and so kindles fires before which he basks.
3 *Hereturi-koka* The scorching effect of fire on the knees of man is seen.
4 *Mahuru* The earth has now acquired warmth, as also have herbage and trees.
5 *Whiringa-nuku* The earth has now become quite warm.
6 *Whiringa-rangi* It has now become summer, and the sun has acquired strength.
7 *Hakihea* Birds are now sitting in their nests.
8 *Kohi-tatea* Fruits have now set, and man eats of the new food products of the season.
9 *Hui-tanguru* The root of Ruhi now rests upon the earth.
10 *Poutu-te-rangi* The crops are now taken up.
11 *Paenga-whawha* All stakes are now stacked at the borders of the plantations.
12 *Haratua* Crops have now been stored in the store pits. The tasks of man are finished.

In different parts of New Zealand, different names were used by different tribes for the 12 or 13 different lunar months.

As month number 1 above suggests, the Maori year was deemed to commence in May or June, whereas in island Polynesia it began in December. The reason for this six-month discrepancy is not known. The actual starting date of the year in New Zealand varied from year to year, and depended on the occurrence of the first new moon after the reappearance of the Pleiades or Rigel in the morning sky.

Throughout Polynesia, each month was composed of thirty named 'nights of the moon', though within New Zealand the names themselves varied slightly from region to region. Despite this, when the individual lists of names are compared from island to island within Polynesia, many of the Maori names recur – even in distant Hawaii, at the far northern apex of the 'Polynesian Triangle'. This linguistic feature indicates that the names for the nights of the moon and other shared Polynesian astronomical terms were part of the ancestral vocabulary found in the Society Islands–Marquesas region more than two thousand years ago, for it was voyagers from this primary dispersal area who were largely responsible for the subsequent occupation of most other Polynesian island groups (with the notable exceptions of Samoa and Tonga, which were settled earlier).

Given the calendar systems at their disposal, the Polynesians were able to pinpoint particular dates during the year with great precision, by simply specifying the month and the actual night of the moon. This was critical among the Maori, for example, in identifying those particular days when the *kumara* had to be planted.

If the Polynesian calendar looks familiar, from a Eurocentric perspective, we cannot say the same of the mythological accounts that were offered to explain the various astronomical objects and events that were observed. In this regard they followed Aboriginal logic, and Polynesian myths also were passed down by word of mouth, and in songs, charms, chants, and as sayings.

The Maori believed that the universe consisted of twelve separate and distinct heavens, the closest one to the earth's surface being the mythological *Rangi*, or Sky Parent, across whose body the stars and other celestial objects moved. According to one legend, two of the offspring of *Rangi* and *Papa* (the Earth Mother) gave birth to the sun, moon, stars and phosphorescence. This can be most simply expressed as a genealogy (after Best, 1955).

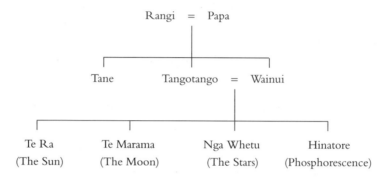

Variations on this genealogy occur throughout New Zealand, though the sun, moon and stars are generally perceived to be siblings.

It was *Tane* who was responsible for the correct placement of his nieces and nephews, in order to bring light to the Earth Mother (his own mother). First he placed phosphorescence on the body of his father, *Rangi*, but darkness prevailed. He then sprinkled the stars along the full length of *Rangi's* body, but it was only when he positioned the moon on *Rangi's* stomach that dim light began to filter through to his mother. Finally, he installed the sun on *Rangi's* chest, and brought full daylight to the world. There is a poignant old Maori saying about the nieces and nephews of *Tane*: 'The sun, moon, and stars all live peacefully together without quarrelling. Evil is unknown to them. When can we become like them?'

The term *Ra* was used for the sun throughout Polynesia, and also in ancient Egypt (though a cultural link should not be assumed between these two areas). According to the Maori, *Ra* started his life speeding across the sky, leaving the world in darkness for almost 24 hours each day, and had to be persuaded to travel at a more leisurely pace in order to produce day and night. An eclipse of the sun occurred when *Ra* was attacked and eaten (or partially eaten) by demons, which is a very different interpretation to that found in Australia.

In vernacular speech the moon was known to the Maori as *marama*, and different terms were used to distinguish a full moon, waxing moon, waning moon, and crescent moon. The Maori also personified the moon as *Hina* or *Hine-te-iwaiwa*, and associated her with a variety of typically female activities, including menstruation and childbirth. *Hina* served as a de facto patron of Maori women, and she was also recognised by this name in the Cook Islands, Tahiti, the Tuamotus and Hawaii. In Samoa she was known as *Sina*. The Maori also recognised a 'woman in the moon', called *Rona*, and she was thought to be responsible for lunar eclipses by attacking and destroying *marama*. It is interesting that in the Tuamotus *Rona* was a noted cannibal! In New Zealand, *Rona* was sometimes known by the longer name, *Rona-whakamau-tai*, or 'Rona the tide-controller', indicating that the Maori knew of the connection between the two.

In pre-European New Zealand, stars were also assigned mythological attributes, and were looked after by *Mangoroiata* (the Milky Way). Occasionally a star would stray from the group to which it belonged, and when it was pursued and struck by the sun or moon it appeared in the sky as a meteor (*matakokiri*). Very bright exploding meteors (termed 'bolides') were personified as *Rongomai*. Occasionally one of these would escape from the sky and crash to earth as a meteorite. Best (1955) relates how last century

> ... when the Pakakutu *pa* [fortified settlement] at Otaki was being besieged [during the 1830s] *Rongomai* was seen in broad daylight, a fiery form rushing through space. It struck the ground and caused dust to rise.

And near Wellington is a place called Te Hapua o Rongomai, where *Rongomai* is said to have descended to earth at some time in the past. It would be worth launching a systematic meteorite search at this locality.

As in Aboriginal Australia, Orion and the Pleiades had a special place in Polynesian astronomy, and in New Zealand they competed with each other for control of the Maori year (according to one myth, Rigel (in Orion) and the Pleiades both descended from the same parents). Elsewhere in Polynesia, the Pleiadian year was an institution. Collectively, the Pleiades were known to the Maori as *Matariki*, and their role was to keep together and paddle their canoe across the body of *Rangi*, all the time ensuring that humans were supplied with food. In the Cook Islands the Pleiades were also called *Matariki*; in Tahiti *Matari'i*; in Hawaii *Makalii*; and in Samoa *Mata-ali'i*.

The Maori also recognised a constellation they called *Te Ra o Tainui* (the 'Sail of Tainui') which took the form of a canoe, with the Pleiades at the bow and Orion's belt at the stern. The Hyades were an inverted triangular sail, and the (very) distant Southern Cross (*Punga a Tama-rereti*) served as the anchor of this mythical ancestral Tainui canoe.

Naked-eye comets also attracted the attention of the Maori, who generally referred to them as *Auahi-roa* or *Auahi-turoa*, though a number of other terms were occasionally used. One of these was *Puaroa*, and in Samoa the word for comet was *Pusaloa*. According to Maori legend, *Auahi-roa* (literally 'long smoke') descended from the sun, and was sent by his father down to earth to supply the people there with fire. When he reached the earth *Auahi-roa* married, and his wife gave birth to five mythical 'Fire Children'. Thus, fire was often figuratively referred to as *Te Tama a Auahi-roa* (the son of *Auahi-roa*) because it was the offspring of a comet. The appearance of a comet could serve as either a good or bad omen, depending upon the orientation and length of the tail.

Concluding remarks

Polynesian and Aboriginal astronomy was intricately interwoven with mythology, ritual and religion. The heavenly bodies were endowed with human attributes, and fantastic stories about their loves, hates, conflicts and achievements were developed. The sun, moon and stars (particularly certain stars or star groups) had enormous power and influence over life on earth, and their surveillance and study was therefore a matter of some importance. In Polynesia this was accomplished by specialist 'astronomers'. No optical aids of any kind were developed to assist the human eye in these endeavours, and in the absence of writing the results of these studies were passed down from generation to generation by word of mouth and in other institutionalised ways.

While the astronomical achievements of the Polynesians and Aborigines did not play any part in the development of scientific astronomy as we know it today, the study of their perspective provides an interesting cameo on the types of intimate relationship that could exist between humans and their environment in a broader cosmic sense.

As Best (1955) has noted: 'The knowledge ... of this subject is meagre and unsatisfactory, but it is now too late to remedy the deficiency.' While this assessment is probably all too true of Polynesia, there may still be opportunities to capture the last vestiges of different Australian Aboriginal astronomical systems before they too disappear into the mists of time.

Bibliography

Best, E. 1955. *The Astronomical Knowledge of the Maori*. Dominion Museum Monograph no. 3. Wellington: Government Printer.

Best, E. 1959. *The Maori Division of Time*. Dominion Museum Monograph no. 4. Wellington: Government Printer.

Grimble, A. 1931. Gilbertese astronomy and astronomical observances. *Journal of the Polynesian Society* 40, 197–224.

Haynes, R. D. 1992. Aboriginal astronomy. *Australian Journal of Astronomy* 4, 127–40.

Kingsley-Smith, C. 1967. Astronomers in puipuis. Maori star lore. *Southern Stars* 22, 5–10.

MacPherson, P. 1881. Astronomy of the Australian Aborigines. *Journal and Proceedings of the Royal Society of New South Wales* 15, 71–80.

Maegraith, B. G. 1932. The astronomy of the Aranda and Luritja tribes. *Proceedings of the Royal Society of South Australia* 56, 19–26.

Mountford, C. P. 1956. *Records of the American–Australian Scientific Expedition to Arnhem Land. Volume 1: Art, Myth and Symbolism*. Melbourne University Press.

F. RICHARD STEPHENSON

Modern Uses of Ancient Astronomy

For the purpose of this chapter, the term 'ancient' will be generalised to mean any time from remote antiquity to the introduction of the telescope in AD 1609. Although certain civilisations made considerable progress in astronomy during this long interval, the accuracy of observation was severely limited by the low acuity of the unaided eye.

Despite the ravages of time, vast numbers of celestial observations are preserved in history from the pre-telescopic period. Some of these records are of little more than historical interest. However, many others have proved of great value in modern scientific studies – notably in astrophysics, solar system studies and geophysics. The application of early astronomical records has developed into a major discipline in recent years. This has become known as Applied Historical Astronomy, to distinguish it from the study of the history of astronomy for its own sake. It is the purpose of the present chapter to outline some of the aims and achievements of this field of research.

It should be emphasised that Applied Historical Astronomy deals almost exclusively with written records. Through these, the present-day investigator has at least some direct contact with the astronomers of antiquity, no matter how brief or obscure an early text might be. Virtually all viable observations prove to be of unusual events; few routine measurements made by the ancients – such as the position of a planet at a certain moment – are of value today. Celestial events recorded in ancient history which have made a significant contribution to modern science are remarkably diverse. They include eclipses of both the sun and moon (Fig. 110), giant stellar outbursts known as supernovae, comets, sunspots, meteor showers and meteorites, and the aurora borealis.

The considerable time-span covered by ancient and medieval observations offers several major advantages over even the most accurate modern measurements. A notable example is the investigation of long-term trends – for instance, in solar activity or the earth's rotation – which cannot be discerned in the relatively short period covered by telescopic observations. The study of very rare phenomena such as supernova explosions is another important aspect of the usefulness of early data.

The reasons why ancient peoples made and recorded celestial observations are

110 Photograph taken about 20 minutes after the start of the lunar eclipse of 9 December 1992, showing the moon gradually entering the dark shadow of the earth. Some 40 minutes later the eclipse became total. Lunar eclipses were frequently recorded in antiquity, often with careful measurements of the time of occurrence. (D.G. McCartan, Department of Physics, University of Newcastle upon Tyne)

varied. Undoubtedly, astrology played a major role. Without the impetus provided by this pseudo-science, many of the observations which have proved of such consequence today might never have been made or reported. The need of early astronomers to maintain a reliable calendar or to make observational tests of their predictions of future events (such as eclipses) has also provided us with many careful measurements. In addition, chroniclers often reported some of the more spectacular celestial phenomena which they witnessed or heard about. As a result, they have preserved numerous valuable observations, even though most of their descriptions lack technical details.

Ideally, ancient observations should be treated purely as scientific data when applying them to problems in current research. However, the analysis of early astronomical records is often complicated by questions of interpretation – a circumstance which can be both challenging and frustrating. Sometimes the precise nature of the phenomenon described is in doubt. Since the cause of many celestial events was not understood in ancient times, their classification did not necessarily follow the strict rules which apply today.

Even if the nature of a certain event recorded in history is clear, dating may be problematical. The dates of the majority of celestial phenomena are accurately reported and can usually be readily converted to the Julian or Gregorian calendar. However, often there are real difficulties. Precise dates are particularly desirable when attempting to synchronise independent records of the same phenomenon from different parts of the world; this is especially true of cometary apparitions. In addition, for a number of very early records of eclipses the uncertainty in the date of occurrence is so large that an unambiguous identification cannot be made.

Before discussing the scientific applications of pre-telescopic records, some consideration of the various historical sources is necessary.

Historical sources

Only a few early civilisations have contributed significant numbers of observations which are viable today. Perhaps rather surprisingly, there appear to be no useful survivals from Egypt in the dynastic period, India or Mesoamerica, for example. Nevertheless, four cultures have bequeathed us astronomical records in abundance, despite the undoubted loss of much material down the centuries. These are Babylon, China (including its cultural satellites of Korea and Japan), Europe and the medieval Arab dominions. Without their contribution, the science of Applied Historical Astronomy might never have developed significantly.

Virtually no reliable astronomical records from any part of the world are older than the eighth century BC. Prior to this time, apart from an isolated series of sightings of the planet Venus during the reign of the Babylonian king Ammiṣaduqa (1702–1682 BC) (see p. 42), little more than occasional allusions to eclipses are extant. Dating of these events, which are mainly preserved in Chinese and Babylonian texts, presents considerable problems. Today it is impossible to assess the extent to which celestial observations were made by any culture at this archaic period or to obtain any idea of their variety.

Towards the end of the eighth century BC we at last encounter the earliest systematic observations. These originate from both Babylon and China and most dates are accurately reported. At no subsequent period in history do we find any major gaps in the combined record of celestial phenomena from different parts of the world. However, it should be noted that between about 50 BC (after which extant Babylonian observations cease) and AD 800 (when regular European and Arab observations began) nearly all the surviving astronomical records come from China alone.

Babylon

The vast majority of celestial observations which are preserved from the centuries prior to the Christian Era originate from Babylon. These are recorded on the Late Babylonian astronomical texts, fragmentary clay tablets which – with very few exceptions – date from between 700 and 50 BC. Virtually all of the surviving texts are now in the British Museum. Most extant observations are cited in astronomical diaries, which originally contained an almost day-to-day record of skywatching for perhaps more than eight centuries. However, despite the extent of the surviving Late Babylonian archive (some 1,500 tablets in all) only about 10% of the original data has survived. During the

Hellenistic period (fourth century BC onwards) the astronomers of Babylon copied many of the observations from past diaries in summary tables devoted to specific phenomena (e.g. lunar eclipses) and other secondary texts. Some of these copies are still extant and they form a valuable supplement to the very incomplete material in the diaries.

Photographs of the earlier diaries – along with transliterations and translations – have been recently published by A.J. Sachs and H. Hunger; this work is continuing with later diaries, which extend down to about 50 BC, and will eventually include subsidiary texts.

The Babylonian astronomers had a special interest in cyclic phenomena, such as eclipses and lunar and planetary movements, and the texts are a prolific source of observations of this type. However, although occasional comets (including Halley's) and bright meteors are also reported, there is not a single reference to a temporary star or sunspot, and only one possible allusion to the aurora.

China (including Japan and Korea)

蜀之東梁楚之南宋星上長盈尺凡四十六日而滅
北斗天機紫微垣三公貫索星長丈餘至天市垣巴
色白長五尺直西北後經文昌斗魁南掃太陽又掃
自八月庚辰彗出井二十四度四十分如南河大星

Between about 720 and 480 BC the earliest surviving series of solar eclipse observations (thirty-six in all) from any part of the world was recorded in China. Along with one or two comets and meteors, these events are noted in the ancient chronicle known as the *Chunqiu* ('Spring and Autumn Annals'). Most dates in this work are precisely given, a feature common to practically all later Chinese astronomical records. However, not until about 200 BC are systematic observations of a wide variety of celestial phenomena reported in Chinese history. These records, which extend almost uninterrupted down to relatively recent times, exist in vast numbers. Most are to be found in special astronomical treatises in the official dynastic histories. Although such works no longer exist in their original form, they have been copied and recopied many times.

Unlike the Babylonians, the astronomers of China were interested in almost all celestial phenomena visible to the unaided eye – whether they appeared to be periodic or not. With only isolated exceptions, Chinese observations of comets (Fig. 111), sunspots and supernovae are unrivalled in any other part of the world before the European Renaissance. Reports of eclipses, meteors and aurorae are also numerous.

111 Part of a page from the astronomical treatise of the *Yuanshi*, the official history of the Yuan dynasty in China (AD 1279–1368), containing a detailed record of Halley's Comet in AD 1301.

Around AD 700 Japanese and Korean astronomers independently began making similar observations to those practised by their Chinese counterparts and these continued for many centuries afterwards. Histories of Japan and Korea – almost invariably written in Classical Chinese – are replete with such material.

Europe

Many eclipses and several comets are recorded in the Greek and Latin Classics but the dates and descriptions are often obscure. Not until around AD 800 (after the Dark Ages) do we begin to find significant numbers of observations of celestial phenomena from Europe. Until the Renaissance these are mainly reported in monastic and town chronicles, whose compilers were often in the habit of including accounts of the more spectacular celestial events along with mundane matters. Large numbers of annals have been published in their original language (usually Latin) and thus are readily accessible.

Eclipses, comets, meteor showers and displays of the aurora borealis are often referred to in medieval European chronicles. Descriptions are often vivid, while dates are usually accurately specified in terms of the Julian calendar. In particular, the solar eclipse records – although generally lacking in technical details such as times – are often of considerable value.

After about AD 1450, Renaissance astronomers began to make careful measurements of the position of comets and supernovae and also timed both solar and lunar eclipses. From the point of view of modern science, perhaps the most valuable observations made in Europe at this period are of the brilliant supernovae of AD 1572 and 1604. These are representative of the very best that astronomers could achieve before the invention of the telescope.

The Arab dominions

Ancient Arab records of celestial phenomena may be found in two quite distinct sources: astronomical treatises mainly covering the period from about AD 800 to 1000, and chronicles extending from around AD 800 to 1500. In general, Arab astronomers showed little interest in sporadic phenomena such as comets and meteors; these were believed to take place in the sublunar region where change was prevalent. Much more concern was shown with the periodic movements of the sun, moon and planets and many carefully timed observations of both solar and lunar eclipses have come down to us.

Muslim chroniclers had no such scruples about reporting events which seemed to show no definite pattern and – like their Christian counterparts in Europe – recorded many comets and meteors, as well as solar and lunar eclipses. Their descriptions tend to be qualitative rather than quantitative, but some records (notably of solar eclipses and meteor showers) are of considerable utility at the present day. It is unfortunate that compared with the huge European corpus, few Arabic chronicles have been published so that many works are inaccessible.

Examples of scientific applications

Investigations of the scientific applications of pre-telescopic observations which have proved particularly fruitful in recent years include the following: i) eclipses and the study of the earth's past rotation; ii) the association of unusually bright stellar outbursts with supernova remnants which are observable today at X-ray, radio and other wavelengths; iii) the long-term orbit of Halley's Comet (the only known periodic comet which is regularly visible to the unaided eye). Other current research is concerned with problems such as iv) long-term solar variability as revealed by sunspots and aurorae and v) variations in the orbits of meteor streams and their links with comets. As examples, I have selected the first two topics and each will be discussed briefly below.

Eclipses and earth's rotation

As was first suggested by Immanuel Kant and William Herschel in the eighteenth century, the period of rotation of the earth is by no means constant. Telescopic observations of occultations of stars by the moon reveal that over the past three centuries the length of the mean solar day has continually fluctuated on time-scales of typically two or three decades. As a result, the length of the mean solar day has sometimes been as much as 4 milliseconds (0.004 seconds) longer or shorter than average. The amount of energy involved in such seemingly small changes is enormous – far greater than the total energy utilised by humans in the same period. Most of the observed variations are probably due to electromagnetic effects linking the liquid core and solid mantle of the earth.

Ancient observations reveal that in addition to short period fluctuations in the length of the day, significant changes are also taking place on the time-scale of millennia. Over a period of a few centuries, these are obscured by the decade variations. It is generally acknowledged that the main cause of long-term changes is the tides raised by the moon and sun in the oceans and seas of the earth.

Calculations indicate that the tides should be responsible for a steady increase in the length of the day by 2.3 milliseconds per century. Hence 1,000 years ago the day would be about 0.023 seconds shorter than it is now due to this cause alone. Despite the smallness of this change, the cumulative effect is so large as to be readily detectable from ancient astronomical observations. This is because an enormous number of days (roughly 365,000), each marginally shorter than at present, has elapsed. It is readily shown that if a clock keeping time by the earth over the past millennium could be compared with an ideal chronometer, it would be rather more than an hour slow. Two thousand years ago the error would be nearly 5 hours, and so on, the error increasing as the square of the elapsed time.

Non-tidal mechanisms, such as the on-going rise of land which was glaciated during the last Ice Age, modify these figures to an uncertain degree. Hence in practice the total accumulated clock error at any time – usually known as ΔT – must be derived from observations alone. For this purpose, time defined by the motion of the sun and moon (known as Terrestrial Time: TT) is used as the standard. At any moment, this differs from time measured relative to the rotation of the earth (Universal Time: UT) by ΔT.

Most ancient observations which can be utilised to study changes in the length of the day are in the form of eclipse timings; very few careful timings of other events are preserved in early literature. Only after the invention of the telescope do observations of occultations become important. Most carefully measured eclipse times recorded in ancient history are from Babylon, China and the Arab lands.

To give just one historical example, I have selected the solar eclipse of 15 April 136 BC (Fig. 112). This was observed to be total in Babylon. A translation based on the preserved accounts in two separate damaged texts which describe the same event is given below:

> Year 175 (SE), intercalary 12th month, day 29. At 24 degrees after sunrise, solar eclipse. When it began on the south-west side, in 18 degrees of day in the morning it became entirely total. Venus, Mercury and the 'Normal Stars' were visible; Jupiter and Mars, which were in their period of disappearance, became visible in its eclipse [. . .]. It threw off (the shadow) from south-west to north-east. (Time interval of) 35 degrees for onset, maximal phase and clearing.
>
> Composite translation by H. Hunger based on
> British Museum texts WA 34034 and WA 45745.

The date according to the Seleucid Era is exactly correct. Time intervals in the above quotation are expressed in *uš*, the interval for the celestial sphere to turn through 1°, or

112 A Babylonian astronomical diary containing a detailed account of the total solar eclipse of 136 BC. (British Museum, WA 45745)

4 minutes. These were probably determined with the aid of a water-clock. It can be seen that the eclipse began 96 minutes after sunrise, became total 72 minutes later and lasted for 140 minutes. It was customary for the Babylonian astronomers to time both solar and lunar eclipses using these same units.

Comparison between the calculated and observed times of the various phases of this eclipse leads to ΔT values in 136 BC of respectively 3.51, 3.38 and 3.42 hours, a mean of 3.44 hours. This result is supported by other roughly contemporaneous Babylonian timings of eclipses, both of moon and sun.

Between about AD 400 and 1300, Chinese astronomers timed many solar and lunar eclipses to the nearest quarter or half hour (Fig. 113). Over a period of two centuries – from about AD 800 to 1000 – Muslim astronomers in Baghdad and later Cairo made similar determinations which were probably accurate to the nearest five minutes or so. The Chinese used water-clocks but their Arab counterparts preferred to determine time indirectly by measuring altitudes of the sun, moon or certain bright stars and then converting to local time.

Untimed observations of total solar eclipses can be used to set limits to the value of ΔT on selected dates since the track of totality across the earth's surface is usually very narrow. The method of computation may be outlined as follows. Modern calculations which make no allowance for the variation in the length of the day result in tracks of totality consistently lying to the west of those places where the sun was seen to be completely obscured. Provided an observation is reliable, the difference in longitude corresponding to each edge of the track yields a range of values for ΔT. European observations recorded in chronicles are particularly valuable for this purpose – as are several Chinese and Arab records.

Combining the various ancient measurements and converting from clock errors to changes in the length of the mean solar day, it is possible to deduce the graph shown in Fig. 114. This diagram, recently derived by F. R. Stephenson and L. V. Morrison, shows that the average rate of increase in the length of the day over the past 2,700 years (compared with the value at AD 1800 as

113 An account of several solar eclipses, including the total solar eclipse of AD 1275, recorded in the astronomical treatise of the *Songshi*, the official history of the Song dynasty in China (AD 960–1279). The appearance of stars is noted, and the text also records that the hens and ducks all went to roost on account of the darkness.

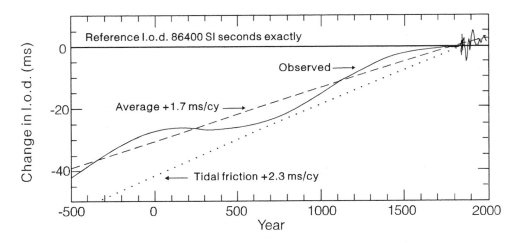

114 Plot of changes in the length of the day since 500 BC as determined by F. R. Stephenson and L.V. Morrison from ancient records of solar and lunar eclipses and telescopic observations of occultations. Fluctuations such as those noted in recent centuries are too rapid to be detected from ancient data, which reveal only the general trend. (F. R. Stephenson and L.V. Morrison, in *Philosophical Transactions of the Royal Society*, series A, vol. 351 (1995), p. 197, fig. 7)

standard) is 1.7 milliseconds per century. This is considerably less than the tidal figure, and marked long-term fluctuations are also evident. Interpretation of results such as these is of considerable importance in modern geophysics.

Historical supernovae

Supernovae are believed to signal either the complete disruption of a compact star which is one of the components of a close binary system (a Type I supernova) or the collapse of a massive star to a neutron star (a Type II event). These huge celestial explosions can be detected at vast distances from us. Although in recent decades many supernovae have been observed telescopically in external galaxies (some of which are very remote) no similar occurrence is definitely known to have been witnessed in our own galaxy since the invention of the telescope. Historical records provide evidence that several outbursts of this nature took place in the Milky Way between AD 1006 and 1604. Hence, for information about the very nearest supernova explosions we have to rely on observations made with the unaided eye.

Temporary stars which rival the brightest stars in the sky appear on average every few decades (the last occurring in Cygnus in 1975). However, most of these are relatively minor explosions known as novae, which remain visible for a relatively short time and then vanish virtually without trace. Supernovae are much rarer events and typically remain in view for many months. A supernova leaves behind it an expanding cloud of gas and dust which acts as a powerful source of radiowaves,

X-rays, etc. for many thousands of years – long after the original outburst has faded from sight. Investigations of supernovae are currently of great importance in astrophysics. In particular, knowledge of the precise time for which a supernova remnant has been expanding since the original explosion, and also information on the maximum brightness and subsequent fading of the outburst, help considerably in our understanding of the supernova process.

In this section, I shall briefly relate the history of the five supernovae which are known to have appeared in our galaxy since AD 1000. Other similar events probably occurred during the previous millennium but the records are rather concise and comparatively little is known about the outbursts.

Around the beginning of May in AD 1006, European, Arab and East Asian observers independently noted a brilliant new star appearing in the southern sky. At the monastery of St Gallen in Switzerland, where it was only barely above the horizon, it was said to be dazzling the eyes and beyond (i.e. to the south of) all the other constellations. It was seen there altogether for three months. At the monastery of Benevento in Italy the star was also described as very brilliant. Arab observers kept the new star in view for several months and both in Egypt and Iraq its light was said to rival the moon in brightness. In particular, the astronomer ʿAlī ibn Riḍwān measured its position as the fifteenth degree of the zodiacal sign of Scorpio.

In China the star was observed for several years, each year setting with the sun in September and reappearing at dawn three months later. Chinese astronomers reported that at first the star, which was yellow in colour, shone so brightly that one could see things clearly by its light and they measured the position as the third degree of the lunar mansion *Di*. Unfortunately there is little information on its changing brightness with time.

The new star of AD 1006 is thus the most brilliant supernova on record and its duration of visibility exceeds that of any other similar objects. Combining the various estimates of position (including the fact that the star could be seen as far north as St Gallen) enables the location to be deduced as within about 1° or 2° of the star β in the constellation of Lupus. This site is extremely close to that of a relatively young (age roughly 1,000 years) and nearby (distance from us some 3,000 light years) supernova remnant which is listed as 1459–41 in a survey by the Parkes radiotelescope in Australia.

Less than 50 years afterwards appeared another supernova, which although much fainter, has become far more well known. This is the new star of AD 1054, which formed the now famous Crab Nebula in Taurus. Although it was noticed in Constantinople, only Chinese and Japanese accounts provide useful information regarding its brightness and position in the sky. The Chinese Astronomer Royal gave the full period of visibility as more than 21 months (from 4 July in 1054 to 17 April in 1056). The pale red star was said to be 'visible in the daytime, like Venus' for a total of 23 days, after which it seems to have gradually faded. Both Chinese and Japanese astronomers noted the position of this 'guest star' as close to *Tianguan* (ζ Tauri).

The estimated position of the supernova lies very close to the site of the Crab Nebula, which is also one of the most powerful sources of X-rays and radiowaves in the entire sky. Some 6,000 light years away, it is unique among the historical supernovae in containing a pulsar.

In the autumn of AD 1181 was seen what is probably the faintest of the historical supernovae. This appeared close to the star ε in Cassiopeia and was detected only by Chinese and Japanese astronomers (Fig. 115). In South China, the star was discovered on 6 August and it remained visible for 185 days. In North China – then a separate empire – it was first sighted on 11 August and kept in view for 156 days. Japanese astronomers noticed it on 7 August but it is not known how long they continued to observe it. One Japanese account, which describes the colour as purple, seems to suggest that the star was no brighter than the planet Saturn.

Apart from its 6-month period of visibility, the best evidence that the star was indeed a supernova comes from its very close proximity to the strong radio source 3C 58 (source number 58 in the third Cambridge catalogue) – a known supernova remnant. This remnant is of similar form to the Crab Nebula but is rather further away from us (more than 8,000 light years). Astronomers have yet to detect a pulsar associated with it.

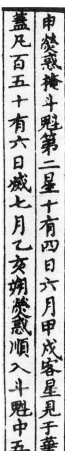

After the new star of AD 1181 faded from view, there is nothing in extant historical records to suggest that any further supernovae were seen for nearly four centuries. Subsequently, two brilliant new stars appeared within the space of about 30 years. The records of both the supernovae of AD 1572 and 1604 by European astronomers are remarkable, both for the accurate positional information which they yield and also the careful estimates of light variation.

The supernova which appeared in November of 1572 became temporarily as bright as Venus and was visible in daylight for a short time. Several European astronomers, including the great Danish observer Tycho Brahe (who was then aged only twenty-five), Michael Maestlin and Thomas Digges measured its position with extreme care relative to the surrounding stars of Cassiopeia. By comparison, East Asian observations are extremely crude and show no advance on those made several centuries previously. The location of the supernova derived from the European measurements lies well within the radio source 3C 10 and only about 3'

115 Part of a page from the astronomical treatise of the *Jinshi*, the official history of the Jin dynasty in China (AD 1115–1234), containing a record of the supernova of AD 1181.

of arc from its centre (Fig. 116). This supernova remnant is located about 7,000 light years from the solar system.

Tycho Brahe, who followed this orange-coloured star for 16 months, made several estimates of its declining brightness at widely spaced intervals. These enable a good light curve to be constructed, which is quite typical of those exhibited by Type I supernovae observed telescopically in external galaxies in recent decades.

Johannes Kepler and other European observers achieved similar success in recording

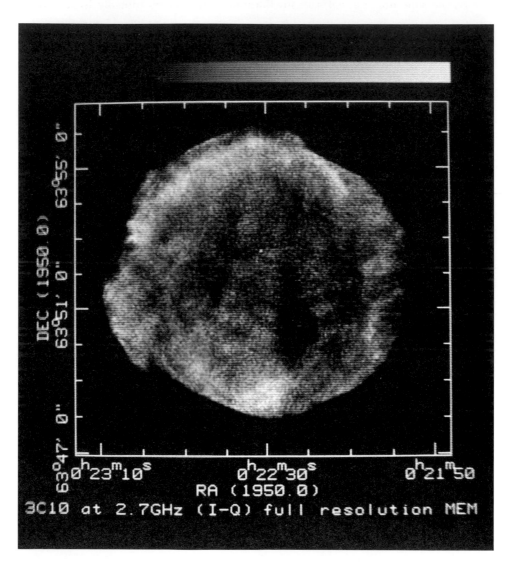

116 A radio image of 3C10, the remnant of Tycho's supernova of AD 1572, observed at 2.7 GHz with the Cambridge 5 km telescope. (D.A. Green, Mullard Radio Astronomy Observatory, Cavendish Laboratory, Cambridge)

the position and changing brightness of the supernova of 1604. By contrast, East Asian positional observations are of poor quality. The star, which was orange in colour and rather brighter than Jupiter, appeared in the southern constellation of Ophiuchus in October of 1604 and remained visible for a total of 12 months. However, the position measured independently by Kepler and David Fabricius lies within about 1' of arc of the centre of the supernova remnant 3C 358. This remnant is about 15,000 light years away, towards the central region of our galaxy.

As the supernovae faded, various European astronomers (notably Kepler) made many comparisons between its brightness and that of nearby stars. Remarkably, the court astronomers of Korea made similar observations which when included with the European data show that the light variation was again characteristic of a Type I supernova.

Conclusion

Although many ancient astronomical records have already been thoroughly researched, their potential to enrich various branches of modern science has by no means been exhausted. Without the observations which the astronomers of antiquity left behind, we would be unaware of many changes occurring both in the far reaches of space and on the earth itself. Present-day astronomers have good reason to be grateful to their poorly equipped but diligent predecessors.

Bibliography

Clark, D. H. and Stephenson, F. R. 1977. *The Historical Supernovae*. Oxford: Pergamon Press.

Stephenson, F. R. and Clark, D. H. 1978. *Applications of Early Astronomical Records*. Bristol: Hilger.

Stephenson, F. R. and Morrison, L.V. 1995. Long-term fluctuations in the Earth's rotation: 700 BC to AD 1990. *Philosophical Transactions of the Royal Society of London* (A) 351, 165–202.

Stephenson, F. R. 1996. *Historical Eclipses and Earth's Rotation*. Cambridge University Press.

Glossary

Some terms are effectively defined by authors in the context of their
own chapters, and these are cross-referenced to the appropriate page.

acronychal rising The day, shortly before opposition, when a star or outer planet is first visible in the east shortly after sunset.

altitude The angular distance of a celestial body above the horizon.

analemma A mathematical procedure for reducing a configuration on a sphere to a plane, thereby simplifying the solution of problems in spherical astronomy. Small circles on the sphere (such as the solar day-circle) are essentially folded about their diameters into a working plane (such as that of the meridian), and problems (such as the determination of time from solar altitude) can then be solved geometrically or trigonometrically in two dimensions.

anomaly 1) An irregularity or inequality in the motion of a celestial body, or the effects thereof, reflecting a departure from uniform motion, typically reflected in inequalities of time or angular motion due to the effects of the variable velocity of the celestial body. 2) The angular distance of a celestial body from its perigee or perihelion.

anomalistic cycle A whole number of synodic months which approximates a whole number of anomalistic months, e.g. 251 synodic months ≡ 269 anomalistic months.

anomalistic month/return in anomaly The interval in which the moon returns to its initial velocity, and thus to its initial position relative to its perigee (or apogee).

lunar/solar anomaly The inequality in the moon's/sun's motion due to its variable velocity resulting (principally) from the eccentricity of its orbit and its mean (angular) distance from perigee.

synodic anomaly An inequality due to the apparent position of a celestial body relative to the sun, encountered most conspicuously in planetary theory, where it is responsible for the stations and retrogradations of the planets. Also called 'second anomaly.'

zodiacal (sidereal) anomaly An inequality in the motion of a celestial body corresponding to its position in the zodiac. It is encountered in Babylonian lunar theory in the variable progress of the sun and moon at syzygy, where the inequality is largely, but not entirely, due to solar anomaly. It is also reflected in both Babylonian and Ptolemaic planetary theories and their derivatives. Also called 'first anomaly'.

aphelion The place in the orbit of a planet which is furthest from the sun.

apogee The place in the orbit of a celestial body which is furthest from the earth.

apsides The two points in an eccentric or elliptical orbit at which a body is at the greatest and the least distance from the body about which it orbits.

archaeoastronomy Interdisciplinary study of the practice of astronomy by ancient indigenous cultures of the world employing both the written and especially the unwritten evidence, e.g. alignments of buildings, iconography.

arcus visionis The minimum difference in altitude (solar depression) between the sun and a celestial body crossing the horizon, which is necessary for that body to be visible shortly before sunrise or after sunset. Ptolemy (*Almagest* 13, 7) gives the following values for the arcus visionis of each planet: Saturn 11°, Jupiter 10°, Mars 11.5°, Venus 5°, and Mercury 10°.

argument The independent variable of a function.

azimuth The horizontal bearing of a celestial object, measured clockwise from north.

body tally To count by making reference to the parts of the body, e.g. fingers, toes, joints.

calendar round A time cycle in widespread use in Mesoamerica comprising the lowest common multiple of *tzol kin* and *Haab*; thus $52 \times 365 = 73 \times 260$. Also called the 52-year cycle.

conjunction The moment when two celestial bodies have the same ecliptic (or polar) longitude. With respect to the sun and planets, at superior conjunction the sun is between the planet and observer, while at inferior conjunction an inner planet is between the sun and observer.

cosmical setting The day, shortly after opposition, when a star or outer planet is last visible in the west shortly before sunrise.

declination The angular distance of a celestial body perpendicular to the celestial equator.

deferent In epicyclic models, the circle which carries the epicycle; see pp. 73–4 and Figs 23 and 29.

direct motion Apparent motion of a celestial body from west to east in the direction of increasing longitude.

eccentric See p. 23 and Figs 64 and 65.

  lunar eccentricity See p. 79 and Fig. 25.

  solar eccentricity See pp. 77–8 and 184.

ecliptic The apparent path of the sun among the stars, and thus the line near which eclipses of the sun and moon take place (whence its name). Also, since ancient times the reference line for the measurement of celestial longitudes, either from a point fixed relative to the stars as among the Babylonians and Indians, or from the intersection of the ecliptic and celestial equator since Ptolemy.

  obliquity of the ecliptic The angle, slightly less than 24°, between the ecliptic and the celestial equator. Its modern value (1950) is 23°26'45", and it varies slowly with time. Ptolemy used the value 23°51'20", which he attributed to Eratosthenes and Hipparchus. Ulugh Beg found it to be 23°30'17".

elongation The angular distance in longitude between two celestial bodies as seen from the earth.

  synodic elongation The elongation of a celestial body with respect to the sun.

ephemeris A list or table of calculated astronomical events at regular intervals.

epicycle See pp. 73–4 and Figs 22 3, 28–9 and 64–5.

epoch 1) A date chosen for reference purposes in quoting astronomical data. 2) A location at a given time.

equation The correction applied to a celestial body's mean motion to obtain its true motion.

equation of time The difference between apparent (solar) time and clock time.

equator (celestial) The projection of the earth's equator onto the celestial sphere.

equinoctial hour One twenty-fourth part of the day, equal to a seasonal hour at the equinoxes.

equinox(es), adj equinoctial The instants when the sun is at one of the intersections of the ecliptic and the equator, and thus by extension these points of intersection. In this sense the spring or vernal equinox, also known as the first Point of Aries, has since (at least) Ptolemy served as the reference point for the measurement of celestial longitudes in Western and Islamic astronomy. (It now lies in the constellation Pisces.) At the equinoxes day and night would be equal, were it not for the small effects of refraction.

evection, lunar The largest inequality in the moon's motion due to the perturbing effect of the sun; discovered by Ptolemy and first explained by Newton.

genethlialogy The 'science' of nativities, making prognostications concerning the life of the native on the basis (primarily) of computed positions of the planets in the zodiac at the time of birth.

Goal Year See p. 51.

Haab A whole day approximation to the Mayan seasonal year of 365 days. It consisted of 18 months of 20 days plus 5 days. (Called *xihuitl* by the Aztecs.)

heliacal rising First pre-dawn appearance of a celestial body after a period during which it has been lost in the glare of the sun.

hierophany A manifestation of the sacred via phenomena involving the sky or the earth, as, for example, a conjunction of planets witnessed in the environment of a ceremonial centre to celebrate the birth of the gods (as at Palenque).

horoscope A record of the computed longitudes of the sun, moon and planets for a given date (usually a birthdate, see **genethlialogy**), as well as certain astrologically significant points of the ecliptic such as the point rising on the eastern horizon.

house In astrology the heavenly sphere was divided (in various ways) into 12 segments or 'houses' by great circles joining the north and south points of the horizon.

hypothesis A geometrical or physical representation of the arrangement of the heavens or of the motion of a heavenly body, also called a model.

inequality See **anomaly**

Laṅkā In Indian astronomy, the name of the point where the prime meridian crosses the equator.

latitude, celestial The angular distance of a celestial body perpendicular to the ecliptic.

libration See pp. 223–6.

long count Five-digit vigesimal (base 20) count of number of days since the most recent primitive Mayan creation (12 August 3114 BC). Comparable in essence to a Julian day-count.

longitude, celestial The angular distance of a celestial body from Aries 0° (sidereal or tropical), measured in degrees eastward along the ecliptic.

longitude, sidereal Distance measured along the ecliptic (path of the sun) relative to the stars, as in the Babylonian and Indian reference systems, rather than relative to the intersection of the ecliptic and (precessing) equator, as has been the Western and Islamic convention since Ptolemy.

lunar mansions In Islamic astronomy: the 28 divisions of the belt around the ecliptic that is the background to the motion of the moon, which spends one day in each mansion. Also a set of 28 groups of stars approximating these divisions.

major standstill The northerly and southerly declination limits that the moon reaches each month themselves vary, over the lunar node cycle of 18.6 years. The major standstill is the time when these monthly limits reach their outermost extent, equivalent to declinations of $\pm(\epsilon + i)$, where ϵ is the obliquity of the ecliptic and i is the inclination of the moon's orbit to the ecliptic. (*Note*: In order to obtain the apparent declination of the moon, a further adjustment must be made for parallax, since we stand at the earth's surface, not at its centre. Other smaller corrections are also needed for exact work but are irrelevant in the context of prehistoric lunar astronomy.)

mean sun Geocentrically, a body moving uniformly along the ecliptic with a period of one year, differing from the motion of the true sun by the solar equation; heliocentrically, the centre of the earth's orbit eccentric to the true sun.

meridian, celestial The great circle on the celestial sphere which passes through the local zenith and both celestial poles.

minor standstill The time when the monthly limits of the moon's motions reach their innermost extent, equivalent to declinations of $\pm(\epsilon - i)$.

nativity See **genethlialogy**

node The point(s) at which the orbit of the moon or a planet cuts the plane of the ecliptic; south to north (ascending node), north to south (descending node). In the case of the moon the nodes move retrograde around the ecliptic in 18.6 years.

obliquity See **ecliptic**

octant An arc of one-eighth the circumference of a circle or 45°; a distance of ±45° or ±135° between the moon or a planet and the sun or either end of the apsidal line.

occultation The covering of a planet or star by the disc of the moon.

opposition The position of the moon or a superior planet when its elongation from the sun is 180°.

parallax The difference between the position of an object (e.g. the moon) as computed from a model assuming it is seen from

the centre of the earth and its position as seen from a particular location on the surface of the earth.

perigee The position in the orbit of a celestial body closest to the earth.

perihelion The position in orbit of a planet when closest to the sun.

phase The apparent changes in shape of the moon over the course of a month. Synodic phases: for the planets their appearances and disappearances, stations and (for the superior planets) acronychal risings.

precession Changes in the direction of the earth's axis of rotation cause the intersection of the equator and the ecliptic to move westward through the fixed stars with time, causing the longitudes of the fixed stars relative to the vernal equinox slowly to increase with time. This motion, known as the precession of the equinoxes, was discovered by Hipparchus, who estimated its rate as at least 36" of arc per year or 1° per century. Its present value is roughly 50.2" per year or 1° per 72 years.

proper motion The individual movement of a star on the celestial sphere.

quadrature The position of the moon or a planet when its elongation from the sun is ± 90°.

retrograde motion/retrogradation/retrogression Apparent sidereal motion of a planet from east to west. Planets, while mostly maintaining west to east (direct) motion relative to the stars, periodically appear to reverse direction and move from east to west. The points at which such retrograde motion begins and ends are known as first and second station(ary point)s; they are near opposition for outer planets and near inferior conjunction for inner planets.

right ascension The angular distance of a celestial body from the vernal equinox, measured eastward along the equator. In the period covered by this volume right ascension was measured in degrees, not in hours, minutes and seconds of time as is customary today.

seasonal hour One-twelfth part of the day or night, of variable length according to the season; sometimes called a 'civil' hour.

sexagesimal system Place-value notation based on the number 60 (Babylonian in origin). See also pp. 51 and 77.

sidereal period Interval between successive passages of the sun, moon or planet by the same star.

sidereal year The period in which the earth completes one revolution around the sun relative to the stars (365.25636... days).

solstice(s), adj. solstitial The instants at which the sun is at its extreme declination (*c.* ±23.5°).

station/stationary point See **retrograde motion**

synodic period/synodic month Interval between successive conjunctions of a celestial body with the sun or between successive phases of the same kind. For the moon, this is the month of the phases; for Venus, the interval between successive heliacal rising events. The mean synodic month is 29.53 days.

syzygy The moment when the moon is in conjunction with or in opposition to the sun.

tithi One thirtieth of a synodic month. In Babylonian astronomy the mean tithi is used (called a 'day'). In Indian astronomy, where the term originates, the true tithi is used, corresponding to an increase of 12° in the elongation of the moon from the sun.

transit The passage of a celestial body across the observer's meridian or from one astrological place to another.

trepidation An erroneous theory that the intersection of the ecliptic and the equator oscillates over a given arc with respect to the fixed stars; the theory survived until the time of Copernicus but was rejected by Tycho Brahe.

tropical year The interval between successive passages of the mean sun across the vernal equinox (365.2422... days).

tzol kin 'Count of the days', a 260-day period comprising 13 coefficients and 20 named days unique to Mesoamerica. (Called *tonalpohualli* by the Aztecs.)

vigesimal system Place-value notation based on the number 20 (Mesoamerican).

zenith The observer's overhead point (altitude 90°).

zīj See pp. 150–52.

zodiacal signs A scale placed upon the ecliptic comprising 12 equal segments of 30° called signs, thus dividing the whole ecliptic into 360°. The zodiac was introduced by the Babylonians, for whom it was sidereally fixed with the vernal equinox at 10° or 8° of Aries (called by them *hun.ga* = the hireling). Since Ptolemy its zero point in tropical systems has been the vernal equinox, with each sign named for a nearby Babylonian or Greek constellation; in sidereal systems it is fixed with respect to the fixed stars. Due to precession the constellations for which the zodiacal signs are named have moved eastwards roughly one sign since Ptolemy's time. The zodiacal signs are: 1) Aries; 2) Taurus; 3) Gemini; 4) Cancer; 5) Leo; 6) Virgo; 7) Libra; 8) Scorpio; 9) Sagittarius; 10) Capricorn; 11) Aquarius; 12) Pisces.

The names of the planets

This table inevitably simplifies the data since, for instance, in Egyptian, Sumerian and Babylonian the names of the planets can be written in several different ways at different periods. Of the Chinese names, group a) was in use about 300 BC and group b) from about 200 BC onwards. The sequence of the planets also varies in different traditions.

	☉	☾	☿	♀	♂	♃	♄
Latin	Sol	Luna	Mercurius	Venus	Mars	Jupiter	Saturnus
Greek	Helios	Selene	Hermes	Aphrodite	Ares	Zeus	Kronos
Egyptian	Rᶜ	Iᶜḥ	Sbg	Sb3 d3	Ḥr-dšr	Ḥr-wpš-t3wy	Ḥr-k3-pt
Sumerian	ᵈutu	ᵈsuen	udu-idim-gu₄-ud	ᵈnin-si₄-an-na	ᵈsi-mu-ud	mul-sag-me-gar	udu-idim-sag-uš
Babylonian	Šamaš	Sin	Šiḫṭu	Delebat	Ṣalbatānu	Nēberu	Kayamānu
Sanskrit	Sūrya	Candra	Budha	Śukra	Bhauma	Bṛhaspati	Śani
Arabic	al Shams	al-Qamar	ᶜUṭārid	al-Zuhara	al-Mirrīkh	al-Mushtarī	Zuḥal
Chinese a)	Ri	Yue	Chenxing	Taibe	Yinghuo	Suixing	Zhenxing
Chinese b)	Tai Yang	Tai Yin	Shuixing	Jinxing	Huoxing	Muxing	Tuxing

Index

Illustration numbers are given in **bold** *type*